Plasmonic Photocatalysts

Plasmonic Photocatalysts

Guest Editor
Ewa Kowalska

Basel • Beijing • Wuhan • Barcelona • Belgrade • Novi Sad • Cluj • Manchester

Guest Editor
Ewa Kowalska
Faculty of Chemistry
Jagiellonian University
Kraków
Poland

Editorial Office
MDPI AG
Grosspeteranlage 5
4052 Basel, Switzerland

This is a reprint of the Special Issue, published open access by the journal *Catalysts* (ISSN 2073-4344), freely accessible at: www.mdpi.com/journal/catalysts/special_issues/plasmonic_photocatal.

For citation purposes, cite each article independently as indicated on the article page online and using the guide below:

Lastname, A.A.; Lastname, B.B. Article Title. *Journal Name* **Year**, *Volume Number*, Page Range.

ISBN 978-3-7258-3398-6 (Hbk)
ISBN 978-3-7258-3397-9 (PDF)
https://doi.org/10.3390/books978-3-7258-3397-9

Cover image courtesy of Ewa Kowalska

© 2025 by the authors. Articles in this book are Open Access and distributed under the Creative Commons Attribution (CC BY) license. The book as a whole is distributed by MDPI under the terms and conditions of the Creative Commons Attribution-NonCommercial-NoDerivs (CC BY-NC-ND) license (https://creativecommons.org/licenses/by-nc-nd/4.0/).

Contents

About the Editor vii

Ewa Kowalska
Plasmonic Photocatalysts
Reprinted from: *Catalysts* **2021**, *11*, 410, https://doi.org/10.3390/catal11040410 1

Maya Endo-Kimura, Bariş Karabiyik, Kunlei Wang, Zhishun Wei, Bunsho Ohtani and Agata Markowska-Szczupak et al.
Vis-Responsive Copper-Modified Titania for Decomposition of Organic Compounds and Microorganisms
Reprinted from: *Catalysts* **2020**, *10*, 1194, https://doi.org/10.3390/catal10101194 6

Zuzanna Bielan, Agnieszka Sulowska, Szymon Dudziak, Katarzyna Siuzdak, Jacek Ryl and Anna Zielińska-Jurek
Defective TiO_2 Core-Shell Magnetic Photocatalyst Modified with Plasmonic Nanoparticles for Visible Light-Induced Photocatalytic Activity
Reprinted from: *Catalysts* **2020**, *10*, 672, https://doi.org/10.3390/catal10060672 32

Shin-ichi Naya, Musashi Fujishima and Hiroaki Tada
Synthesis of Au–Ag Alloy Nanoparticle-Incorporated AgBr Crystals
Reprinted from: *Catalysts* **2019**, *9*, 745, https://doi.org/10.3390/catal9090745 52

Jose I. Garcia-Peiro, Javier Bonet-Aleta, Carlos J. Bueno-Alejo and Jose L. Hueso
Recent Advances in the Design and Photocatalytic Enhanced Performance of Gold Plasmonic Nanostructures Decorated with Non-Titania Based Semiconductor Hetero-Nanoarchitectures
Reprinted from: *Catalysts* **2020**, *10*, 1459, https://doi.org/10.3390/catal10121459 59

Tharishinny Raja-Mogan, Bunsho Ohtani and Ewa Kowalska
Photonic Crystals for Plasmonic Photocatalysis
Reprinted from: *Catalysts* **2020**, *10*, 827, https://doi.org/10.3390/catal10080827 86

Hung Ji Huang, Jeffrey Chi-Sheng Wu, Hai-Pang Chiang, Yuan-Fong Chou Chau, Yung-Sheng Lin and Yen Han Wang et al.
Review of Experimental Setups for Plasmonic Photocatalytic Reactions
Reprinted from: *Catalysts* **2019**, *10*, 46, https://doi.org/10.3390/catal10010046 106

Go Kawamura and Atsunori Matsuda
Synthesis of Plasmonic Photocatalysts for Water Splitting
Reprinted from: *Catalysts* **2019**, *9*, 982, https://doi.org/10.3390/catal9120982 131

Maya Endo-Kimura and Ewa Kowalska
Plasmonic Photocatalysts for Microbiological Applications
Reprinted from: *Catalysts* **2020**, *10*, 824, https://doi.org/10.3390/catal10080824 144

Zheng Gong, Jialong Ji and Jingang Wang
Photocatalytic Reversible Reactions Driven by Localized Surface Plasmon Resonance
Reprinted from: *Catalysts* **2019**, *9*, 193, https://doi.org/10.3390/catal9020193 165

Jingang Wang, Xinxin Wang and Xijiao Mu
Plasmonic Photocatalysts Monitored by Tip-Enhanced Raman Spectroscopy
Reprinted from: *Catalysts* **2019**, *9*, 109, https://doi.org/10.3390/catal9020109 178

Zhishun Wei, Marcin Janczarek, Kunlei Wang, Shuaizhi Zheng and Ewa Kowalska
Morphology-Governed Performance of Plasmonic Photocatalysts
Reprinted from: *Catalysts* **2020**, *10*, 1070, https://doi.org/10.3390/catal10091070 . . . . **197**

About the Editor

Ewa Kowalska

Ewa Kowalska received her PhD with honors in chemical technology from Gdansk University of Technology in 2004. After completing JSPS (2005–2007) and GCOE (2007–2009) postdoctoral fellowships in Japan, and Marie Sklodowska-Curie (MSC) fellowships in France (2002–2003) and Germany (2009–2012), she worked at the Institute for Catalysis (ICAT), Hokkaido University, as an associate professor and a leader of a research cluster on plasmonic photocatalysis (2012–2023). In 2022, she was appointed professor at the Faculty of Chemistry, Jagiellonian University. Her research interests focus on environmental protection, AOPs, solar energy conversion, plasmonic photocatalysis, nanoscience, "green" chemistry, and vis-responsive materials for the degradation of chemical and microbiological pollutants. She has published over 120 papers, has led research grants from different founding agencies (e.g., Bill & Melinda Gates Foundation, CONCERT Japan, European Commission (MSC), NCN, NAWA), and has served as editor of *Materials Science in Semiconductor Processing* (MSSP), *Micro & Nano Letters* (MNL) and *Energy, Water and Air Catalysis Research* (EWA Cat. Res.).

Editorial

Plasmonic Photocatalysts

Ewa Kowalska

Institute for Catalysis (ICAT), Hokkaido University, Sapporo 001-0021, Japan; kowalska@cat.hokudai.ac.jp

Citation: Kowalska, E. Plasmonic Photocatalysts. *Catalysts* **2021**, *11*, 410. https://doi.org/10.3390/catal 11040410

Received: 19 March 2021
Accepted: 22 March 2021
Published: 24 March 2021

Publisher's Note: MDPI stays neutral with regard to jurisdictional claims in published maps and institutional affiliations.

Copyright: © 2021 by the author. Licensee MDPI, Basel, Switzerland. This article is an open access article distributed under the terms and conditions of the Creative Commons Attribution (CC BY) license (https:// creativecommons.org/licenses/by/ 4.0/).

Plasmonic photocatalysts, i.e., photocatalysts using plasmonic properties to gain activity under visible-light (vis) irradiation, have been intensively studied in recent years for various possible applications, including environmental purification and energy conversion. Most typical plasmonic photocatalysts are composed of wide-bandgap semiconductor, e.g., titanium (IV) oxide, and deposits of noble metals (NM). Although, NM-modified titania photocatalysts have already been extensively investigated for more than 40 years as NMs work as an electron sink inhibiting charge carriers' recombination [1–4], the use of plasmonic properties for photocatalysis could be considered as new. It is thought that the first report, proving that plasmon resonance of gold was responsible for vis response, was published by Tian and Tatsuma in 2005 [5]. Since then more and more studies have been published [6–10], and the new term of "plasmonic photocatalysis" and "plasmonic photocatalysts" have been proposed [11]. It seems that the plasmonic photocatalysis is still a "hot" topic of research on the vis-responsive materials, as observed by a high number of scientific papers published yearly (Figure 1), including highly cited and "hot" papers (46 and 3, respectively, according to Web of Science: 18 March 2021). It should be pointed out that many reports have not been even included in Figure 1, as other terms have been commonly used, e.g., "visible-light-induced photocatalysis through surface plasmon excitation of NM . . . ", "gold-modified titania with vis activity", "vis response of silver-modified titanium (IV) oxide", etc. Not only research papers on plasmonic photocatalysis have been published, but also reviews and book chapters. Although, plasmonic photocatalysts have proven to be active under broad range of irradiation (UV/vis/IR) and for various applications, including wastewater and water treatment [12,13], air purification [14,15], antimicrobial materials [16–20], solar energy conversion [21,22], synthesis of organic compounds [23], and CO_2 conversion [24], the activity under solar radiation (vis/NIR) is still low for broader commercialization. Therefore, various studies have been performed on activity and stability improvements of plasmonic photocatalysts, as presented in the Special Issue of *Catalysts* on "Plasmonic Photocatalysts".

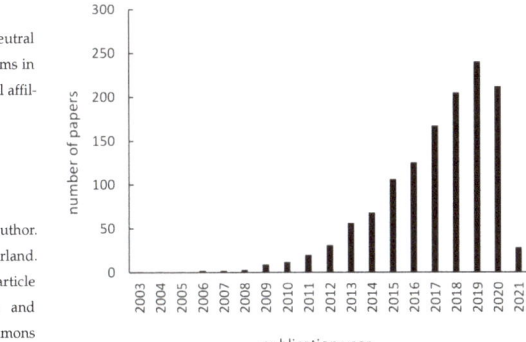

Figure 1. Number of papers published annually on plasmonic photocatalysis searched in Web of Science (18 March 2021) using "plasmon* photocatalyst*" and "plasmon* catalys*".

The Special Issue is a collection of 11 papers, including three research papers and eight reviews; in total, five manuscripts have been selected as feature papers. Original papers have presented different photocatalysts, considering both semiconductor type and plasmonic metal, i.e., Au–Ag alloy incorporated AgBr crystals [25], commercial titania samples modified with Cu deposits [26] and self-synthesized titania with defects decorated with mono- and bi-metallic (Pt/Cu) nanoparticles (NPs) [27]. The study by Naya et al. has shown how to efficiently form Au–Ag alloy NP of 5-nm mean size incorporated in micrometer-sized AgBr crystals by photochemical method [25]. It has been found that an increase in Au content in the alloy results in a drastic decrease of the local electric field enhancement (calculated by finite-difference time-domain; FDTD) and bathochromic shift of the localized surface plasmon resonance (LSPR) peak. Finally, it has been proposed that Au–Ag@AgBr material is highly promising because of high efficiency of sunlight harvesting.

Bielan et al. have proposed defective titania modified with Pt and Cu NPs of core/interlayer/shell structure with magnetic core (Fe_3O_4) for easy separation and recycling of photocatalysts after photocatalytic reactions [27]. It has been shown that defective titania (titanium vacancies; d-TiO_2) allows efficient light harvesting, and Pt-modified d-TiO_2 exhibits the highest activity under vis irradiation. It should be pointed out that defects might accelerate NM aggregation due to the seed-mediated growth on the titania surface, and thus causing a decrease in metal and semiconductor interface, resulting in worsening of UV/vis activity. Importantly, it has been proven that the formation of Fe_3O_4@SiO_2/d-TiO_2-Pt/Cu structures allows an effective separation of the obtained magnetic photocatalysts.

The third research manuscript by Endo-Kimura et al. has discussed the modification of seven commercial titania photocatalysts of different properties (composition: anatase/rutile and particle/crystallite sizes) with Cu NPs by photo-deposition method performed in the presence or absence of methanol as a hole scavenger, i.e., during methanol dehydrogenation and water oxidation, respectively [26]. It should be underlined that though zero-valent copper has been deposited on the surface of titania (indicated by violet color of samples), copper has been oxidized in air, resulting in the co-presence of its different forms. Interestingly, the prolonged irradiation (from 1 to 5 h) during preparation of samples has caused copper aggregation, and thus being detrimental for vis activity. Moreover, it has been concluded that oxidized forms of copper are more active than plasmonic one during oxidative decomposition of acetic acid under vis irradiation. Furthermore, it has been confirmed that Cu/TiO_2 samples show high antimicrobial effect, mainly due to the intrinsic ("dark") activity of copper species.

The review papers have presented various aspects of plasmonic photocatalysts, including nano-architecture design, experimental systems and various applications. For example, Wang et al. have discussed time-dependent measurements of tip-enhanced resonance Raman spectroscopy (TERRS) [28]. It has been shown that thermal electrons might dissociate organic compounds (Malachite Green), i.e., "plasma scissors" as an effective tool for molecular dissociation. The importance of plasmon in the field of heterogeneous photocatalysis has been underlined, such as allowing hydrogen to carry out dissociation reactions on Au NPs at room temperature. Additionally, theoretical calculations combined with real-time observations of plasmon-induced chemical reactions have been discussed, e.g., for the dissociation of S-S bonds on the surface of plasmon-induced Ag and Cu.

Gong et al. have reviewed LSPR-driven photocatalytic reactions, such as dissociations of hydrogen, water, nitrogen, and carbon dioxide [29]. First, interesting introduction on the synthesis of 4,4'-dimercaptoazobenzene (DMAB) from 4-nitrobenzenethiol (4NBT) using surface-enhanced Raman scattering (SERS) technology has been presented. Then, the photosynthetic and degradation processes of 4NBT have been discussed, including the impact of various factors, such as laser wavelength, reaction time, substrate, and pH value. Finally, the competition between degradation and photosynthetic pathways for this reaction by SERS technology on Au-film substrate over a nanosphere has been presented. It has been concluded that plasmon–exciton interaction might increase the efficiency of chemical reactions driven by plasmonic excitation.

A very interesting review has been prepared by Kawamura and Matsuda on the photocatalytic water splitting [30]. Although plasmonic enhancement for water splitting has been investigated broadly, the mechanisms have not been clarified and understood yet. Authors have pointed out that additionally, the yield of water splitting should be improved for commercial application. They have presented various types of plasmonic materials and semiconductors with different morphologies. The typical plasmonic materials (NM), but also other metals (e.g., Al) and some transition metal nitrides, being attractive alternatives to the precious and expensive Au and Ag, have been discussed in detail. Interesting aspects on the cost reduction and scale-up of plasmonic photocatalyst fabrication have also been addressed.

Huang et al. have proposed that unclarified mechanism of plasmonic photocatalysis is caused by the complexity of plasmonic optical responses and complicated microchemical processes during photocatalytic reactions [31]. Accordingly, they have presented various experimental methods used in the research on plasmonic photocatalysis and provided alternative mechanisms, based on the theoretical modeling. The five most important advantages of plasmonic effect in photocatalytic reactions have been presented, i.e., (i) efficient light harvesting, (ii) intensive light scattering, (iii) strengthening of electromagnetic field, (iv) generation of hot carriers, and (v) plasmonic heating.

The application of plasmonic photocatalysts for microbial studies have been discussed by Endo-Kimura and Kowalska [32]. It should be remembered that in the case of microorganisms, intrinsic properties of plasmonic photocatalysts (in the dark) should be considered, both for well-known antimicrobial agents as silver and copper, but also for wide-bandgap semiconductors. Accordingly, this review has presented various microbiological applications of plasmonic photocatalysis, including antibacterial, antifungal, and antiviral properties, but also a novel research on the delivery of drugs and cancer treatment. It should be pointed out that the mechanism of plasmonic photocatalysts' interactions with microorganisms has not been understood well, and contradictory data have also been published. Accordingly, the further research is necessary on the key-factor properties and the mechanisms of inactivation of microorganisms and the treatment of cancer cells.

For efficient light harvesting the application of photonic crystals (PC) as a support for NMs have been discussed by Raja-Mogan et al. [33]. It is important to mention that though PCs might possess vis response due to photonic bandgap (PBG) and slow photon effects at vis/IR range of solar spectrum, it should not result in vis-activity of wide-bandgap semiconductors. In contrast, PCs-based plasmonic photocatalysts are promising candidates for high photocatalytic performance, as proven in discussed papers. However, it should be pointed out that the synthesis of both PCs and PCs-based plasmonic photocatalysts is challenging, involving multi-step procedures, and also the preservation of PC structure during activity testing must be ensured (not to lose photonic properties). It has been concluded that the best method to obtain highly active photocatalyst is to tune the photo-absorption properties, i.e., to match the PBG edge (slow photon effect) with LSPR wavelengths by the change in the void diameter of inverse opal PC and the size of NMNPs, respectively.

Similarly, the reviews by Wei et al. [34] and Garcia-Peiro et al. [35] have underlined the importance of the morphology. For example, faceted nanoparticles, aerogels, nanowires, nanotubes, and super-nano structures of semiconductors have shown enhanced photocatalytic activity and high stability [34]. Additionally, NM deposition on special parts of supports, such as facets, might accelerate the activity. It should be pointed out the though bi-, and ternary-metal-modifications have been suggested for enhanced light harvesting, the co-interactions between NMs have also caused a decrease in activity. Interestingly, the difficulties in photocatalysts comparison have been pointed out, as different irradiation sources, photoreactors, tested compounds, and reaction conditions are commonly used, and quantum yields have been estimated only in several studies. Another problem is connected with the use of solar radiation or solar simulators (also containing UV), and thus excitation of semiconductor should be rather considered than plasmonic photocatalysis. Additionally, during mechanism study the activity tests should not be performed for dye discoloration, as photocatalyst sensitization by dye might be the predominant mechanism of their degradation rather than vis-activity caused by plasmonic photocatalysts.

Interesting review has been prepared by Garcia-Peiro et al. since they have looked for an alternative to famous titania to obtain photocatalysts of better performance [35]. They have discussed the correlation between the morphology and photocatalytic activities of gold-modified non-titania-based semiconductors, including Au-CeO$_2$, Au-Cu$_2$O/Au-CuS, and Au-CdS. It has been confirmed that the morphology is key-factor of activity, and that the less-conventional design might result in breakthroughs in plasmonic photocatalysis. Furthermore, the anisotropic metals are highly attractive for efficient light harvesting. They have also emphasized the need to establish the control and clear definition of hetero-nanoarchitectures at the nanoscale level for fast progress in research, especially for biomedical applications.

It is obvious that excellent research has been performed worldwide on plasmonic photocatalysts for various types of materials, reactions systems, and applications. It might be concluded that plasmonic photocatalysis is a very active field of research, with promising results for future commercialization. On the other hand, further research is highly needed, especially to clarify the mechanism, standardization of research practices and to find the key-factors of photocatalytic activity.

Finally: I would like to thank all authors for their valuable contributions, without which this special issue would not have been possible. I would like to express my sincerest thanks to the editorial team of *Catalysts* (especially Ms. Janine Li) for their kind support, advices, and fast responses.

Conflicts of Interest: The author declares no conflict of interest.

References

1. Kraeutler, B.; Bard, A.J. Heterogeneous photocatalytic preparation of supported catalysts. Photodeposition of platinum on TiO$_2$ powder and other substrates. *J. Am. Chem. Soc.* **1978**, *100*, 4317–4318. [CrossRef]
2. Ohtani, B.; Kakimoto, M.; Nishimoto, S.; Kagiya, T. Photocatalytic reaction of neat alcohols by metal-loaded titanium(IV) oxide particles. *J. Phys. Chem. A Chem.* **1993**, *70*, 265–272. [CrossRef]
3. Herrmann, J.-M.; Disdier, J.; Pichat, P. Photoassisted platinum deposition on TiO$_2$ powder using various platinum complexes. *J. Phys. Chem.* **1986**, *90*, 6028–6034. [CrossRef]
4. Kowalska, E.; Remita, H.; Colbeau-Justin, C.; Hupka, J.; Belloni, J. Modification of titanium dioxide with platinum ions and clusters: Application in photocatalysis. *J. Phys. Chem. C* **2008**, *112*, 1124–1131. [CrossRef]
5. Tian, Y.; Tatsuma, T. Mechanisms and Applications of Plasmon-Induced Charge Separation at TiO$_2$ Films Loaded with Gold Nanoparticles. *J. Am. Chem. Soc.* **2005**, *127*, 7632–7637. [CrossRef] [PubMed]
6. Kowalska, E.; Abe, R.; Ohtani, B. Visible light-induced photocatalytic reaction of gold-modified titanium(IV) oxide particles: Action spectrum analysis. *Chem. Commun.* **2009**, 241–243. [CrossRef] [PubMed]
7. Kominami, H.; Tanaka, A.; Hashimoto, K. Mineralization of organic acids in aqueous suspension of gold nanoparticles supported on cerium(IV) oxide powder under visible light irradiation. *Chem. Commun.* **2010**, *46*, 1287–1289. [CrossRef] [PubMed]
8. Verbruggen, S.W.; Keulemans, M.; Goris, B.; Blommaerts, N.; Bals, S.; Martens, J.A.; Lenaerts, S. Plasmonic 'rainbow' photocatalyst with broadband solar light response for environmental applications. *Appl. Catal. B Environ.* **2016**, *188*, 147–153. [CrossRef]
9. Mukherjee, S.; Libisch, F.; Large, N.; Neumann, O.; Brown, L.V.; Cheng, J.; Lassiter, J.B.; Carter, E.A.; Nordlander, P.; Halas, N.J. Hot Electrons Do the Impossible: Plasmon-Induced Dissociation of H$_2$ on Au. *Nano Lett.* **2013**, *13*, 240–247. [CrossRef]
10. Mizeikis, V.; Kowalska, E.; Juodkazis, S. Resonant, localization, enhancement, and polarization o optical fields in nano-scale interface regions for photo-catalytic application. *J. Nanosci. Nanotechnol.* **2011**, *11*, 2814–2822. [CrossRef] [PubMed]
11. Awazu, K.; Fujimaki, M.; Rockstuhl, C.; Tominaga, J.; Murakami, H.; Ohki, Y.; Yoshida, N.; Watanabe, T. A plasmonic photocatalyst consisting of sliver nanoparticles embedded in titanium dioxide. *J. Am. Chem. Soc.* **2008**, *130*, 1676–1680. [CrossRef]
12. Kowalska, E.; Janczarek, M.; Rosa, L.; Juodkazi, S.; Ohtani, B. Mono- and bi-metallic plasmonic photocatalysts for degradation of organic compounds under UV and visible light irradiation. *Catal. Today* **2014**, *230*, 131–137. [CrossRef]
13. Wang, X.P.; Tang, Y.X.; Chen, Z.; Lim, T.T. Highly stable heterostructured Ag-AgBr/TiO$_2$ composite: A bifunctional visible-light active photocatalyst for destruction of ibuprofen and bacteria. *J. Mater. Chem.* **2012**, *22*, 23149–23158. [CrossRef]
14. Wysocka, I.; Markowska-Szczupak, A.; Szweda, P.; Ryl, J.; Endo-Kimura, M.; Kowalska, E.; Nowaczyk, G.; Zielinska-Jurek, A. Gas-phase removal of indoor volatile organic compounds and airborne microorganisms over mono- and bimetal-modified (Pt, Cu, Ag) titanium(IV) oxide nanocomposites. *Indoor Air* **2019**, *29*, 979–992. [CrossRef] [PubMed]
15. Verbruggen, S.W. TiO$_2$ photocatalysis for the degradation of pollutants in gas phase: From morphological design to plasmonic enhancement. *J. Photoch. Photobio. C* **2015**, *24*, 64–82. [CrossRef]

16. Rtimi, S.; Baghriche, O.; Sanjines, R.; Pulgarin, C.; Bensimon, M.; Kiwi, J. TiON and TiON-Ag sputtered surfaces leading to bacterial inactivation under indoor actinic light. *J. Photoch. Photobio. A* **2013**, *256*, 52–63. [CrossRef]
17. Endo, M.; Janczarek, M.; Wei, Z.; Wang, K.; Markowska-Szczupak, A.; Ohtani, B.; Kowalska, E. Bactericidal properties of plasmonic photocatalysts composed of noble-metal nanoparticles on faceted ana-tase titania. *J. Nanosci. Nanotechnol.* **2019**, *19*, 442–452. [CrossRef] [PubMed]
18. Endo, M.; Wei, Z.S.; Wang, K.L.; Karabiyik, B.; Yoshiiri, K.; Rokicka, P.; Ohtani, B.; Markowska-Szczupak, A.; Kowalska, E. Noble metal-modified titania with visible-light activity for the decomposition of microorganisms. *Beilstein J. Nanotech.* **2018**, *9*, 829–841. [CrossRef]
19. Wei, Z.; Endo, M.; Wang, K.; Charbit, E.; Markowska-Szczupak, A.; Ohtani, B.; Kowalska, E. Noble metal-modified octahedral anatase titania particles with enhanced activity for decomposition of chemical and microbiological pollutants. *Chem. Eng. J.* **2017**, *318*, 121–134. [CrossRef] [PubMed]
20. Kowalska, E.; Wei, Z.; Karabiyik, B.; Herissan, A.; Janczarek, M.; Endo, M.; Markowska-Szczupak, A.; Remita, H.; Ohtani, B. Silver-modified titania with enhanced photocatalytic and antimicrobial properties under UV and visible light irradiation. *Catal. Today* **2015**, *252*, 136–142. [CrossRef]
21. Solarska, R.; Bienkowski, K.; Zoladek, S.; Majcher, A.; Stefaniuk, T.; Kulesza, P.J.; Augustynski, J. Enhanced Water Splitting at Thin Film Tungsten Trioxide Photoanodes Bearing Plasmonic Gold-Polyoxometalate Particles. *Angew. Chem. Int. Ed.* **2014**, *53*, 14196–14200. [CrossRef] [PubMed]
22. Warren, S.C.; Thimsen, E. Plasmonic solar water splitting. *Energy Environ. Sci.* **2012**, *5*, 5133–5146. [CrossRef]
23. Wang, C.L.; Astruc, D. Nanogold plasmonic photocatalysis for organic synthesis and clean energy conversion. *Chem. Soc. Rev.* **2014**, *43*, 7188–7216. [CrossRef] [PubMed]
24. Wang, C.J.; Ranasingha, O.; Natesakhawat, S.; Ohodnicki, P.R.; Andio, M.; Lewis, J.P.; Matranga, C. Visible light plasmonic heating of Au-ZnO for the catalytic reduction of CO_2. *Nanoscale* **2013**, *5*, 6968–6974. [CrossRef]
25. Naya, S.-I.; Fujishima, M.; Tada, H. Synthesis of Au–Ag Alloy Nanoparticle-Incorporated AgBr. *Crystals* **2019**, *9*, 745. [CrossRef]
26. Endo-Kimura, M.; Karabiyik, B.; Wang, K.; Wei, Z.; Ohtani, B.; Markowska-Szczupak, A.; Kowalska, E. Vis-responsive copper-modified titania for decomposition of organic compounds and microorganisms. *Crystals* **2020**, *10*, 1194. [CrossRef]
27. Bielan, Z.; Sulowska, A.; Dudziak, S.; Siuzdak, K.; Ryl, J.; Zielinska-Jurek, A. Defective TiO_2 core-shell magnetic photocatalyst modified with plasmonic nanoparticles for visible light-induced photocatalytic activity. *Catalysts* **2020**, *10*, 672. [CrossRef]
28. Wang, J.; Wang, X.; Mu, X. Plasmonic Photocatalysts Monitored by Tip-Enhanced Raman Spectroscopy. *Catalysts* **2019**, *9*, 109. [CrossRef]
29. Gong, Z.; Ji, J.; Wang, J. Photocatalytic Reversible Reactions Driven by Localized Surface Plasmon Resonance. *Catalysts* **2019**, *9*, 193. [CrossRef]
30. Kawamura, G.; Matsuda, A. Synthesis of Plasmonic Photocatalysts for Water Splitting. *Catalysts* **2019**, *9*, 982. [CrossRef]
31. Huang, H.J.; Wu, J.C.-S.; Chiang, H.-P.; Chou Chau, Y.-F.; Lin, Y.-S.; Wang, Y.H.; Chen, P.-J. Review of Experimental Setups for Plasmonic Photocatalytic Reactions. *Catalysts* **2020**, *10*, 46. [CrossRef]
32. Endo-Kimura, M.; Kowalska, E. Plasmonic Photocatalysts for Microbiological Applications. *Catalysts* **2020**, *10*, 824. [CrossRef]
33. Raja-Mogan, T.; Ohtani, B.; Kowalska, E. Photonic crystals for plasmonic photocatalysis. *Catalysts* **2020**, *10*, 827. [CrossRef]
34. Wei, Z.; Janczarek, M.; Wang, K.; Zheng, S.; Kowalska, E. Morphology-governed performance of plasmonic photocatalysts. *Catalysts* **2020**, *10*, 1070. [CrossRef]
35. Garcia-Peiro, J.I.; Bonet-Aleta, J.; Bueno-Alejo, C.J.; Hueso, J.L. Recent Advances in the Design and Photocatalytic Enhanced Performance of Gold Plasmonic Nanostructures Decorated with Non-Titania Based Semiconductor Hetero-Nanoarchitectures. *Catalysts* **2020**, *10*, 1459. [CrossRef]

Article

Vis-Responsive Copper-Modified Titania for Decomposition of Organic Compounds and Microorganisms

Maya Endo-Kimura [1], Bariş Karabiyik [1], Kunlei Wang [1,2], Zhishun Wei [3], Bunsho Ohtani [1], Agata Markowska-Szczupak [4] and Ewa Kowalska [1,*]

1. Institute for Catalysis, Hokkaido University, N21, W10, Sapporo 001-0021, Japan; m_endo@cat.hokudai.ac.jp (M.E.-K.); bakarabiyik@gmail.com (B.K.); kunlei@cat.hokudai.ac.jp (K.W.); ohtani@cat.hokudai.ac.jp (B.O.)
2. Northwest Research Institute, Co., Ltd. of C.R.E.C., Lanzhou 730099, China
3. Hubei Provincial Key Laboratory of Green Materials for Light Industry, Hubei University of Technology, Wuhan 430068, China; wei.zhishun@hbut.edu.cn
4. Department of Chemical and Process Engineering, West Pomeranian University of Technology in Szczecin, Piastów 42, 71-065 Szczecin, Poland; Agata.Markowska@zut.edu.pl
* Correspondence: kowalska@cat.hokudai.ac.jp; Tel.: +81-11-706-9130

Received: 19 September 2020; Accepted: 14 October 2020; Published: 16 October 2020

Abstract: Seven commercial titania (titanium(IV) oxide; TiO_2) powders with different structural properties and crystalline compositions (anatase/rutile) were modified with copper by two variants of a photodeposition method, i.e., methanol dehydrogenation and water oxidation. The samples were characterized by diffuse reflectance spectroscopy (DRS), X-ray powder diffraction (XRD), scanning electron microscopy (SEM), energy-dispersive X-ray spectroscopy (EDS) and X-ray photoelectron spectroscopy (XPS). Although zero-valent copper was deposited on the surface of titania, oxidized forms of copper, post-formed in ambient conditions, were also detected in dried samples. All samples could absorb visible light (vis), due to localized surface plasmon resonance (LSPR) of zero-valent copper and by other copper species, including Cu_2O, CuO and Cu_xO (x:1-2). The photocatalytic activities of samples were investigated under both ultraviolet (UV) and visible light irradiation (>450 nm) for oxidative decomposition of acetic acid. It was found that titania modification with copper significantly enhanced the photocatalytic activity, especially for anatase samples. The prolonged irradiation (from 1 to 5 h) during samples' preparation resulted in aggregation of copper deposits, thus being detrimental for vis activity. It is proposed that oxidized forms of copper are more active under vis irradiation than plasmonic one. Antimicrobial properties against bacteria (*Escherichia coli*) and fungi (*Aspergillus niger*) under vis irradiation and in the dark confirmed that Cu/TiO_2 exhibits a high antibacterial effect, mainly due to the intrinsic activity of copper species.

Keywords: copper-modified titania; plasmonic photocatalysts; heterogeneous photocatalysis; bactericidal activity; antifungal effect

1. Introduction

Titania (TiO_2) is one of the most widely investigated semiconductor photocatalysts because it is abundant, cost-effective, highly active, and environmentally friendly. Furthermore, titania has high photocatalytic activity and good chemical and thermal stability [1–3]. Therefore, it has been broadly used for water and wastewater treatment, air purification and energy conversion [1,4–8]. The crystalline form (i.e., anatase, rutile and brookite), the surface properties (e.g., specific surface area, crystallinity and crystallite size) and the morphology of titania photocatalysts govern the photocatalytic performance. For example, detailed study of commercial titania P25 photocatalyst, which contains anatase, rutile and non-crystallite phase, has revealed that the superiority of anatase or rutile depends on the photocatalytic reaction system, i.e., anatase is more active for oxidative decomposition of organic compounds, whereas rutile is more efficient for methanol dehydrogenation [9]. Although titania photocatalysts are considered as highly active, the charge carriers' recombination (typical for all semiconductors) results in lower quantum yields of photocatalytic reaction than expected 100%. Accordingly, various studies on activity enhancement have been performed, including also the morphology architecture, and thus titania in the form of nanowires, nanoflakes, nanotubes, nanorods, mesocrystals, mesoporous networks (aerogels) and as faceted nanoparticles have been intensively investigated [10–15]. For example, titania nanorods have shown higher activity than both nanoparticles and nanospheres in the degradation of dye, probably due to the nanoporous structure, which might allow an effective transport of the reactant molecules to the active sites [13].

There are two main mechanisms of organic compounds' decomposition during photocatalytic degradation, i.e., (i) direct reaction of a substrate with photo-generated charge carriers on the photocatalyst surface, and (ii) indirect reaction with reactive oxygen species (ROS), generated from oxygen and water. In situ Fourier transform infrared (FT-IR) study has elucidated that 4-chlorophenol degradation occurs both on the surface of titania, and by the reductive dechlorination proceeds by superoxide anion radical (O_2^-) [16], and thus it is proposed that the affinity of titania surface to a substrate is crucial for the effectiveness of photocatalytic degradation [16,17]. Recently, the complete decomposition of pollutants, such as organic compounds and textile dyes, has been reported for a variety of modified titania photocatalysts under ultraviolet (UV) irradiation [18–20].

However, as titania absorbs only UV light and the solar spectrum contains just 3–4% of UV, only a small portion of the incident solar energy is utilized by conventional titania photocatalysts [21], thus limiting their reactivity. Consequently, there is great interest in modification of titania to extend the onset of absorption into the visible (vis) region of solar spectrum. The methods of titania modification include: (i) doping with non-metals [22–25], (ii) coupling with narrower-bandgap semiconductors [26–29], (iii) preparation of oxygen-deficient titania ("self-doped") [30], and (iv) various kinds of surface modifications, e.g., with nanoparticles/nanoclusters of metals [31–35], metal complexes [36,37], anions (e.g., surface fluorination [38]), metal cations and their oxides (e.g., Co, Cr, Cu, Fe, Mo, V, and W [39], Li and rare-earth elements [40]), carbon-containing compounds (e.g., alcohols [41], glucose [42], graphene [43]). It might be concluded that the surface modification has been more intensively studied and recommended since doping might generate recombination centers in the crystal structure (dopants as recombination centers), resulting in a decrease in photocatalytic activity under UV irradiation [44].

Since the pioneering work of Bard in 1978 [31], noble metals have played the leading role among available surface modifiers since not only do they shift photoactivation towards vis response, but they also enhance photocatalytic activity under UV irradiation. Noble-metal-loading provides chemically active sites, and thus metal deposits lower the activation barrier and prevent charge carriers' recombination by formation of a Schottky barrier (resulting from higher work function of metals than electron affinity of TiO_2) increasing the transfer of photoexcited electrons to the substrates under UV illumination. Although activity enhancement for titania (and other semiconductors) under UV irradiation has been well known for about 50 years, activation of wide-bandgap semiconductors at vis range by noble metals has been investigated in the last decade [44–46]. Despite the precise mechanism of vis activity remaining unclarified (whether through energy transfer, electron transfer or plasmonic heating), it is apparent that surface plasmon resonance (SPR) of noble metals is responsible for this phenomenon [47]. SPR, the collective oscillation of the valence electrons, is induced when the frequency of photons matches the frequency of the surface electrons. The plasmon wavelength depends on the nature of the metal and the properties of metallic deposits, such as the size and the shape [48,49]. Hence, the entire solar spectrum might be used by the proper design of the composition and the morphology of plasmonic nanostructures [48,50–54]. Light below the plasmon frequency is reflected (screening of electric field of light) and above is transmitted (electrons cannot respond fast enough) [55]. Moreover, the energetic charge carriers might release the energy by converting it to local thermal energy or by transferring it to the surroundings [56,57].

In addition to the photocatalytic properties of titania for degradation of organic compounds and energy conversion, titania has also been proposed as an antimicrobial agent after an initial report by Matsunaga et al. in 1985 [58]. Since then, many studies have been performed and titania has been shown to inhibit bacteria, fungi, viruses and protozoa. Accordingly, titania might be used as an effective disinfectant for air, water and various surfaces (e.g., tables, walls and medical devices) [59–65]. The recent study has mainly focused on solar disinfection (SODIS), which indicates that titania (and other semiconductor photocatalysts) significantly enhance the overall disinfection rate [66]. Although many reports on SODIS have been published, including mechanistic studies, e.g., the combined effect of heating and generation of ROS under UV irradiation [67], a lack of knowledge about the mechanism of titania-based photocatalysis under vis conditions remains. Thus, effective design of photoactive materials with optimal antimicrobial activity under solar radiation continues to prove elusive [68].

Copper (Cu) is considered as an excellent environmental purifier due to its attractive cost (in comparison to other noble metals), low toxicity for humans and high antimicrobial activity. It has been proposed that the bactericidal mechanisms of copper involve adsorption of Cu ions on the surface of bacteria, and then (1) the surface proteins are denatured [69], and/or (2) adsorbed Cu ions induce oxidative stress [70]. Furthermore, it has been reported that Cu_2O has higher activity in inactivation of bacteria than CuO and Ag [69]. On the other hand, Cu nanoparticles (NPs) have lower toxicity than Ag NPs against some species of fungi [71]. It has also been revealed that copper-modified titania photocatalysts show high antimicrobial efficiency [72,73]. It is believed that modification of titania with small amount of an inexpensive noble metal (Cu) would result in preparation of environmentally friendly materials with high levels of activity. However, it is also known that the properties of photocatalysts govern the overall activity. Accordingly, in this study, Cu NPs have been deposited on various commercial titania photocatalysts under different preparation conditions, and investigated for the photocatalytic decomposition of the organic compounds (acetic acid and methanol) and microorganisms (bacteria and fungi) in order to clarify the property-governed activity.

2. Results and Discussion

2.1. Preparation of Cu-Modified Titania

In this study, seven commercial titania samples, i.e., ST-01, FP6, P25, ST-41, TIO-6, ST-G1 and TIO-5, have been used, and their properties are shown in Table 1.

Table 1. The characteristics of commercial titania samples.

Titania	Crystalline Form [I] (A%) [II]	Crystallite Size [III] (nm)	SSA [III] ($m^2 g^{-1}$)	ETs [III] ($\mu mol\ g^{-1}$)
ST-01	A (100)	8	298	84
FP6	A/R (92.5)	15	103.7 [IV]	154 [IV]
P25	A/R (84.8)	28	59	50
ST-41	A/R (99.7)	208	11	25
TIO-6	R (0)	15	100	242
ST-G1	R/A (1.9)	205	5.7	50
TIO-5	R/A (8.6)	570	3	14

[I] A: anatase, R: rutile, A/R and R/A: mixture with majority of anatase and rutile, respectively. [II] The content of anatase (without consideration of non-crystalline phase, i.e., [A]% + [R]% = 100%). [III] data reported previously (except for FP6) in [74] and [IV] FP6 in [75], SSA—specific surface area, ETs—density of defective sites (electron traps).

Copper was photodeposited on titania in two reaction systems, i.e., during dehydrogenation of methanol, and water oxidation. The reactions occurring during Cu deposition can be presented by Equations (1)–(2) and (4)–(5), respectively, and summarized by Equations (3) and (6), as shown below:

(i) during methanol dehydrogenation (Equations (1)–(3)):

$$Cu^{2+} + e^- \rightarrow Cu^+ + e^- \rightarrow Cu^0 \qquad (1)$$

$$CH_3OH + 2h^+ \rightarrow HCHO + 2H^+ \qquad (2)$$

$$Cu^{2+} + CH_3OH \rightarrow Cu^0 + HCHO + 2H^+ \qquad (3)$$

(ii) during water oxidation (Equations (4)–(6)):

$$Cu^{2+} + e^- \rightarrow Cu^+ + e^- \rightarrow Cu^0 \qquad (4)$$

$$H_2O + 2h^+ \rightarrow 1/2O_2 + 2H^+ \qquad (5)$$

$$Cu^{2+} + H_2O \rightarrow Cu^0 + 1/2O_2 + 2H^+ \qquad (6)$$

In the case of methanol dehydrogenation, the formation of metallic deposits on the titania surface (here zero-valent Cu) results in the evolution of hydrogen (H_2), as Cu (similar to other noble metals) works as a co-catalyst on which molecular hydrogen is formed (negligible activity of titania without co-catalyst for alcohol dehydrogenation), as follows (Equations (7) and (8), and summarized as Equation (9)):

$$2H^+ + 2e^- \rightarrow H_2 \qquad (7)$$

$$CH_3OH + 2h^+ \rightarrow HCHO + 2H^+ \qquad (8)$$

$$CH_3OH \rightarrow H_2 + HCHO \qquad (9)$$

The obtained results of methanol dehydrogenation and water oxidation during Cu NPs formation are shown in Figure 1. It was found that the ST-01 sample (fine anatase) was the most active, reaching an approximately 10-fold higher rate of H_2 evolution than other titania samples, especially during the first hour of irradiation (Figure 1a). Almost all samples (except TIO-6) show good linearity in H_2 generation after 30 min of irradiation (induction time for formation of Cu NPs), indicating that 30 min could be sufficient for complete deposition of Cu in methanol solution. On the other hand, the efficiency of water oxidation is low, and only slightly linear evolution of oxygen was observed for TIO-6 sample (fine rutile), as shown in Figure 1b. It is thought that water oxidation is not effective, probably because copper is less noble than platinum and gold, and thus easily oxidizable in water and air. Indeed, there are no other reports on Cu for photocatalytic water oxidation.

Almost linear evolution of hydrogen has been observed for all samples (except TIO-6; probably due to the largest content of electron traps slowing down the copper deposition) during 1 h irradiation (Figure 1a) in methanol. Prolonged irradiation (up to 5 h) has also been applied to check how and if the properties of obtained samples differ, and the data of H_2-evolution rate are shown in Figure 1c. Similar to 1 h photodeposition, the ST-01 sample shows the highest efficiency for methanol dehydrogenation, which is not surprising as large specific surface area has been reported to be a key-factor of hydrogen evolution for anatase-based titania, e.g., in the case of gold- [44], silver- [76] and platinum- [77] modified titania samples. Although quite good linearity in H_2 evolution was observed after 30 min of irradiation (Figure 1a), the data for prolonged irradiation indicate that at least 1 h of irradiation is necessary to ensure complete deposition of copper (linear evolution of H_2 starting at about 1 h).

Moreover, the efficiencies of anatase are higher than rutile for H_2 evolution (methanol dehydrogenation), whereas rutile-containing photocatalysts are more active than anatase ones for O_2 evolution, probably due to the smaller Ti–Ti distance in rutile than that in anatase, leading to the formation of a surface structures, such as Ti–OO–Ti [78], and thus enhanced evolution of O_2. Hence, the photocatalytic activity during Cu-photodeposition corresponded to the findings by Buchalska et al. showing that rutile is an active phase in reduction reactions (by photogenerated electrons), whereas anatase in oxidation reactions (by photogenerated holes) [79].

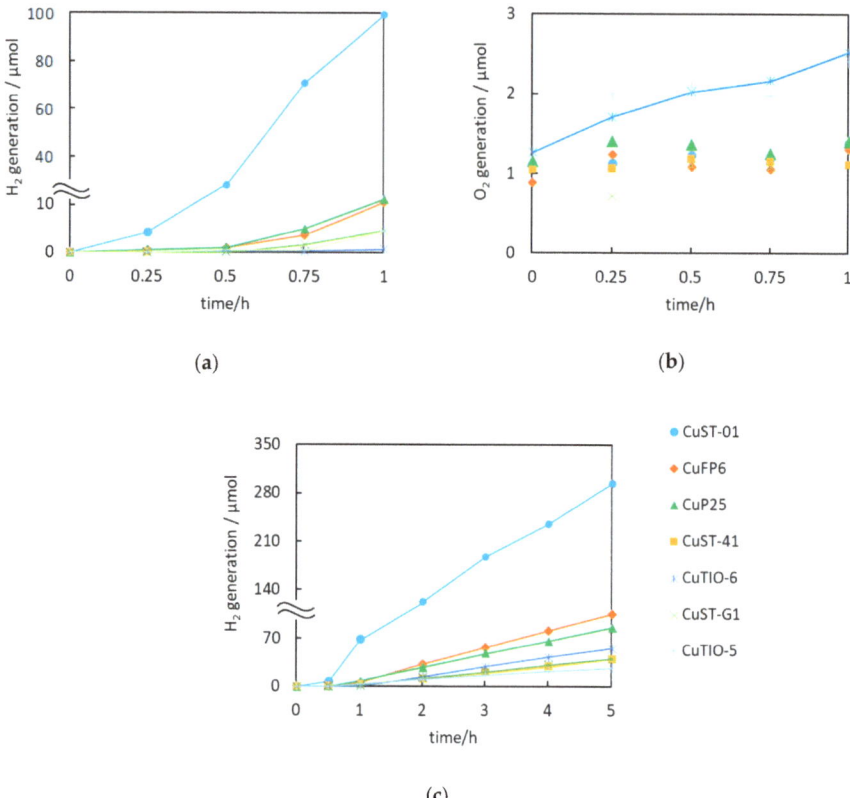

Figure 1. Hydrogen and oxygen evolution during the photodeposition of copper on titania: (**a**) in methanol/water for 1 h, (**b**) in water for 1 h, (**c**) in methanol/water for 5 h.

2.2. Photoabsorption Properties (Ultraviolet-Visible (UV-Vis) Diffuse Reflectance Spectroscopy)

The modification of titania with copper causes its coloration, i.e., color change from white into light green-blue for anatase-rich samples of fine particles (ST-01, FP6 and P25), light green for the fine rutile sample (TIO-6) and light grey for titania samples with large particles (ST-41, ST-G1 and TIO-5), as shown in the insets of Figure 2. The greyish color of large titania samples indicates the polydispersity of deposited copper NPs, similar to reported gold-modified titania samples (Although usually violet/pink samples are obtained, photodeposition of gold on large rutile results in greyish coloration because of high polydispersity of gold deposits, i.e., both size and shape of NPs, nanorods and irregularly-shaped deposits, and thus broad localized-surface plasmon resonance (LSPR) peak [44]). It has been proposed that noble metals are mainly deposited on surface defects [80], and since fine titania contains large content of defects (Table 1, ETs [77]), fine noble-metal NPs are formed and uniformly deposited on the titania surface [44]. In contrast, in the case of well-crystallized large titania samples, a low content of defects (Table 1, [77]) results in aggregation of noble-metal NPs, and thus larger NPs and with various shapes are finally deposited on the surface of large titania samples [44]. The green-blue color of Cu-modified titania has commonly been reported, indicating the possible presence of mixed-oxidation state of copper (Cu(II), Cu(I), Cu(0) and Cu_xO for x = 1–2) [72,81–83].

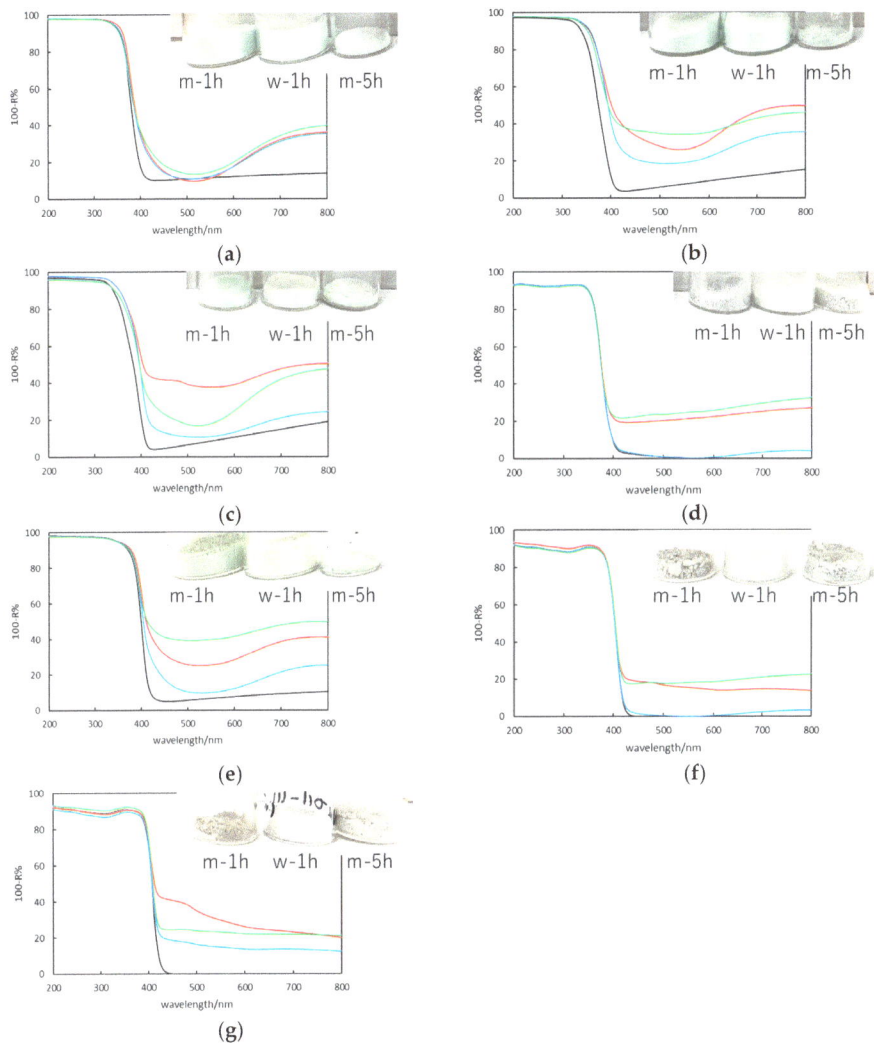

Figure 2. Diffuse reflectance (DR) spectra with respective photographs of copper-modified titania (with diffuse reflectance spectroscopy (DRS) for bare titania as reference): (**a**) ST-01, (**b**) FP6, (**c**) P25, (**d**) ST-41, (**e**) TIO-6, (**f**) ST-G1 and (**g**) TIO-5 (black: bare, red: m-1h, blue: w-1h, green: m-5h).

Indeed, the absorption spectra confirm the possibility of co-existence of different copper forms, as shown in Figure 2. The typical absorption of zero-valent Cu (due to LSPR), interband absorption of Cu_2O, and intrinsic exciton CuO and/or d-d transition of Cu^{2+} are expected at about 570, 500–600, and 600–800 nm, respectively [72,84,85]. Moreover, the clear absorption in the range of 400–500, especially for fine titania samples (Figure 2a–c,e) might be caused by the interfacial charge transfer (IFCT) from the valence band (VB) of TiO_2 to Cu_xO clusters [86]. Interestingly, very similar photoabsorption properties have been obtained for ST-01-based samples, independently on the preparation conditions (Figure 2a), i.e., with clear interfacial charge transfer (Cu_xO/TiO_2) between 400 and 500 nm and intrinsic exciton of CuO and/or d-d transition of Cu^{2+}, suggesting that oxidized forms of copper are predominant in these samples. Furthermore, the CuST-01w-1h sample exhibits the lowest absorption at 570 nm, and thus the lack of zero-valent copper in this sample is expected as a result of its preparation in the presence of oxygen (O_2 evolution). It has been reported that even in deaerated methanol solution, although zero-valent copper is formed and deposited on titania (violet coloration of suspension during photodeposition), copper is further oxidized easily during sample washing and drying, forming mixed-oxidation-state deposits, possibly the metallic core and oxidized shell [81,82]. Interestingly, the absorption spectra of Cu/FP6 samples indicate the possibility of co-presence of zero-valent copper because of stronger absorption at about 570 nm than that for a bare titania sample, especially for CuFP6m-5h. Unfortunately, no clear LSPR peak could be seen, probably because of peak damping caused by interband transition of CuO [49]. In contrast, in the case of Cu/P25 samples, shorter irradiation in reductive conditions (m) results in stronger photoabsorption at LSPR range, as shown in Figure 2c. In the case of large anatase (ST-41), samples prepared in methanol have almost the same photoabsorption properties until 600 nm, and prolonged irradiation results in stronger "absorption", which could be caused by light scattering on larger NPs (Figure 2d). It is expected that prolonged photodeposition might result in aggregation of copper deposits since it is carried out under continuous stirring allowing an easy contact between deposits (similar aggregation has been observed for gold-modified titania during long-term irradiation, especially for large titania samples (unpublished data)). Indeed, all samples prepared during 5 h irradiation (green lines) show a continuous increase in photoabsorption with an increase in wavelengths, indicating enhanced scattering on larger copper deposits, which is clearly seen especially for large titania samples. In the case of fine rutile samples (TIO-6), similar spectra to that by fine anatase samples have been obtained, indicating the co-presence of various forms of copper, predominantly in oxidized state (Figure 2e). In the case of large rutile samples (ST-G1 and TIO-5), the photoabsorption spectra correlate with those of large anatase, and no clear absorption peaks could be determined, well-corresponding with greyish color of samples (absorption at whole vis range). Interestingly, in the case of P25 and ST-G1 samples modified with copper during 1 h irradiation in methanol (red lines), clear peaks at about 470 nm could be seen, corresponding to the Cu_2O presence, as also confirmed by X-ray diffraction (XRD) and X-ray photoelectron spectroscopy (XPS) analysis (discussed further).

2.3. X-Ray Diffraction

The presence of copper has been confirmed in all samples by X-ray diffraction, as summarized in Table 2 and shown in Figure 3. In the case of fine titania particles, titania peaks are very broad and thus overlapping with copper ones. Accordingly, Rietveld refinement has been used for peaks' separation. It should be pointed out that titania peaks do not change after copper photodeposition (i.e., shift, broadening), confirming that surface modification does not influence the crystalline properties of titania. Although for ST-01 and P25 samples clear peaks of copper could be hardly seen, XRD patterns for other samples, especially those prepared in methanol, prove the copper presence (as zero-valent or Cu_2O), as shown in Figure 3b,d–g. As expected, usually larger content of crystalline cuprous oxide has been obtained in samples prepared during water oxidation than that during methanol dehydrogenation. Moreover, prolonger irradiation in methanol causes the preparation of samples with larger content of zero-valent copper, with the exception of very fine titania (FP6 and TIO-6), which could be caused by formation of very fine Cu NPs, and thus easily oxidizable (whole NPs) in air, whereas larger Cu NPs are only partly oxidized (surface as a shell) keeping zero-valent core. The larger content of deposited copper, and especially zero-valent copper during longer irradiation time might be caused by: (i) its more efficient reduction, and thus complete deposition of copper on titania, and (ii) aggregation of copper NPs, causing partial stabilization of zero-valent copper inside these aggregates. Moreover, the largest content of zero-valent copper has been obtained for large titania NPs, i.e., ST-41, ST-G1 and TIO-5 (and thus with large copper NPs) and for the finest titania (ST-01), which could confirm the stabilization of zero-valent copper either as a core of larger particles ($Cu@Cu_xO$) or densely surrounded by fine titania (similar to zero-valent copper placed in titania aerogel nanostructure [82]), respectively. Moreover, it has been confirmed (according to diffuse reflectance (DR) spectra) that larger crystals have been formed on larger titania NPs due to the lower content of surface defects, as already found for gold-modified titania samples [44].

Table 2. The crystalline properties of Cu-modified titania by XRD measurement.

Titania		Crystalline Content/%				Crystallite Size/nm			
		Anatase	Rutile	Cu	Cu$_2$O	Anatase	Rutile	Cu	Cu$_2$O
ST-01	m-1h	98.0	0	1.3	0.7	8.5	ND	8.6	8.6
	w-1h	99.7	0	0.01	0.3	8.5	ND	ND	0.7
	m-5h	96.9	0	2.4	0.7	8.3	ND	1.9	0.7
FP6	m-1h	92.0	7.0	0.2	0.8	12.5	12.0	12.3	9.7
	w-1h	90.9	7.0	1.7	0.5	12.6	11.1	11.3	15.6
	m-5h	93.9	5.8	0.1	0.2	12.3	14.7	9.3	9.2
P25	m-1h	81.6	16.7	0.4	1.4	23.2	38.8	2.4	2.9
	w-1h	82.9	16.1	0.1	0.9	22.9	41.2	10.2	11.2
	m-5h	83.6	15.5	0.8	0.15	22.9	44.6	10.1	4.6
ST-41	m-1h	95.7	3.3	0.5	0.5	80.8	12.0	40.3	34.9
	w-1h	98.3	0.9	0.4	0.4	78.3	57.3	14.9	17.0
	m-5h	96.8	1.6	1.0	0.7	81.1	36.9	68.2	32.6
TIO-6	m-1h	0	97.5	2.5	0	ND	18.8	8.2	ND
	w-1h	0	97.5	2.5	0	ND	19.3	4.2	ND
	m-5h	0.8	97.3	0.1	1.8	7.5	19.0	22.2	0.7
ST-G1	m-1h	0.9	97.6	0.2	1.3	20.7	87.1	46.3	71.7
	w-1h	1.2	98.6	0.02	0.1	46.7	93.8	ND	74.1
	m-5h	0.2	98.1	0.9	0.9	20.0	95.5	51.4	3.6
TIO-5	m-1h	8.9	90.0	0	1.1	152.1	168.6	5.2	42.7
	w-1h	4.3	94.2	0.1	1.5	147.1	157.2	58.3	13.3
	m-5h	8.9	89.2	0.8	1.1	169.5	197.6	36.8	66.6

It should be pointed out that although cupric oxide has not been detected, the oxidation of zero-valent copper and cuprous oxide is highly possible (considering DRS and XPS data, and previous studies). It is thought that CuO might predominantly exist as an amorphous form rather than a crystalline one since samples have not been calcined after preparation. In the case of large titania particles, clear Cu and Cu$_2$O peak could be observed in XRD patterns, as shown in Figure 3d,f,g. Interestingly, the clear photoabsorption peak of Cu$_2$O at about 470 nm for CuP25m-1h and CuTIO-5m-1h (red lines in Figure 2c,g) correlates well with its largest content in those samples.

2.4. Scanning Electron Microscopy

The morphology of samples has been investigated by scanning electron microscopy (SEM) observation. It has been confirmed that the properties of copper NPs depend on the properties of TiO$_2$ particles (Figure 4a–d), as suggested from DR spectra. Indeed, although Cu NPs on fine TiO$_2$ particles (ST-01, FP6, TIO-6) are hardly detected due to nano-sized nature (Figure 4a), relatively large Cu NPs are observed in the case of prolonged irradiation (CuTIO-6m-5h, Figure 4b), possibly due to the aggregation of Cu NPs during long-term stirring. In contrast, clear Cu deposits are seen on large TiO$_2$ particles (ST-41, ST-G1 and TIO-5), as shown for CuTIO-5m-1h and 5h sample in Figure 4c,d. Copper NPs have aggregated forming large particles (>100 nm) on the surface of large titania crystals (TIO-5). Moreover, although a large number of fine Cu NPs (of several nanometers) are clearly observed in CuTIO-5m-1h sample (in addition to large NPs), fine NPs have almost disappeared in the CuTIO-5m-5h sample, confirming that longer stirring during sample preparation results in NP aggregation. Energy-dispersive X-ray spectra of Cu-modified titania samples are shown (with corresponding SEM images) in Figure 4c,d for CuST-01m-5h and CuST-41m-5h samples, respectively. Although, CuSO$_4$ has been used as a copper source, sulfur has not been detected in analyzed samples, confirming efficient sample washing after copper photodeposition. The estimated content of copper on fine (ST-01) and large (ST-41) anatase samples prepared during 5 h irradiation in methanol differs slightly from expected value (2 wt%), reaching 1.22 wt% and 2.60 wt%, respectively.

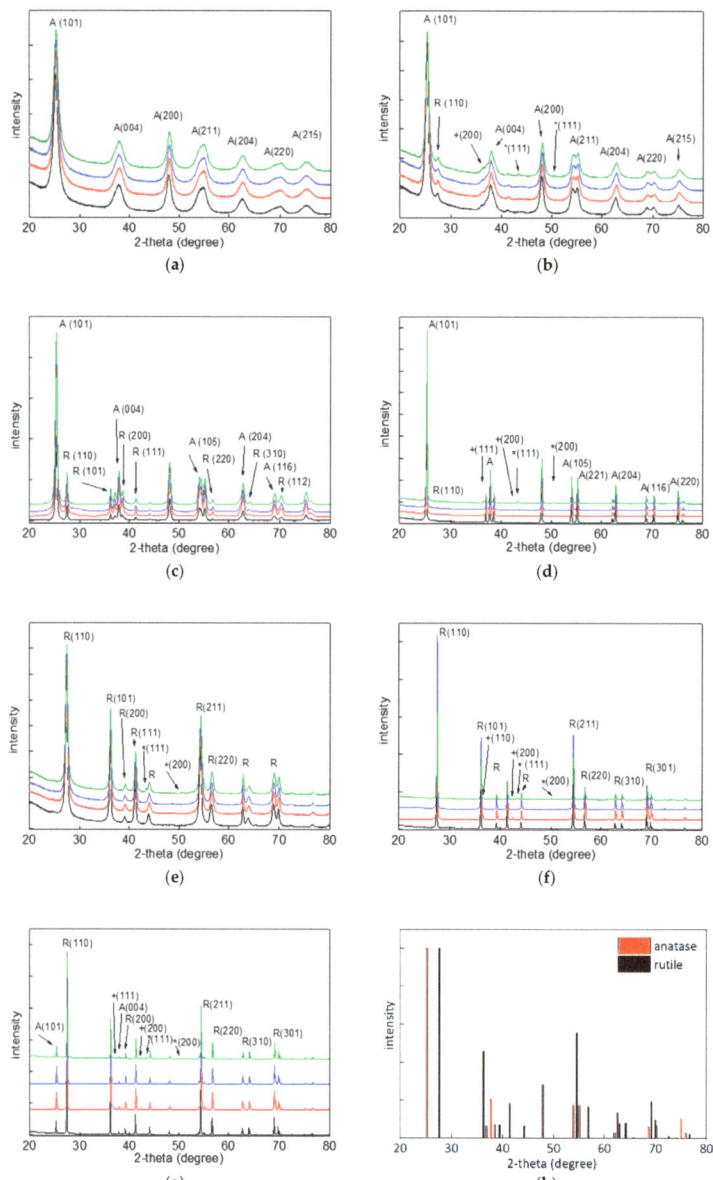

Figure 3. X-ray diffraction (XRD) patterns of: (a-g) bare and copper-modified TiO_2: (**a**) ST-01, (**b**) FP6, (**c**) P25, (**d**) ST-41, (**e**) TIO-6, (**f**) ST-G1 and (**g**) TIO-5 (black: bare, red: m-1h, blue: w-1h, green: m-5h); A and R indicate anatase and rutile, respectively; Symbols (* and +) indicate Cu and Cu_2O, respectively; (**h**) standard XRD patterns of anatase (red) and rutile (black).

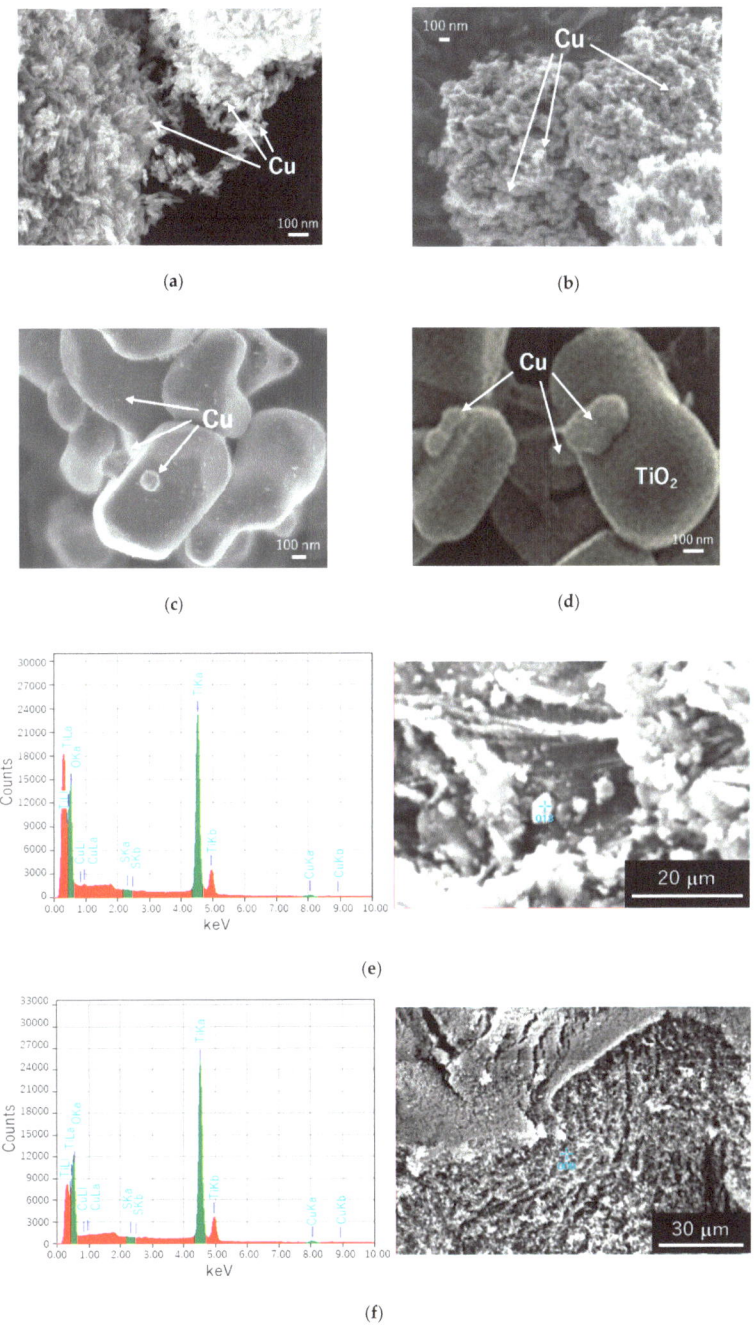

Figure 4. (**a–d**) Scanning electron microscopy (SEM) images of: (**a**) CuTIO-6m-1h, (**b**) CuTIO-6m-5h, (**c**) CuTIO-5m-1h and (**d**) CuTIO-5m-5h. (**e**,**f**) Energy-dispersive X-ray spectroscopy (EDS) and corresponding SEM images of (**e**) CuST-01m-5h and (**f**) CuST-41m-5h.

2.5. X-Ray Photoelectron Spectroscopy

For detailed analysis of surface composition of samples, including copper forms, XPS analysis has been performed, and summarized data are shown in Table 3. The oxygen to titanium ratio in all samples has exceeded that in chemical formula (2.0 for TiO_2), as typical for titania samples due to adsorbed water and carbon dioxide. Moreover, the content of reduced form of titanium (Ti^{3+}) has increased after copper deposition, as a result of reductive conditions during copper photodeposition, i.e., from 0.5% and 1.4% (bare titania) to 3.3–4.9% and 3.0–3.4% for P25 and TIO-5, respectively. Copper has been detected in all analyzed samples, and its content (Cu:Ti) exceeds that used for sample preparation (2 wt%, i.e., 0.025 of Cu(at):Ti(at)), which is reasonable considering that the surface of titania must be enriched with copper. Moreover, it has been confirmed that prolonged irradiation results in more efficient deposition of copper on titania (m-1h vs. m-5h). The Cu $3p_{3/2}$ peak could be divided to at least four oxidation forms, i.e., zero-valent copper, Cu_2O, CuO and $Cu(OH)_2$ with binding energies of about 930, 932, 933 and 934 eV, respectively. For P25 samples, XPS results agree well with DRS data (Figure 2c), i.e., the higher content of zero-valent copper for samples prepared in methanol (corresponding to more intense absorption at LSPR range) than prepared in water, and high content of cuprous oxide in CuP25m-1h sample (clear absorption peak at about 470 nm). Interestingly, samples prepared in water (both P25 and TIO-5) contain the smallest content of zero-valent copper, suggesting that reduction of copper during water oxidation might not been completed (Equation 4). Summarizing, XPS data confirm the co-existence of different forms of copper in the samples prepared by photodeposition methods.

Table 3. Surface composition of samples determined by X-ray photoelectron spectroscopy (XPS) analysis.

Samples		C 1s	O 1s	Ti $2p_{3/2}$	Cu $2p_{3/2}$	Cu/Ti	Ti $2p_{3/2}$ (%)		Cu $2p_{3/2}$ (%)			
							Ti^{3+}	Ti^{4+}	Cu	Cu_2O	CuO	$Cu(OH)_2$
P25	bare	47.1	39.7	13.2	-	-	0.5	99.5	-	-	-	-
	m-1h	39.9	39.8	18.9	1.4	0.075	3.6	96.4	7.9	79.2	12.4	0.5
	w-1h	49.2	35.2	14.7	0.9	0.059	3.3	96.7	1.5	28.3	69.6	0.6
	m-5h	47.8	36.1	14.7	1.5	0.099	4.9	95.1	18.8	70.0	8.3	2.9
TIO-5	bare	47.0	40.1	12.9	-	-	1.4	98.6	-	-	-	-
	m-1h	49.7	34.8	13.3	3.0	0.180	3.4	96.6	8.5	25.6	59.4	6.5
	w-1h	54.3	32.4	11.3	2.0	0.220	3.0	97.0	0.3	71.9	3.4	24.4
	m-5h	49.8	34.6	13.1	2.6	0.195	3.0	97.0	8.1	15.9	43.5	32.5

2.6. Photocatalytic Activity

The oxidative decomposition of acetic acid to CO_2 and H_2O on the surface of irradiated titania is a common reaction used for photocatalytic activity testing [44]. Although bare TiO_2 is also able to decompose acetic acid (black bars in Figure 5a), the reaction rate is limited due to charge carriers' recombination. Accordingly, noble metals in both zero-valent and oxide forms have been used to hinder this recombination via scavenging of photogenerated electrons and heterojunction between two oxides (p-n junction or/and Z-scheme mechanism), respectively [81,82,87–90]. Indeed, modification with copper enhances the photocatalytic activity, especially for anatase samples, by 2–3 times, as shown in Figure 5a. Interestingly, shorter irradiation during copper photodeposition results in better photocatalytic performance (red vs. green bars), due to the unwanted aggregation of copper deposits during prolonged irradiation, and thus worse distribution of copper on titania (fewer active sites for reaction).

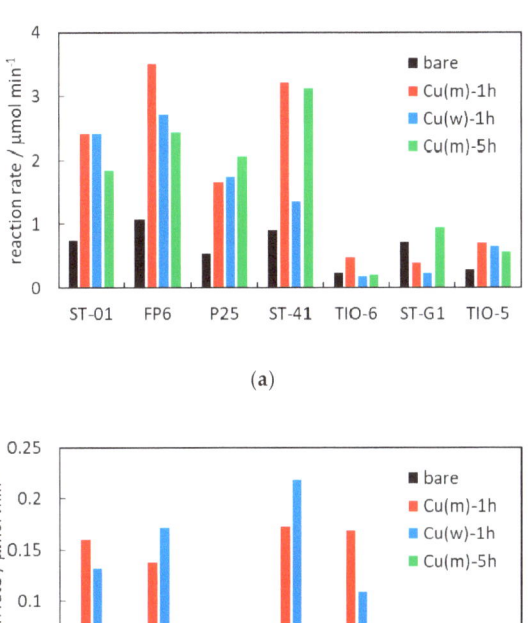

Figure 5. Comparison of photocatalytic efficiency for acetic acid decomposition under: (**a**) ultraviolet/visible (UV/vis) and (**b**) vis irradiation.

It should be pointed out that mainly oxidized forms of copper have been detected in present samples, and thus the activity enhancement by type II heterojunction or Z-scheme mechanism for Cu_xO and TiO_2 might be considered as predominant mechanism. Interestingly, the Cu_2O/TiO_2 photocatalysts, prepared by the grinding of commercial Cu_2O and different titania samples, have shown similar activity enhancement, i.e., 4–5 times for samples containing of anatase and only slight enhancement for those with rutile [87]. It has been proposed that due to more negative potential of the conduction band (CB) of rutile than anatase [79], type II heterojunction for rutile/Cu_2O and Z-scheme mechanism for anatase/Cu_2O are most probable [87]. The latter is preferential, resulting in the generation of charges with stronger redox potential (electrons in CB of Cu_2O and positive holes in VB of TiO_2). It has been proved that photocatalytic activity for the oxidation reactions depends directly on the oxidation potential of photogenerated holes, and anatase is a stronger oxidant than rutile, due to the more positive position of the VB [79]. Accordingly, lower activities of rutile might be simply explained by less positive potential of its VB. In this regard, it has been considered that the more negative potential of CB of rutile might cause a type II heterojunction with not so good redox properties (electrons in CB of titania and holes in VB of cuprous oxide), and thus not so high improvement of photocatalytic activity. It should be pointed that although the content of zero-valent copper is low, the function of copper as an electron scavenger should not be omitted since it is known that even a very low content of noble metals might cause significant enhancement of activity, e.g., 0.05 wt% of Au [90]. Therefore, participation of all forms of copper might be responsible for activity enhancement, e.g., by cascade mechanism or Z-scheme with co-electron scavenging by zero-valent copper, etc.

The photocatalytic activity under vis ($\lambda > 450$ nm) irradiation has also been evaluated, and obtained data are shown in Figure 5b. Although acetic acid decomposition under vis irradiation is much slower than that under UV (by about one order), the vis-activity of copper-modified titania has been proven. In contrast to UV activity, prolonged irradiation during samples' preparation (green bars) has resulted in a significant decrease in activity, especially for fine titania samples. Accordingly, it is proposed that aggregation of copper NPs is highly detrimental for fine titania samples. It should be reminded that titania absorbs only UV light, whereas under vis irradiation, only copper species (except FP6 titania with some defects) might absorb photons, and thus larger particles (aggregates) mean less species absorbing photons and lower interface between titania and copper, and thus lower probability for efficient charge carriers' transfer. Although, various mechanisms of oxidative decomposition of organic compounds on copper-modified titania under vis irradiation have already been proposed, including (i) plasmonic excitation (for zero-valent copper) [82], (ii) type II heterojunction with an electron transfer from the CB of copper oxides (Cu_2O/CuO) to CB of titania [87], and (iii) interfacial charge transfer with excitation of electrons from VB of titania into CB of Cu_xO [86], it is highly probable that those mechanisms might proceed simultaneously, depending on the photocatalyst's properties, i.e., especially the form of copper and its content.

It should be pointed out that the activity under vis irradiation for copper-modified titania differs significantly from that for gold-modified titania samples, where the highest activity has been obtained for large rutile samples (the same titania samples as used in this study) with aggregated gold deposits [44,91,92]. Additionally, other studies have also confirmed that for plasmonic photocatalysis, an increase in the polydispersity of noble metals result in higher activity as a result of efficient light harvesting (broad LSPR peak and strong field enhancement) [93,94]. Therefore, it might be postulated that here the vis activity of copper-modified titania is mainly caused by copper oxides, which correlates well with the activity of Cu_2O/TiO_2 photocatalysts (except ST-41 samples), prepared by grinding, i.e., Cu_2O/TIO-6 > Cu_2O/ST-01 > $Cu_2O/P25$ > Cu_2O/ST-41 > Cu_2O/large rutile [87]. The highest activity of copper-modified large anatase (ST-41) samples, especially that prepared in water, should be caused by the presence of CuO (Figure 2d), which has recently been reported as highly active under vis irradiation [95]. Moreover, only TIO-6 samples exhibit high activity among the rutile group. Therefore, in the case of acetic acid decomposition under vis irradiation, the presence of CuO (as confirmed in DRS spectra) is beneficial for the activity. Accordingly, it might be concluded that for vis activity the copper oxides might be more active than plasmonic photocatalysts. However, it should be pointed out that in the case of easily oxidizable metal (Cu), only slight content of zero-valent copper has been kept in the samples (after their surface oxidation in air), and thus the preparation of samples with stabilized copper should be performed, which is now under progress (e.g., Cu(core)/TiO_2(shell) photocatalysts), to conclude which forms of copper in Cu-modified TiO_2 are the most recommended.

2.7. Antibacterial Activity

The bactericidal activities against Gram-negative *Escherchia coli* of copper-modified titania photocatalysts have been evaluated under vis irradiation and in the dark (Figure 6). Although bare anatase samples do not exhibit high activity, bare rutile samples are shown to be active (Figure 6e,f), which might be explained by the narrower bandgap of rutile, and thus slight content of ETs inside bandgap (self-doped titania; Table 1) resulting in slight vis response. Recently, high antibacterial activity of samples containing rutile against various bacteria (e.g., *E. coli*, *Bacillus subtilis*, and *Staphylococcus aureus*) under fluorescent irradiation has been confirmed by Werapun and Pechwang [96]. Moreover, it has been reported that rutile particles (200 nm) induce hydrogen peroxide formation and oxidative DNA damage for human bronchial epithelial cells in the dark condition, whereas anatase particles (200 nm) do not [97]. Similarly, Vargas and Rodríguez-Páez concluded that rutile NPs possess higher antibacterial efficiency than anatase NPs in the dark [98].

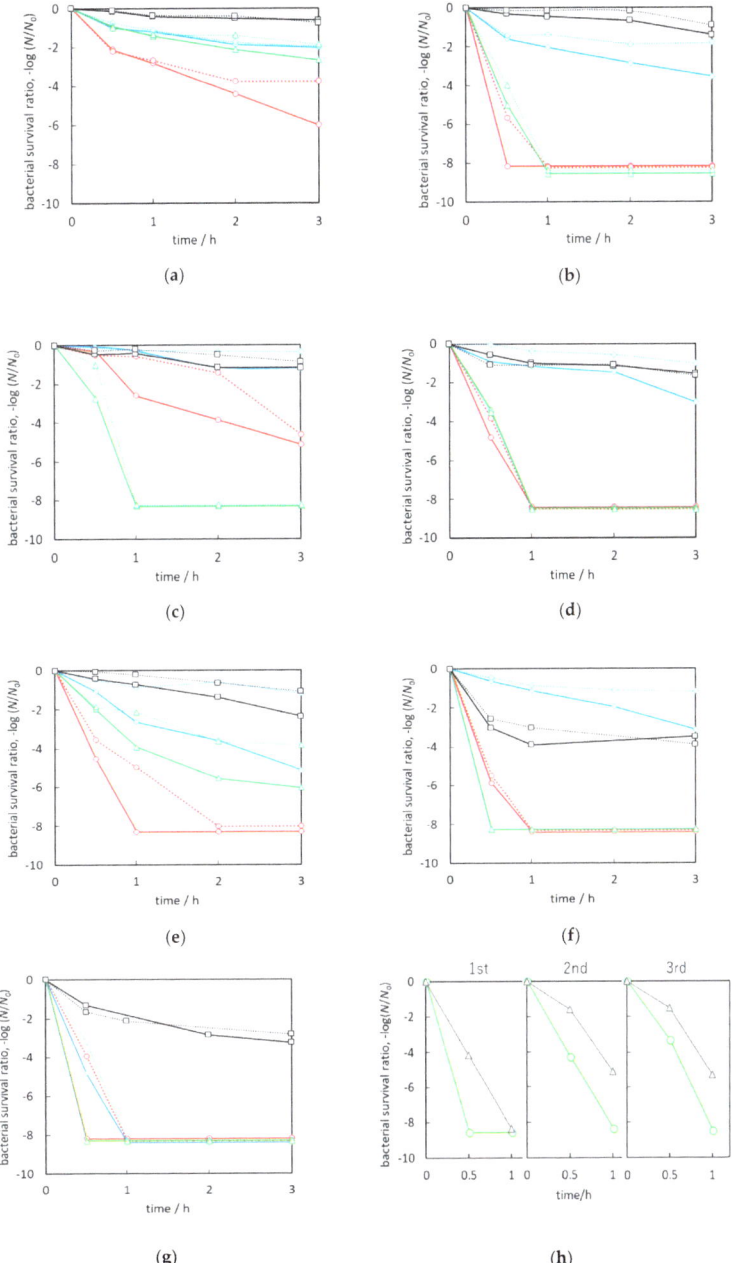

Figure 6. Bactericidal activities of bare and copper modified titania: (**a**) ST-01, (**b**) FP6, (**c**) P25, (**d**) ST-41, (**e**) TIO-6, (**f**) ST-G1 and (**g**) TIO-5 (black: bare, red: m-1h, blue: w-1h, green: m-5h) under visible light (solid line) and in the dark (dashed line); (**h**) the activity of CuTIO-5m-1h photocatalyst during three successive experiments under vis irradiation (green) and in the dark (grey).

It has been found that the properties of photocatalysts influence significantly antibacterial activity. For example, photocatalysts with fine particles (CuST-01 and CuTIO-6) show relatively lower activity than respective titania (anatase or rutile) but with larger particles, e.g., CuST-41 and CuTIO-5, respectively. The bactericidal activity of the CuP25m-5h sample is much higher than other P25-containing samples, possibly due to higher copper content (Cu/Ti ratio) on its surface (Table 3). Moreover, the comparison between large anatase and rutile indicates that rutile samples (CuST-G1 and CuTIO-5) exhibit higher activity than the anatase one (CuST-41). Across all samples, the samples formed in methanol possess much higher bactericidal activity than those in water, possible due to much larger content of zero-valent copper (Table 3). However, interestingly, CuTIO-5w-1h sample shows also high activity, comparable to those by samples prepared in methanol (CuTIO-5m-1h and CuTIO-5m-5h), probably due to the presence of Cu_2O (Figure 3g), known as the most active form of copper against bacteria. On the other hand, the light-induced activity is mainly observed in m-1h samples, but not in m-5h samples, probably due to aggregation of copper NPs, and a decrease in the interface between copper and bacteria. Importantly, high difference between the activities during decomposition of acetic acid and bacteria have been noticed, since the bactericidal activity is highly dependent on the oxidation state of copper, e.g., the optimal ratio proposed for Cu_xO clusters ($Cu_2O/CuO = 1.3$) is critical for the antimicrobial activity [72].

Interestingly, the samples prepared by grinding Cu_2O with titania show much higher activity enhancement during vis irradiation (even 4–7 orders in magnitude higher activity than that in the dark) [87], whereas the samples, prepared in this study by photodeposition of copper, exhibit only at most two orders of magnitude higher vis-activity than that in the dark (e.g., CuFP6m-1h during first 30 min; red lines in Figure 6b). Accordingly, it might be concluded that co-existence of other copper species, i.e., zero-valent copper and CuO, result in high dark activity of these samples.

Moreover, although the bactericidal activity of Cu_2O/ST-01 (ground sample) has been much higher than CuST-01 (prepared here), other titania samples (P25, ST-41 and TIO-6) exhibit similar activity under vis irradiation (depending on the synthesis conditions) [87], and thus the bactericidal activity of Cu-modified titania might be attributed not only to the initial content of Cu_2O (Table 2), but also the charge-transfer during the irradiation, including plasmonic photocatalysis [99].

The possible mechanism of antimicrobial action of copper-modified titania under vis irradiation has been explained by Wang et al. [100], where generated reactive oxygen species, such as O_2^-, $^{\bullet}OH$ or H_2O_2, have been proposed to cause an oxidative stress and cell disruption. Additionally, part of the copper might be released from the photocatalyst surface, and adsorb on the bacteria surface (or cross bacterial membrane). However, the toxic mechanism of copper has not been clarified yet. It has been proposed that the excess copper ions might interact with functional group of enzymes, required for defense against oxidative damage or other copper/zinc proteins (Cu ions compete with zinc for important binding sites). The generation of reactive hydroxyl radicals in a Fenton-type reaction has also been proposed as one of the possible mechanisms of bacteria inactivation [101].

The stability of the photocatalyst (CuTIO-5m-1h) has been investigated during three successive experiments under vis irradiation and in the dark, and obtained results are shown in Figure 6h. The activity decreases after the first experiment, especially in the dark, but almost the same performance has been observed during the second and third experiments. Accordingly, it might be concluded that even when initial activity has been decreased (probably due to the change in the surface composition of copper: $Cu/Cu_2O/CuO$ as a result of direct redox reaction with bacteria), the photocatalyst might be efficiently reused, maintaining a high performance, i.e., complete bacteria inactivation during 1 h vis irradiation.

Summing up, copper-modified titania photocatalysts have proven to be a promising candidate for preparation of antibacterial materials or active coatings under vis irradiation. In comparison with other Cu-modified titania prepared by different methods (e.g., sol-gel [102] and impregnation [72] method), $CuTiO_2$ samples inactivated bacteria much faster. Moreover, the dark activity of the samples was significant, indicating the possibility of efficient use in an indoor environment. Nevertheless, antibacterial activity against only one type of model bacteria is limited in scope, and thus more studies against different bacteria, especially antibiotic resistant (ABR) ones should be carried out.

2.8. Antifungal Activity

The filamentous or mould fungi are eukaryotic microorganisms that have a more complex morphology of cells and cell-walls structures than prokaryotic bacteria. Due to this difference, fungi are generally more resistant to the antimicrobial agents, drugs [103] and photocatalytic process [104]. That is caused by high chitin content (around 15% of cell wall dry mass) in cell walls, which protects fungi against mechanistic and oxidative damage. A relatively thick layer of polysaccharide (chitin) cover is also resistant to oxidative stress due to a glucan layer, mainly composed of β-1,3-glucan. Extracellular layer of fungal cell wall is formed by glycoproteins which are very sensitive to oxidative damage (around 15% of cell wall dry mass). Pigments, such as dark-brown melanin (in a fungal cell wall) protect them against some environmental factors e.g., UV radiation and ionizing radiation [105]. The importance of the cell wall for fungal cells and their spore survival during the photocatalytic process has been proven by Kühn et al. [106]. Additionally, in our previous study, it has been shown that antimicrobial activity depends on the type of noble metal, e.g., silver-modified titania activated under vis irradiation shows the highest bactericidal activity, whereas gold-modified titania are the most active against fungi [107]. Therefore, since the bactericidal materials do not necessarily inhibit the fungal growth, the fungistatic performance of these Cu-modified titania is now described. The recent study has shown that copper is micronutrient for fungal growth and proliferation, but might also generate toxicity and even cell-death [108].

The antifungal activity of anatase samples (CuST-01m-1h, CuFP6m-1h and CuST-41m-1h) was evaluated against *Aspergillus niger*, under both dark and vis irradiation conditions, by the agar disc diffusion method (Figure 7). No inhibition zones were observed and fungi were overgrown over all the plate's surface, excluding only the paper discs, Figure 7b–d. The only exception was for the control sample (0.9% NaCl solution) for which mold covered also paper discs (Figure 7a). In other words, Cu-modified titania in concentration of 1 g/L suppressed the fungal growth on the surface without obvious differences between vis and dark conditions. Therefore, slight fungistactic activity of Cu-modified titania was not caused by photocatalytic activity but rather by the antifungal properties of copper. Moreover, no clear differences among copper-modified titania samples could be noticed. In addition, as reported previously [107], the lack of inhibition zones (around paper discs) indicate the stability of metallic deposits on the titania surface (no leakage).

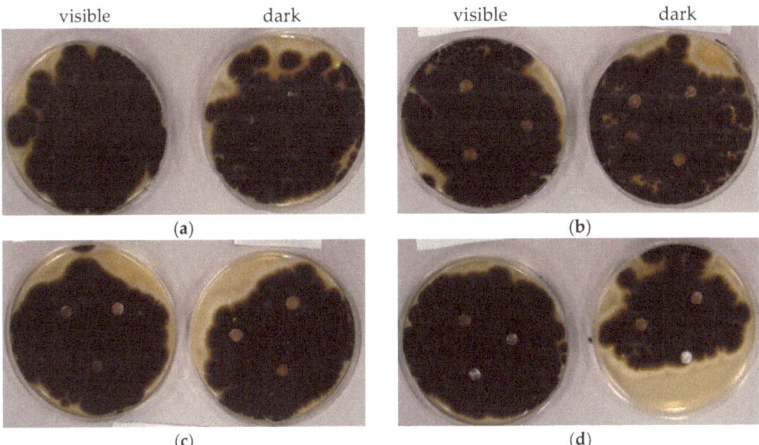

Figure 7. The photographs of *Aspergillus niger* colony on Malt Extract Agar (MEA) plates: (**a**) 0.9% NaCl (control), (**b**) CuST-01m-1h, (**c**) CuFP6m-1h, (**d**) CuST-41m-1h, incubated under vis (left) and in the dark (right).

Chen et al. showed that growth of *A. niger* on wood surfaces coated with TiO_2 film was completely inhibited after 20 days under UV light of 365 nm wavelength [109]. Despite such a long time of exposure, fungal spores have remined viable and their re-growth has been observed when irradiation was stopped. Taking into account the results obtained in this study, it has been concluded that vis irradiation is insufficient to activate copper-modified TiO_2 during the first period of fungal growth. A previous study, upon nitrogen-doped TiO_2 activated by indoor light, demonstrated a significant effect in the first 48 h of incubation, when the germination phase of spores takes place [110], and this is followed by the elongation of white (i.e., melanin-free) hyphae [111]. Although significant differences in the initiation of melanin production exist among filamentous fungi, for *Aspergillus* species, this is characterized by darkening of the hyphae and formation of mature conidiophores. Melanin is a powerful free-radical scavenger, acting as a sponge for intracellular free radicals generated when environmental factors place the cell under oxidative stress [112]. Thus, there seems little rationale for prolongation of irradiation during these vis irradiation tests. The real challenge is to find a photocatalyst, which is suitable to stop production of melanin within 24 h by as many fungal species as achievable. For further analysis, the other method should be also applied, for example sporulation tests (next paragraph). The mechanism deserves further investigation, because the effects might help to optimize the methodology for antifungal susceptibility testing and developing photocatalytic self-cleaning building materials or coatings, air cleaners, and filters used to remove the microorganism from indoors aerosol and surfaces, especially where UV-A exposure is not possible (under visible light).

The fungistatic activity against *A. niger* has also been investigated by the spore-counting method under fluorescence light and in the dark (Figure 8). The CuFP6m-1h sample significantly suppresses the generation of spores under both conditions, but the mycelium has been still observed (Figure 8b), indicating that only sporulation is inhibited. A pristine FP6 sample also exhibits slight inhibition of sporulation, since fluorescence light contains a small portion of UV. It should be pointed out that the number of spores in the dark for both samples is smaller than that under irradiation, possibly due to the stimulation of fungal growth and sporulation by light [110,113–115]. Therefore, it is proposed that fungal growth might exceed the photocatalytic fungistatic effect [88]. Moreover, some proteins and sugars on fungal cells can be easily adsorbed on titania (independent on the irradiation) [116–119].

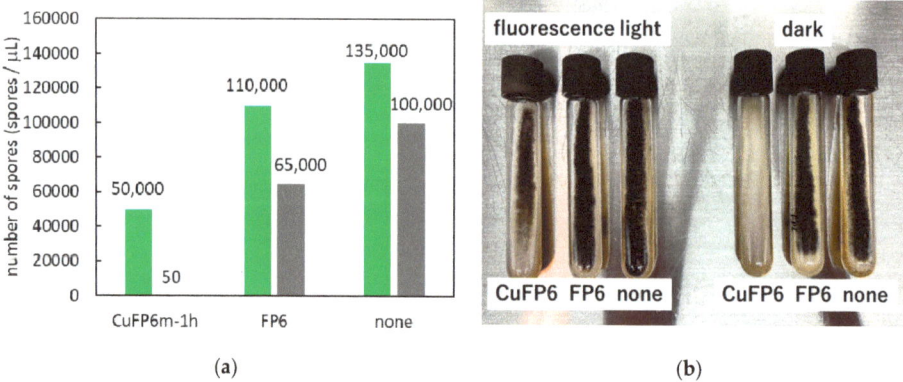

Figure 8. Sporulation tests: (**a**) number of spores after 3-day growth of *A. niger* under florescence light (green bars) and in the dark (grey bars); and (**b**) representative photographs of slants for CuFP6m-1h, FP6 and in the absence of photocatalyst.

Additionally, the presence of melanin in the spore protects the fungal cells against ROS [112] and light. The increased inhibition of sporulation and melanization observed for CuFP6m-1h can be explained by the possibility of increased copper uptake through the action of one of the family of methionine-rich copper transport proteins. These transmembrane proteins are expected to be mostly similar to those already characterized for *A. fumigatus* and *A. nidulans* [108]. Moreover, as mentioned above, mycelium growth is not inhibited and this is in accordance with a previous study which showed *A. niger* to be an efficient organism for bioremediation of copper-rich waste streams [120], where removal of copper was accompanied by a growth in mycelium but without growth of conidia or spores.

3. Materials and Methods

3.1. Preparation of Cu-Modified Titania

Seven commercial titania samples, i.e., ST-01 (Ishihara Shangyo Kaisha, Ltd., Osaka, Japan), FP6 (Showa Denko Ceramics Co. Ltd., Tokyo, Japan), P25 (AEROXIDE® TiO_2 P25 produced by Nippon Aerosil Co., Ltd., Tokyo, Japan), ST-41 (Ishihara Shangyo Kaisha, Ltd., Osaka, Japan), TIO-6 (Catalysis Society of Japan, Tokyo, Japan), ST-G1 (Showa Denko Ceramics Co. Ltd., Tokyo, Japan) and TIO-5 (Catalysis Society of Japan, Tokyo, Japan), were used without pretreatment for the surface modification with copper (Table 1). We mixed 600 mg of TiO_2 with 0.94 mL of 0.2 M $CuSO_4 \cdot 5H_2O$ (99.9%, FUJIFILM Wako Pure Chemical Corporation, Osaka, Japan) aqueous solution in a 50 mL Pyrex-glass tube containing of 26.3 mL of 50 vol% methanol-water (Milli-Q) solution. Samples were also prepared in the absence of alcohol, i.e., in 26.3 mL of Milli-Q water. The suspension was deaerated with argon pre-bubbling, and the tube was tightly closed with a rubber septum and wrapped in parafilm. The samples were irradiated under ultraviolet light, using home-built high-pressure Hg lamp ($\lambda > 220$ nm, output power of 400 W, light intensity of about 20 mW cm^{-2}) in a thermostatic water bath at 25 °C under continuous stirring (500 rpm) for 1 or 5 h. Before and during irradiation, 0.2 mL of the gas sample was taken from the tube with a syringe and the amount of evolved H_2 or O_2 (in the case of methanol/water and water, respectively) was detected by gas chromatography with a thermal conductivity detector (GC-TCD; GC-8A, SHIMADZU CORPORATION, Kyoto, Japan). After irradiation, the samples were collected by centrifugation (3000 rpm), washed twice by resuspending in the methanol (followed by centrifugation), then washed twice by resuspending in water (followed by centrifugation), and finally dried overnight at 120 °C. Then, samples were ground in an agate mortar. The code names of samples indicate the titania type (ST-01, FP6, P25, ST-41, TIO-6, ST-G1 and TIO-5), environment (water (w) or methanol/water (m)) and duration (1 h or 5 h) of irradiation, e.g., the CuST-01m-1h sample was prepared by deposition of copper on ST-01 titania in a methanol/water solution during 1 h irradiation.

3.2. Characterization

Photoabsorption properties of samples were analyzed by diffuse reflectance spectroscopy (DRS; JASCO V-670 equipped with a PIN-757 integrating sphere, JASCO, LTD., Pfungstadt, Germany). Barium sulfate was used as a reference for DRS analysis. For the characterization of crystal structure, X-ray powder diffraction (XRD; Rigaku intelligent XRD SmartLab with a Cu target, Rigaku, LTD., Tokyo, Japan) was operated (accelerating voltage: 40kV, emission current: 30 mA). The samples were investigated between 10° and 90° at 1°/min scan speed and scan step of 0.008°. The crystal structures were analyzed with Rigaku PDXL software (Version 2.6.1.2, Rigaku, LTD., Tokyo, Japan, 2007–2015). Crystallite sizes of anatase, rutile, Cu and Cu_2O were estimated by the Scherrer equation, using the Joint Committee on Powder Diffraction Standards (JCPDS) card numbers of 5757, 7818, 31057, 8728, respectively. The surface properties of samples and the oxidation states of elements were analyzed by X-ray photoelectron spectroscopy (XPS; JPC-9010MC, JEOL, Tokyo, Japan). At first, samples were attached to carbon tape on a sample holder and dried under a vacuum. The Mg X-ray source, 10 V accelerating voltage and 10 mA emission current under high vacuum were set as operation conditions. The binding energies were examined between 1000 eV and 0 eV and for specific ranges depending on the binding energies of elements. The data were evaluated using JEOL SpecSurf Software (Version 1.7.3.9, JEOL, Tokyo, Japan, 2000). All graphs were charge-corrected with a carbon 1s peak to 284.7 eV. The morphology of the samples was analyzed by field emission-scanning electron microscopy (FE-SEM; JSM-7400F, JEOL, Tokyo, Japan) under a high vacuum. The samples were spread on carbon paste and dried under a vacuum overnight. Images were acquired in a wide range of magnifications in secondary electron imaging mode (SEI). Samples were also analyzed by energy-dispersive X-ray spectroscopy (EDS, JSM-6360 LA Analytical SEM, JEOL, Tokyo, Japan) under high vacuum to investigate the chemical composition of copper-modified titania.

3.3. Photocatalytic Activity Tests

The photocatalytic activity of samples was investigated for oxidative decomposition of acetic acid under UV/vis and vis irradiation. Fifty milligrams of sample were dispersed in 5 mL of 5 vol% acetic acid in a Pyrex-glass test tube. The suspension was irradiated with a home-made Hg lamp ($\lambda > 220$ nm) or Xe lamp (output power 450 W, light intensity about 20 mW cm^{-2}, $\lambda > 450$ nm; water IR filter, cold mirror and cut-off filter Y48) in a thermostat water bath at 25 °C under continuous stirring. The generated carbon dioxide was measured by GC-TCD or gas chromatography with flame ionization detector (GC-FID; SHIMADZU CORPORATION, Kyoto, Japan), respectively.

3.4. Bactericidal Tests

The bactericidal activity was evaluated under vis irradiation ($\lambda > 450$ nm) using *Escherichia coli* K12 (ATCC 29425, ATCC, Manassas, VA, USA), as described elsewhere [107]. In brief, a 50 mg of sample was suspended in 5 mL of *E. coli* suspension (about $1\sim5 \times 10^8$ cells/mL) in a Pyrex-glass test tube. The suspension was irradiated with Xe lamp ($\lambda > 450$ nm) or kept in the dark (control experiment) under continuous stirring at 25 °C. During irradiation (i.e., at 0.5, 1, 2 and 3 h), a portion of suspension was withdrawn, diluted and inoculated on the Plate Count Agar (Becton, Dickinson and Company, Franklin Lakes, NJ, USA) medium. Agar plates were incubated at 37 °C overnight, and then the bacterial colonies were counted to calculate the colony-forming unit/mL (CFU/mL). The bacterial survival ratio was computed, and presented as $-\log (N_t/N_0)$, where N_0 is the initial number of bacteria, and N_t is the number of bacteria after time t. For the reusability test, the titania sample (CuTIO-5m-1h) was collected by leaving it to stand, washing it with sterile Milli-Q water, and drying it at 120 °C. Then, the same experiment was repeated twice.

3.5. Antifungal Tests

3.5.1. Disc Diffusion Test

The antifungal tests were performed with mould fungus, *Aspergillus niger*, isolated from damp basement air (Zachodniopomorski Uniwersytet Technologiczny (ZUT) collection, Szczecin, Poland). The antifungal activity of the photocatalysts was tested using the disc diffusion method [121]. Culture plates were prepared with 20 mL of Malt Extract Agar (MEA; Merck, Darmstadt, Germany). Sterilized culture media were poured into Petri dishes and, after solidification of the medium, 0.25 mL of fungal suspension was spread on the plate using a spreader. The concentration of microorganisms was about 10^6 CFU/mL. The sterile paper discs (Whatman No.1, diameter 5 mm) impregnated with 10 µL/disc photocatalyst suspension in concentration of 1 g/L (10 µL/disc) were placed at different locations on the culture plates. The control discs were impregnated with saline solution (0.9% NaCl). The plates were incubated both in the dark and under indoor fluorescence light (intensity of about 120 lx) for 72 h. The temperature was maintained at about 25 °C.

3.5.2. Spore-Counting Test

CuFP6m-1h and bare FP6 samples (10 g/L) in MEA were autoclaved and the agar slants were prepared. Spores of *A. niger* (1000 cells/µL) were suspended in 8.5 g/L NaCl aqueous solution, 20 µL of spore suspension was inoculated on the agar slants and irradiated with fluorescence light (intensity of about 120 lx) or kept in the dark for 3 days (about 23 °C). Generated spores were collected with 0.05% Triton X-100 in NaCl solution by vortex mixing and counted under an optical microscope.

4. Conclusions

Surface modification of titania with copper results in the preparation of highly active photocatalysts under both UV and vis irradiation, due to the inhibition of charge-carrier recombination and vis absorption by copper species (Cu, Cu_2O, CuO and Cu_xO), respectively. The conditions during samples' preparation, i.e., time and medium (reaction system) and titania properties (size and crystalline composition) have governed the properties of formed Cu NPs, such as size, composition and oxidation state, and thus resultant activities. Since copper is easily oxidizable, it is thought that even though zero-valent Cu NPs are formed during methanol dehydrogenation, these NPs are easily oxidized in air, and thus co-mixed copper deposits have been obtained in all samples. The aggregation of copper during prolonged deposition results in the formation of copper deposits with a larger content of zero-valent copper (as a core). In the case of copper photodeposition during water oxidation, the least content of zero-valent copper was obtained, probably due to insufficient copper cations' reduction. It should be pointed out that even though the same method of copper deposition was used, as well as the same copper source and its content, a difference in titania kind resulted in preparation of samples with significantly different properties, and thus overall activities, e.g., one order for acetic acid decomposition and few orders for bacteria inactivation.

The modification of titania with copper causes a significant increase in the activity under UV irradiation, especially for anatase samples, suggesting the enhancement according to a Z-scheme mechanism between two types of semiconductor, i.e., copper oxides and anatase. Copper-modified samples have also been active under vis irradiation, but those activities do not correlate with the content of zero-valent copper, and thus it is thought that the p-n junction between two oxides (with electron transfer from CB of Cu_2O to CB of TiO_2) rather than plasmonic photocatalysis is the main driving force of vis response. Moreover, modified samples exhibit antimicrobial activity under both vis irradiation and in the dark. Although vis-irradiation has increased activity only slightly (predominant activity in the dark resulting from intrinsic properties of Cu, mainly Cu_2O), the overall antibacterial effect is very high, causing complete inactivation of bacteria within 0.5–1 h for rutile-containing samples. Unfortunately, the mixed-oxidation state of copper in all samples makes it difficult to establish the key-factor properties for high photocatalytic activity. Despite that, it is believed that copper-modified titania are promising photocatalysts for broad environmental application because of low price, availability and high activity for decomposition of both organic compounds and microorganisms.

Author Contributions: Conceptualization, E.K., writing—original draft preparation, M.E.-K. and B.K., writing-revision and editing, M.E.-K., K.W., Z.W., A.M.-S., B.O. and E.K., sample preparation, B.K. and K.W., sample characterization and photoactivity tests, M.E.-K., K.W. and B.K., microbiological tests, M.E.-K., B.K. and A.M.-S., data analysis, M.E.-K., K.W., Z.W., A.M.-S. and E.K., supervision, E.K., funding acquisition, E.K. and B.O. All authors have read and agreed to the published version of the manuscript.

Funding: This work was financially supported by a Grand Challenges Explorations Grant (GCE RB, OPP1060234) from Bill and Melinda Gates Foundation, and "Yugo-Sohatsu Kenkyu" for an Integrated Research Consortium on Chemical Sciences (IRCCS) project from the Ministry of Education and Culture, Sport, Science and Technology-Japan (MEXT). The APC was funded by Z.W., A.M.-S. and E.K.

Conflicts of Interest: The authors declare no conflict of interest.

References

1. Hoffmann, M.R.; Martin, S.T.; Choi, W.; Bahnemann, D.W. Environmental applications of photocatalysis. *Chem. Rev.* **1995**, *95*, 69–96. [CrossRef]
2. Wang, Y.; Huang, Y.; Ho, W.; Zhang, L.; Zou, Z.; Lee, S. Biomolecule-controlled hydrothermal synthesis of C-N-S-tridoped TiO_2 nanocrystalline photocatalysts for NO removal under simulated solar light irradiation. *J. Hazard. Mater.* **2009**, *169*, 77–87. [CrossRef] [PubMed]

3. Su, C.; Tseng, C.M.; Chen, L.F.; You, B.H.; Hsu, B.C.; Chen, S.S. Sol-hydrothermal preparation and photocatalysis of titanium dioxide. *Thin Solid Films* **2006**, *498*, 259–265. [CrossRef]
4. Martin, S.T.; Lee, A.T.; Hoffmann, M.R. Chemical mechanism of inorganic oxidants in the TiO_2/UV process: Increased rates of degradation of chlorinated hydrocarbons. *Environ. Sci. Technol.* **1995**, *29*, 2567–2573. [CrossRef]
5. Verbruggen, S.W. TiO_2 photocatalysis for the degradation of pollutants in gas phase: From morphological design to plasmonic enhancement. *J. Photochem. Photobiol. C Photochem. Rev.* **2015**, *24*, 64–82. [CrossRef]
6. Fujishima, A.; Honda, K. Electrochemical photolysis of water at a semiconductor electrode. *Nature* **1972**, *238*, 37–38. [CrossRef]
7. Fujishima, A.; Rao, T.N.; Tryk, D.A. Titanium dioxide photocatalysis. *J. Photochem. Photobiol. C Photochem. Rev.* **2000**, *1*, 1–21. [CrossRef]
8. Byrne, C.; Subramanian, G.; Pillai, S.C. Recent advances in photocatalysis for environmental applications. *J. Environ. Chem. Eng.* **2018**, *6*, 3531–3555. [CrossRef]
9. Ohtani, B.; Prieto-Mahaney, O.O.; Li, D.; Abe, R. What is degussa (Evonic) P25? Crystalline composition analysis, reconstruction from isolated pure particles and photocatalytic activity test. *J. Photochem. Photobiol. A Chem.* **2010**, *216*, 179–182. [CrossRef]
10. Crampton, A.S.; Cai, L.; Janvelyan, N.; Zheng, X.; Friend, C.M. Methanol photo-oxidation on rutile TiO_2 nanowires: Probing reaction pathways on complex materials. *J. Phys. Chem. C* **2017**, *121*, 9910–9919. [CrossRef]
11. Liu, Y.; Chen, L.; Hu, J.; Li, J.; Richards, R. TiO_2 nanoflakes modified with gold nanoparticles as photocatalysts with high activity and durability under near UV irradiation. *J. Phys. Chem. C* **2010**, *114*, 1641–1645. [CrossRef]
12. Anitha, V.C.; Hamnabard, N.; Banerjee, A.N.; Dillip, G.R.; Joo, S.W. Enhanced electrochemical performance of morphology-controlled titania-reduced graphene oxide nanostructures fabricated via a combined anodization-hydrothermal process. *RSC Adv.* **2016**, *6*, 12571–12583. [CrossRef]
13. Huang, X.; Meng, L.; Du, M.; Li, Y. TiO_2 nanorods: Hydrothermal fabrication and photocatalytic activities. *J. Mater. Sci. Mater. Electron.* **2016**, *27*, 7222–7226. [CrossRef]
14. Bian, Z.; Tachikawa, T.; Zhang, P.; Fujitsuka, M.; Majima, T. Au/TiO_2 superstructure-based plasmonic photocatalysts exhibiting efficient charge separation and unprecedented activity. *J. Am. Chem. Soc.* **2014**, *136*, 458–465. [CrossRef] [PubMed]
15. Abbas, W.A.; Ramadan, M.; Faid, A.Y.; Abdellah, A.M.; Ouf, A.; Moustafa, N.; Allam, N.K. Photoactive catalysts for effective water microbial purification: Morphology-activity relationship. *Environ. Nanotechnol. Monit. Manag.* **2018**, *10*, 87–93. [CrossRef]
16. Stafford, U.; Gray, K.A.; Kamat, P.V.; Varma, A. An in situ diffuse reflectance FTIR investigation of photocatalytic degradation of 4-chlorophenol on a TiO_2 powder surface. *Chem. Phys. Lett.* **1993**, *205*, 55–61. [CrossRef]
17. Smirnova, N.; Fesenko, T.; Zhukovsky, M.; Goworek, J.; Eremenko, A. Photodegradation of stearic acid adsorbed on superhydrophilic TiO_2 surface: In situ FT-IR and LDI study. *Nanoscale Res. Lett.* **2015**, *10*, 1–7. [CrossRef]
18. Mamaghani, A.H.; Haghighat, F.; Lee, C.S. Role of titanium dioxide (TiO_2) structural design/morphology in photocatalytic air purification. *Appl. Catal. B Environ.* **2020**, *269*, 118735. [CrossRef]
19. Ghosh, M.; Lohrasbi, M.; Chuang, S.S.C.; Jana, S.C. Mesoporous titanium dioxide nanofibers with a significantly enhanced photocatalytic activity. *ChemCatChem* **2016**, *8*, 2525–2535. [CrossRef]
20. Al-Mamun, M.R.; Kader, S.; Islam, M.S.; Khan, M.Z.H. Photocatalytic activity improvement and application of UV-TiO_2 photocatalysis in textile wastewater treatment: A review. *J. Environ. Chem. Eng.* **2019**, *7*, 103248. [CrossRef]
21. Ohtani, B. Preparing articles on photocatalysis—Beyond the illusions, misconceptions, and speculation. *Chem. Lett.* **2008**, *37*, 217–229. [CrossRef]
22. Anpo, M.; Takeuchi, M. The design and development of highly reactive titanium oxide photocatalysts operating under visible light irradiation. *J. Catal.* **2003**, *216*, 505–516. [CrossRef]
23. Karvinen, S.; Hirva, P.; Pakkanen, T.A. Ab initio quantum chemical studies of cluster models for doped anatase and rutile TiO_2. *J. Mol. Struct. Theochem.* **2003**, *626*, 271–277. [CrossRef]

24. Zhu, J.; Deng, Z.; Chen, F.; Zhang, J.; Chen, H.; Anpo, M.; Huang, J.; Zhang, L. Hydrothermal doping method for preparation of Cr^{3+}-TiO_2 photocatalysts with concentration gradient distribution of Cr^{3+}. *Appl. Catal. B Environ.* **2006**, *62*, 329–335. [CrossRef]
25. Tada, H.; Mitsui, T.; Kiyonaga, T.; Akita, T.; Tanaka, K. All-solid-state Z-scheme in CdS-Au-TiO_2 three-component nanojunction system. *Nat. Mater.* **2006**, *5*, 782–786. [CrossRef]
26. Ghosh, M.; Liu, J.; Chuang, S.S.C.; Jana, S.C. Fabrication of hierarchical V_2O_5 nanorods on TiO_2 nanofibers and their enhanced photocatalytic activity under visible light. *ChemCatChem* **2018**, *10*, 3305–3318. [CrossRef]
27. Yang, J.; Zhang, J.; Zou, B.; Zhang, H.; Wang, J.; Schubert, U.; Rui, Y. Black SnO_2-TiO_2 Nanocomposites with high dispersion for photocatalytic and photovoltaic applications. *ACS Appl. Nano Mater.* **2020**, *3*, 4265–4273. [CrossRef]
28. Zhang, L.; Yu, W.; Han, C.; Guo, J.; Zhang, Q.; Xie, H.; Shao, Q.; Sun, Z.; Guo, Z. Large scaled synthesis of heterostructured electrospun TiO_2/SnO_2 nanofibers with an enhanced photocatalytic activity. *J. Electrochem. Soc.* **2017**, *164*, H651–H656. [CrossRef]
29. Li, J.J.; Weng, B.; Cai, S.C.; Chen, J.; Jia, H.P.; Xu, Y.J. Efficient promotion of charge transfer and separation in hydrogenated TiO_2/WO_3 with rich surface-oxygen-vacancies for photodecomposition of gaseous toluene. *J. Hazard. Mater.* **2018**, *342*, 661–669. [CrossRef]
30. Chen, X.; Liu, L.; Yu, P.Y.; Mao, S.S. Increasing solar absorption for photocatalysis with black hydrogenated titanium dioxide nanocrystals. *Science* **2011**, *331*, 746–750. [CrossRef]
31. Kraeutler, B.; Bard, A.J. Heterogeneous photocatalytic preparation of supported catalysts. Photodeposition of platinum on TiO_2 powder and other substrates. *J. Am. Chem. Soc.* **1978**, *100*, 4317–4318. [CrossRef]
32. Sclafani, A.; Mozzanega, M.-N.; Pichat, P. Effect of silver deposits on the photocatalytic activity of titanium dioxide samples for the dehydrogenation or oxidation of 2-propanol. *J. Photochem. Photobiol. A Chem.* **1991**, *59*, 181–189. [CrossRef]
33. Disdier, J.; Herrmann, J.-M.; Pichat, P. Platinum/titanium dioxide catalysts. *J. Chem. Soc. Faraday Trans.* **1983**, *79*, 651–660. [CrossRef]
34. Kamat, P.V. Photochemistry on nonreactive and reactive (semiconductor) surfaces. *Chem. Rev.* **1993**, *93*, 267–300. [CrossRef]
35. Wang, C.Y.; Bahnemann, D.W.; Dohrmann, J.K. A novel preparation of iron-doped TiO_2 nanoparticles with enhanced photocatalytic activity. *Chem. Commun.* **2000**, 1539–1540. [CrossRef]
36. Zakeeruddin, S.M.; Nazeeruddin, M.K.; Humphry-Baker, R.; Péchy, P.; Quagliotto, P.; Barolo, C.; Viscardi, G.; Grätzel, M. Design, synthesis, and application of amphiphilic ruthenium polypyridyl photosensitizers in solar cells based on nanocrystalline TiO_2 films. *Langmuir* **2002**, *18*, 952–954. [CrossRef]
37. Bae, E.; Choi, W. Effect of the anchoring group (carboxylate vs phosphonate) in Ru-complex-sensitized TiO_2 on hydrogen production under visible light. *J. Phys. Chem. B* **2006**, *110*, 14792–14799. [CrossRef]
38. Park, H.; Choi, W. Effects of TiO_2 surface fluorination on photocatalytic reactions and photoelectrochemical behaviors. *J. Phys. Chem. B* **2004**, *108*, 4086–4093. [CrossRef]
39. Di Paola, A.; Marcì, G.; Palmisano, L.; Schiavello, M.; Uosaki, K.; Ikeda, S.; Ohtani, B. Preparation of polycrystalline TiO_2 photocatalysts impregnated with various transition metal ions: Characterization and photocatalytic activity for the degradation of 4-nitrophenol. *J. Phys. Chem. B* **2002**, *106*, 637–645. [CrossRef]
40. Reszczyńska, J.; Grzyb, T.; Wei, Z.; Klein, M.; Kowalska, E.; Ohtani, B.; Zaleska-Medynska, A. Photocatalytic activity and luminescence properties of RE^{3+}-TiO_2 nanocrystals prepared by sol-gel and hydrothermal methods. *Appl. Catal. B Environ.* **2016**, *181*, 825–837. [CrossRef]
41. Kusiak-Nejman, E.; Janus, M.; Grzmil, B.; Morawski, A.W. Methylene Blue decomposition under visible light irradiation in the presence of carbon-modified TiO_2 photocatalysts. *J. Photochem. Photobiol. A Chem.* **2011**, *226*, 68–72. [CrossRef]
42. Markowska-Szczupak, A.; Rokicka, P.; Wang, K.; Endo, M.; Morawski, A.; Kowalska, E. Photocatalytic water disinfection under solar irradiation by d-glucose-modified titania. *Catalysts* **2018**, *8*, 316. [CrossRef]
43. Wang, K.; Endo-Kimura, M.; Belchi, R.; Zhang, D.; Habert, A.; Bouclé, J.; Ohtani, B.; Kowalska, E.; Herlin-Boime, N. Carbon/graphene-modified titania with enhanced photocatalytic activity under UV and vis irradiation. *Materials* **2019**, *12*, 4158. [CrossRef] [PubMed]
44. Kowalska, E.; Mahaney, O.O.P.; Abe, R.; Ohtani, B. Visible-light-induced photocatalysis through surface plasmon excitation of gold on titania surfaces. *Phys. Chem. Chem. Phys.* **2010**, *12*, 2344–2355. [CrossRef]

45. Linic, S.; Christopher, P.; Ingram, D.B. Plasmonic-metal nanostructures for efficient conversion of solar to chemical energy. *Nat. Mater.* **2011**, *10*, 911–921. [CrossRef]
46. Tian, Y.; Tatsuma, T. Mechanisms and applications of plasmon-induced charge separation at TiO_2 films loaded with gold nanoparticles. *J. Am. Chem. Soc.* **2005**, *127*, 7632–7637. [CrossRef]
47. Raja-Mogan, T.; Ohtani, B.; Kowalska, E. Photonic crystals for plasmonic photocatalysis. *Catalysts* **2020**, *10*, 827. [CrossRef]
48. Rycenga, M.; Cobley, C.M.; Zeng, J.; Li, W.; Moran, C.H.; Zhang, Q.; Qin, D.; Xia, Y. Controlling the synthesis and assembly of silver nanostructures for plasmonic applications. *Chem. Rev.* **2011**, *111*, 3669–3712. [CrossRef]
49. Kazuma, E.; Yamaguchi, T.; Sakai, N.; Tatsuma, T. Growth behaviour and plasmon resonance properties of photocatalytically deposited Cu nanoparticles. *Nanoscale* **2011**, *3*, 3641–3645. [CrossRef]
50. El-sayed, M.A. Some interesting properties of metals confined in time and nanometer space of different shapes. *Acc. Chem. Res.* **2001**, *34*, 257–264. [CrossRef]
51. Burda, C.; Chen, X.; Narayanan, R.; El-Sayed, M.A. Chemistry and properties of nanocrystals of different shapes. *Chem. Rev.* **2005**, *105*, 1025–1102. [CrossRef] [PubMed]
52. Brus, L. Noble metal nanocrystals: Plasmon electron transfer photochemistry and single molecule raman spectroscopy. *Acc. Chem. Res.* **2008**, *41*, 1742–1749. [CrossRef] [PubMed]
53. Xia, Y.; Xiong, Y.; Lim, B.; Skrabalak, S.E. Shape-controlled synthesis of metal nanocrystals: Simple chemistry meets complex physics? *Angew. Chem. Int. Ed.* **2009**, *48*, 60–103. [CrossRef]
54. Kelly, K.L.; Coronado, E.; Zhao, L.L.; Schatz, G.C. The optical properties of metal nanoparticles: The influence of size, shape, and dielectric environment. *J. Phys. Chem. B* **2003**, *107*, 668–677. [CrossRef]
55. Hou, W.; Cronin, S.B. A review of surface plasmon resonance-enhanced photocatalysis. *Adv. Funct. Mater.* **2013**, *23*, 1612–1619. [CrossRef]
56. Evanoff, D.D.; Chumanov, G. Synthesis and optical properties of silver nanoparticles and arrays. *ChemPhysChem* **2005**, *6*, 1221–1231. [CrossRef]
57. Jain, P.K.; Huang, X.; El-Sayed, I.H.; El-Sayed, M.A. Noble metals on the nanoscale: Optical and photothermal properties and some applications in imaging, sensing, biology, and medicine. *Acc. Chem. Res.* **2008**, *41*, 1578–1586. [CrossRef]
58. Matsunaga, T.; Tomoda, R.; Nakajima, T.; Wake, H. Photoelectrochemical sterilization of microbial cells by semiconductor powders. *FEMS Microbiol. Lett.* **1985**, *29*, 211–214. [CrossRef]
59. Wolfrum, E.J.; Huang, J.; Blake, D.M.; Maness, P.C.; Huang, Z.; Fiest, J.; Jacoby, W.A. Photocatalytic oxidation of bacteria, bacterial and fungal spores, and model biofilm components to carbon dioxide on titanium dioxide-coated surfaces. *Environ. Sci. Technol.* **2002**, *36*, 3412–3419. [CrossRef]
60. Armon, R.; Weitch-Cohen, G.; Bettane, P. Disinfection of Bacillus spp. spores in drinking water by TiO_2 photocatalysis as a model for *Bacillus anthracis*. *Water Sci. Technol. Water Supply* **2004**, *4*, 7–14. [CrossRef]
61. Dillert, R.; Siemon, U.; Bahnemann, D. Photocatalytic disinfection of municipal wastewater. *Chem. Eng. Technol.* **1998**, *21*, 356–358. [CrossRef]
62. Wei, C.; Lin, W.Y.; Zainal, Z.; Williams, N.E.; Zhu, K.; Kruzlc, A.P.; Smith, R.L.; Rajeshwar, K. Bactericidal activity of TiO_2 photocatalyst in aqueous media: Toward a solar-assisted water disinfection system. *Environ. Sci. Technol.* **1994**, *28*, 934–938. [CrossRef] [PubMed]
63. Kim, B.; Kim, D.; Cho, D.; Cho, S. Bactericidal effect of TiO_2 photocatalyst on selected food-borne pathogenic bacteria. *Chemosphere* **2003**, *52*, 277–281. [CrossRef]
64. Wu, P.; Xie, R.; Imlay, K.; Shang, J.K. Monolithic ceramic foams for ultrafast photocatalytic inactivation of bacteria. *J. Am. Ceram. Soc.* **2009**, *92*, 1648–1654. [CrossRef]
65. Keller, V.; Keller, N.; Ledoux, M.J.; Lett, M.C. Biological agent inactivation in a flowing air stream by photocatalysis. *Chem. Commun.* **2005**, 2918–2920. [CrossRef]
66. Porley, V.; Chatzisymeon, E.; Meikap, B.C.; Ghosal, S.; Robertson, N. Field testing of low-cost titania-based photocatalysts for enhanced solar disinfection (SODIS) in rural India. *Environ. Sci. Water Res. Technol.* **2020**, *6*, 809–816. [CrossRef]
67. Hockberger, P.E. A History of ultraviolet photobiology for humans, animals and microorganisms. *Photochem. Photobiol.* **2002**, *76*, 561. [CrossRef]
68. Markowska-Szczupak, A.; Ulfig, K.; Morawski, A.W. The application of titanium dioxide for deactivation of bioparticulates: An overview. *Catal. Today* **2011**, *169*, 249–257. [CrossRef]

69. Sunada, K.; Minoshima, M.; Hashimoto, K. Highly efficient antiviral and antibacterial activities of solid-state cuprous compounds. *J. Hazard. Mater.* **2012**, *235*, 265–270. [CrossRef]
70. Deng, C.H.; Gong, J.L.; Zeng, G.M.; Zhang, P.; Song, B.; Zhang, X.G.; Liu, H.Y.; Huan, S.Y. Graphene sponge decorated with copper nanoparticles as a novel bactericidal filter for inactivation of *Escherichia coli*. *Chemosphere* **2017**, *184*, 347–357. [CrossRef]
71. Jafari, A.; Pourakbar, L.; Farhadi, K.; Mohamadgolizad, L.; Goosta, Y. Biological synthesis of silver nanoparticles and evaluation of antibacterial and antifungal properties of silver and copper nanoparticles. *Turkish J. Biol.* **2015**, *39*, 556–561. [CrossRef]
72. Qiu, X.; Miyauchi, M.; Sunada, K.; Minoshima, M.; Liu, M.; Lu, Y.; Li, D.; Shimodaira, Y.; Hosogi, Y.; Kuroda, Y.; et al. Hybrid Cu_xO/TiO_2 nanocomposites as risk-reduction materials in indoor environments. *ACS Nano* **2012**, *6*, 1609–1618. [CrossRef] [PubMed]
73. Endo, M.; Janczarek, M.; Wei, Z.; Wang, K.; Markowska-Szczupak, A.; Ohtani, B.; Kowalska, E. Bactericidal properties of plasmonic photocatalysts composed of noble metal nanoparticles on faceted anatase titania. *J. Nanosci. Nanotechnol.* **2019**, *19*, 442–452. [CrossRef] [PubMed]
74. Ohtani, B.; Mahaney, O.O.P.; Amano, F.; Murakami, N.; Abe, R. What are titania photocatalysts?-An exploratory correlation of photocatalytic activity with structural and physical properties. *J. Adv. Oxid. Technol.* **2010**, *13*, 247–261. [CrossRef]
75. Nitta, A.; Takashima, M.; Takase, M.; Ohtani, B. Identification and characterization of titania photocatalyst powders using their energy-resolved distribution of electron traps as a fingerprint. *Catal. Today* **2019**, *321*, 2–8. [CrossRef]
76. Kowalska, E.; Wei, Z.; Karabiyik, B.; Herissan, A.; Janczarek, M.; Endo, M.; Markowska-Szczupak, A.; Remita, H.; Ohtani, B. Silver-modified titania with enhanced photocatalytic and antimicrobial properties under UV and visible light irradiation. *Catal. Today* **2015**, *252*, 136–142. [CrossRef]
77. Prieto-Mahaney, O.-O.; Murakami, N.; Abe, R.; Ohtani, B. Correlation between photocatalytic activities and structural and physical properties of titanium(IV) oxide powders. *Chem. Lett.* **2009**, *38*, 238–239. [CrossRef]
78. Kakuma, Y.; Nosaka, A.Y.; Nosaka, Y. Difference in TiO_2 photocatalytic mechanism between rutile and anatase studied by the detection of active oxygen and surface species in water. *Phys. Chem. Chem. Phys.* **2015**, *17*, 18691–18698. [CrossRef]
79. Buchalska, M.; Kobielusz, M.; Matuszek, A.; Pacia, M.; Wojtyła, S.; Macyk, W. On oxygen activation at rutile-and anatase-TiO_2. *ACS Catal.* **2015**, *5*, 7424–7431. [CrossRef]
80. Min, B.K.; Wallace, W.T.; Goodman, D.W. Support effects on the nucleation, growth, and morphology of gold nano-clusters. *Surf. Sci.* **2006**, *600*, L7–L11. [CrossRef]
81. Janczarek, M.; Wei, Z.; Endo, M.; Ohtani, B.; Kowalska, E. Silver- and copper-modified decahedral anatase titania particles as visible light-responsive plasmonic photocatalyst. *J. Photonics Energy* **2017**, *7*, 012008. [CrossRef]
82. DeSario, P.A.; Pietron, J.J.; Brintlinger, T.H.; McEntee, M.; Parker, J.F.; Baturina, O.; Stroud, R.M.; Rolison, D.R. Oxidation-stable plasmonic copper nanoparticles in photocatalytic TiO_2 nanoarchitectures. *Nanoscale* **2017**, *9*, 11720–11729. [CrossRef] [PubMed]
83. Wei, Z.; Janczarek, M.; Endo, M.; Wang, K.; Balčytis, A.; Nitta, A.; Méndez-medrano, M.G.; Colbeau-justin, C.; Juodkazis, S.; Ohtani, B.; et al. Noble metal-modified faceted anatase titania photocatalysts: Octahedron versus decahedron. *Appl. Catal. B Environ.* **2018**, *237*, 574–587. [CrossRef]
84. Banerjee, S.; Chakravorty, D. Optical absorption by nanoparticles of Cu_2O. *Europhys. Lett.* **2000**, *52*, 468–473. [CrossRef]
85. Dias Filho, N.L. Adsorption of Cu(II) and Co(II) complexes on a silica gel surface chemically modified with 2-mercaptoimidazole. *Mikrochim. Acta* **1999**, *130*, 233–240. [CrossRef]
86. Irie, H.; Kamiya, K.; Shibanuma, T.; Miura, S.; Tryk, D.A.; Yokoyama, T.; Hashimoto, K. Visible light-sensitive Cu(II)-grafted TiO_2 photocatalysts: Activities and X-ray absorption fine structure analyses. *J. Phys. Chem. C* **2009**, *113*, 10761–10766. [CrossRef]
87. Janczarek, M.; Endo, M.; Zhang, D.; Wang, K.; Kowalska, E. Enhanced photocatalytic and antimicrobial performance of cuprous oxide/titania: The effect of titania matrix. *Materials* **2018**, *11*, 2069. [CrossRef]
88. Endo-Kimura, M.; Janczarek, M.; Bielan, Z.; Zhang, D.; Wang, K.; Markowska-Szczupak, A.; Kowalska, E. Photocatalytic and antimicrobial properties of Ag_2O/TiO_2 heterojunction. *ChemEngineering* **2019**, *3*, 3. [CrossRef]

89. Ikeda, S.; Sugiyama, N.; Pal, B.; Marcí, G.; Palmisano, L.; Noguchi, H.; Uosaki, K.; Ohtani, B. Photocatalytic activity of transition-metal-loaded titanium(IV) oxide powders suspended in aqueous solutions: Correlation with electron-hole recombination kinetics. *Phys. Chem. Chem. Phys.* **2001**, *3*, 267–273. [CrossRef]
90. Janczarek, M.; Kowalska, E. On the origin of enhanced photocatalytic activity of copper-modified titania in the oxidative reaction systems. *Catalysts* **2017**, *7*, 317. [CrossRef]
91. Kowalska, E.; Abe, R.; Ohtani, B. Visible light-induced photocatalytic reaction of gold-modified titanium(IV) oxide particles: Action spectrum analysis. *Chem. Commun.* **2009**, 241–243. [CrossRef]
92. Kowalska, E.; Rau, S.; Ohtani, B. Plasmonic titania photocatalysts active under UV and visible-light irradiation: Influence of gold amount, size, and shape. *J. Nanotechnol.* **2012**, *2012*, 1–11. [CrossRef]
93. Zielińska-Jurek, A.; Kowalska, E.; Sobczak, J.W.; Lisowski, W.; Ohtani, B.; Zaleska, A. Preparation and characterization of monometallic (Au) and bimetallic (Ag/Au) modified-titania photocatalysts activated by visible light. *Appl. Catal. B Environ.* **2011**, *101*, 504–514. [CrossRef]
94. Wei, Z.; Rosa, L.; Wang, K.; Endo, M.; Juodkazis, S.; Ohtani, B.; Kowalska, E. Size-controlled gold nanoparticles on octahedral anatase particles as efficient plasmonic photocatalyst. *Appl. Catal. B Environ.* **2017**, *206*, 393–405. [CrossRef]
95. Méndez-Medrano, M.G.; Kowalska, E.; Ohtani, B.; Bahena Uribe, D.; Colbeau-Justin, C.; Rau, S.; Rodríguez-López, J.L.; Remita, H. Heterojunction of CuO nanoclusters with TiO_2 for photo-oxidation of organic compounds and for hydrogen production. *J. Chem. Phys.* **2020**, *153*, 034705. [CrossRef]
96. Werapun, U.; Pechwang, J. Synthesis and Antimicrobial Activity of Fe:TiO_2 Particles. *J. Nano Res.* **2019**, *56*, 28–38. [CrossRef]
97. Gurr, J.; Wang, A.S.S.; Chen, C.; Jan, K. Ultrafine titanium dioxide particles in the absence of photoactivation can induce oxidative damage to human bronchial epithelial cells. *Toxicology* **2005**, *213*, 66–73. [CrossRef]
98. Vargas, M.A.; Rodríguez-Páez, J.E. Facile Synthesis of TiO_2 Nanoparticles of different crystalline phases and evaluation of their antibacterial effect under dark conditions against *E. coli*. *J. Clust. Sci.* **2019**, *30*, 379–391. [CrossRef]
99. Endo-Kimura, M.; Kowalska, E. Plasmonic photocatalysts for microbiological applications. *Catalysts* **2020**, *10*, 824. [CrossRef]
100. Wang, M.; Zhao, Q.; Yang, H.; Shi, D.; Qian, J. Photocatalytic antibacterial properties of copper doped TiO_2 prepared by high-energy ball milling. *Ceram. Int.* **2020**, *46*, 16716–16724. [CrossRef]
101. Grass, G.; Rensing, C.; Solioz, M. Metallic copper as an antimicrobial surface. *Appl. Environ. Microbiol.* **2011**, *77*, 1541–1547. [CrossRef] [PubMed]
102. Mathew, S.; Ganguly, P.; Rhatigan, S.; Kumaravel, V.; Byrne, C.; Hinder, S.J.; Bartlett, J.; Nolan, M.; Pillai, S.C. Cu-Doped TiO_2: Visible light assisted photocatalytic antimicrobial activity. *Appl. Sci.* **2018**, *8*, 2067. [CrossRef]
103. Miró-canturri, A.; Ayerbe-algaba, R.; Smani, Y. Drug repurposing for the treatment of bacterial and fungal infections. *Front. Microbiol.* **2019**, *10*, 41.
104. Markowska-Szczupak, A.; Wang, K.; Rokicka, P.; Endo, M.; Wei, Z.; Ohtani, B.; Morawski, A.W.; Kowalska, E. The effect of anatase and rutile crystallites isolated from titania P25 photocatalyst on growth of selected mould fungi. *J. Photochem. Photobiol. B Biol.* **2015**, *151*, 54–62. [CrossRef] [PubMed]
105. Eisenman, H.C.; Casadevall, A. Synthesis and assembly of fungal melanin. *Appl. Microbiol. Biotechnol.* **2015**, *93*, 931–940. [CrossRef]
106. Kühn, K.P.; Chaberny, I.F.; Massholder, K.; Stickler, M.; Benz, V.W.; Sonntag, H.G.; Erdinger, L. Disinfection of surfaces by photocatalytic oxidation with titanium dioxide and UVA light. *Chemosphere* **2003**, *53*, 71–77. [CrossRef]
107. Endo, M.; Wei, Z.; Wang, K.; Karabiyik, B.; Yoshiiri, K.; Rokicka, P.; Ohtani, B.; Markowska-Szczupak, A.; Kowalska, E. Noble metal-modified titania with visible-light activity for the decomposition of microorganisms. *Beilstein J. Nanotechnol.* **2018**, *9*, 829–841. [CrossRef]
108. Antsotegi-Uskola, M.; Markina-Iñarrairaegui, A.; Ugalde, U. New insights into copper homeostasis in filamentous fungi. *Int. Microbiol.* **2020**, *23*, 65–73. [CrossRef]
109. Chen, F.; Yang, X.; Wu, Q. Antifungal capability of TiO_2 coated film on moist wood. *Build. Environ.* **2009**, *44*, 1088–1093. [CrossRef]
110. Markowska-Szczupak, A.; Ulfig, K.; Morawski, A.W. Antifungal effect of titanium dioxide, indoor light and the photocatalytic process in in vitro test on different media. *J. Adv. Oxid. Technol.* **2012**, *15*, 30–33. [CrossRef]

111. Meletiadis, J.; Meis, J.F.G.M.; Mouton, J.W. Analysis of growth characteristics of filamentous fungi in different nutrient media. *J. Clin. Microbiol.* **2001**, *39*, 478–484. [CrossRef] [PubMed]
112. Pombeiro-Sponchiado, S.R.; Sousa, G.S.; Andrade, J.C.R.; Lisboa, H.F.; Gonçalves, R.C.R. Production of melanin pigment by fungi and its biotechnological applications. In *Melanin*; Blumenberg, M., Ed.; InTechOpen: London, UK, 2017.
113. Carlile, M.J. The Photobiology of Fungi. *Annu. Rev. Plant. Physiol* **1965**, *16*, 175–202. [CrossRef]
114. Hill, E.P. Effect of light on growth and sporulation of *Aspergillus ornatus*. *J. Gen. Microbiol* **1976**, *95*, 39–44. [CrossRef] [PubMed]
115. Kopke, K.; Hoff, B.; Bloemendal, S.; Katschorowski, A.; Kamerewerd, J.; Kück, U. Members of the *Penicillium chrysogenum* velvet complex play functionally opposing roles in the regulation of penicillin biosynthesis and conidiation. *Eukaryot. Cell* **2013**, *12*, 299–310. [CrossRef] [PubMed]
116. Raffaini, G.; Ganazzoli, F. Molecular modelling of protein adsorption on the surface of titanium dioxide polymorphs. *Philos. Trans. R. Soc. A Math. Phys. Eng. Sci.* **2012**, *370*, 1444–1462. [CrossRef] [PubMed]
117. Liu, Q.; Zhang, Y.; Laskowski, J.S. The adsorption of polysaccharides onto mineral surfaces: An acid/base interaction. *Int. J. Miner. Process.* **2000**, *60*, 229–245. [CrossRef]
118. Topoglidis, E.; Cass, A.E.G.; Gilardi, G.; Sadeghi, S.; Beaumont, N.; Durrant, J.R. Protein adsorption on nanocrystalline TiO$_2$ films: An immobilization strategy for bioanalytical devices. *Anal. Chem.* **1998**, *70*, 5111–5113. [CrossRef]
119. Ellingsen, J.E. A study on the mechanism of protein adsorption to TiO$_2$. *Biomaterials* **1991**, *12*, 593–596. [CrossRef]
120. Price, M.S.; Classen, J.J.; Payne, G.A. *Aspergillus niger* absorbs copper and zinc from swine wastewater. *Bioresour. Technol.* **2001**, *77*, 41–49. [CrossRef]
121. Markov, S.L.; Vidaković, A.M. Testing methods for antimicrobial activity of TiO$_2$ photocatalyst. *Acta Period. Technol.* **2014**, *45*, 141–152. [CrossRef]

Publisher's Note: MDPI stays neutral with regard to jurisdictional claims in published maps and institutional affiliations.

© 2020 by the authors. Licensee MDPI, Basel, Switzerland. This article is an open access article distributed under the terms and conditions of the Creative Commons Attribution (CC BY) license (http://creativecommons.org/licenses/by/4.0/).

Article

Defective TiO₂ Core-Shell Magnetic Photocatalyst Modified with Plasmonic Nanoparticles for Visible Light-Induced Photocatalytic Activity

Zuzanna Bielan [1,*], Agnieszka Sulowska [1], Szymon Dudziak [1], Katarzyna Siuzdak [2], Jacek Ryl [3] and Anna Zielińska-Jurek [1,*]

[1] Department of Process Engineering and Chemical Technology, Faculty of Chemistry, Gdansk University of Technology (GUT), G. Narutowicza 11/12, 80-233 Gdansk, Poland; Sulowska.as@gmail.com (A.S.); dudziakszy@gmail.com (S.D.)
[2] Physical Aspects of Ecoenergy Department, The Szewalski Institute of Fluid-Flow Machinery Polish Academy of Science, Fiszera 14, 80-231 Gdansk, Poland; katarzyna.siuzdak@imp.gda.pl
[3] Department of Electrochemistry, Corrosion and Materials Engineering, Faculty of Chemistry, Gdansk University of Technology (GUT), G. Narutowicza 11/12, 80-233 Gdansk, Poland; jacek.ryl@pg.edu.pl
* Correspondence: zuzanna.bielan@gmail.com (Z.B.); annjurek@pg.edu.pl (A.Z.-J.)

Received: 19 May 2020; Accepted: 11 June 2020; Published: 15 June 2020

Abstract: In the presented work, for the first time, the metal-modified defective titanium(IV) oxide nanoparticles with well-defined titanium vacancies, was successfully obtained. Introducing platinum and copper nanoparticles (NPs) as surface modifiers of defective d-TiO₂ significantly increased the photocatalytic activity in both UV-Vis and Vis light ranges. Moreover, metal NPs deposition on the magnetic core allowed for the effective separation and reuse of the nanometer-sized photocatalyst from the suspension after the treatment process. The obtained $Fe_3O_4@SiO_2/d\text{-}TiO_2\text{-}Pt/Cu$ photocatalysts were characterized by X-ray diffractometry (XRD) and specific surface area (BET) measurements, UV-Vis diffuse reflectance spectroscopy (DR-UV/Vis), X-ray photoelectron spectroscopy (XPS) and transmission electron microscopy (TEM). Further, the mechanism of phenol degradation and the role of four oxidative species (h^+, e^-, $^\bullet OH$, and $^\bullet O_2^-$) in the studied photocatalytic process were investigated.

Keywords: titanium vacancies; phenol degradation; scavengers; magnetic photocatalysts; platinum-modified defective TiO₂

---

1. Introduction

In recent years, among wastewater treatment and environmental remediation technologies, photocatalysis has gained attention as a promising technique for the degradation of persistent organic pollutants at ambient temperature and pressure [1–4]. Pilot scale-installations for water treatment using photocatalysis are more and more popular among the world [5,6]. The structural and surface properties of photocatalysts significantly influence their physicochemical and photocatalytic properties. In this regard, one of the most important issues in the photocatalytic process is the preparation of well-characterized and highly active photocatalytic material.

Titanium(IV) oxide (TiO₂), the most widely used semiconductor in photocatalysis, is extensively exploited to obtain highly photoactive in UV-Vis range semiconductor material. The TiO₂ nanoparticles differing in size and surface area. Nonetheless, despite different morphology and polymorphic composition, all pristine titanium(IV) oxide particles own wide bandgap energy (Eg), which differs in the range of 3.0–3.2 eV, for rutile and anatase, respectively [7]. In this regard, TiO₂ photoexcitation is possible only with UV irradiation (λ < 388 nm), and therefore the application of solar radiation is highly limited.

Much effort has been done to shift TiO_2 excitation energy to longer wavelengths, especially in visible light range (400–750 nm). Among various methods [8–10], surface modification with noble and semi-noble metals is the most widely used and effective method. Wysocka et al. [11] obtained mono- (Pt, Ag, and Cu) and bimetal- (Cu/Ag, Ag/Pt, and Cu/Pt) modified TiO_2 photocatalysts, where TiO_2 matrix was commercially available ST01 (fine anatase particles). Metal ions were reduced using the chemical ($NaBH_4$ solution) as well as thermal treatment methods. Klein et al. [12] prepared TiO_2-P25 modified with Pt, Pd, Ag, and Au using radiolysis reduction. Obtained photocatalysts were further immobilized on the glass plate and used for toluene removal from the gas phase. Moreover, Janczarek et al. [13] proposed a method of obtaining Ag- and Cu-modified decahedral anatase particles (DAP) by photodeposition. Wei et al. [14] reported that selective deposition of nanometals (Au/Ag nanoarticles) on (001) facets of decahedral anatase particles (DAP), together with octahedral anatase particles (OAP) result in significant photocatalytic process improvement under visible light irradiation.

Another possibility of titania visible light activation is the introduction of intrinsic defects to its crystal structure. Titanium or oxygen vacancies and surface disorders led to changes in electronic and crystal changes, resulting in better electrons and holes separation and even bandgap narrowing [15]. Defected TiO_2 is often evidenced by their color: pale blue for oxygen vacancies (due to d–d transitions of bandgap states) and yellow for titanium vacancies (consumption of free electrons and holes), which also suggest shifting its light absorption to the visible light range [15].

Nevertheless, TiO_2 particle size and shape determine not only the number of active sites but also separation and reusability properties. Commercially available titanium(IV) oxide—P25 forms a stable suspension due to its nanometric sizes, which cause the detached process to be highly expensive and energy-consuming [16,17]. The immobilization of nanoparticles on solid substrates [18] could, in turn, result in a decrease in photocatalytic activity due to significantly reducing specific surface area [19,20]. Alternative way of photocatalyst separation is its deposition on magnetic compound, such as Fe_3O_4 [21,22], $CoFe_2O_4$ [23], $ZnFe_2O_4$ [24], and $BaFe_{12}O_{19}$ [25]. Along with using magnetic materials, the percentage of photocatalyst recovery and the possibility of its reuse significantly increase [26–29]. However, the direct contact of ferrite particles with photocatalyst (e.g., TiO_2) may result in unfavorable electron transfer from TiO_2 into the magnetic compound, causing its transformation and photocorrosion [26]. Thus, their separation with an inert interlayer of silica [26–31] or carbon [32] could effectively prevent the charge carriers recombination. In this regard, the magnetic Fe_3O_4@SiO_2/TiO_2 nanocomposites with a core-shell structure, where the core was Fe_3O_4, the photoactive shell was TiO_2, and silica was used as an inert interlayer, are a relatively new and promising group of composite materials.

The previous studies focused on the preparation and characterization of noble metal NPs modified with different titania matrices [33,34], as well as TiO_2 nanoparticles deposited on various magnetic cores (Fe_3O_4, $CoFe_2O_3$, and $ZnFe_2O_4$) [26–29]. However, in the literature, there is a lack of complex researches on the correlation between structural defects and TiO_2 modification with different materials (metal nanoparticles or other oxides).

In this regard, deeply characterized defective d-TiO_2 with stable titanium vacancies, obtained by a simple hydrothermal method was further modified with Pt and Cu nanoparticles as well as deposited on the magnetite core. The obtained samples were characterized by X-ray diffractometry (XRD), specific surface area (BET) measurements, UV-Vis diffuse reflectance spectroscopy (DR-UV/Vis), X-ray photoelectron spectroscopy (XPS), and transmission electron microscopy (TEM). The photodegradation of phenol as a model organic pollutant in the presence of the obtained photocatalysts was subsequently investigated in the range of UV-Vis and Vis irradiation. Further, the mechanism of phenol degradation and the role of four oxidative species (h^+, e^-, $^\bullet OH$, and $^\bullet O_2^-$) in the studied photocatalytic process were investigated.

2. Results

2.1. Physicochemical Characterization of d-TiO$_2$-Pt/Cu and Magnetic Fe$_3$O$_4$@SiO$_2$/d-TiO$_2$-Pt/Cu Photocatalysts

The preparation of d-TiO$_2$ photocatalysts was based on the hydrothermal method in the oxidative environment (addition of 20–75 mol% of HIO$_3$). The electron paramagnetic resonance (EPR) analysis confirmed the presence of titanium vacancies with g value of 1.995 after calcination. The signal at EPR spectra for bulk Ti^{3+} was not observed for pure TiO$_2$ sample, see in Figure 1.

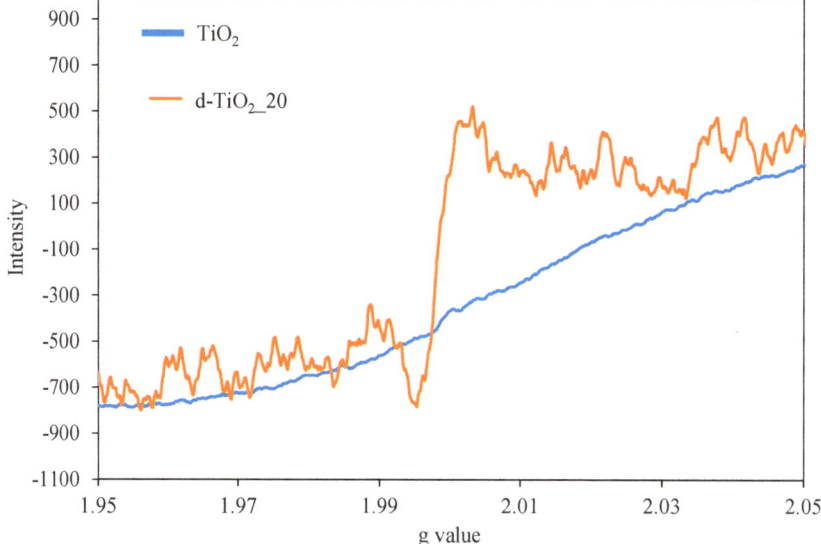

Figure 1. The electron paramagnetic resonance (EPR) spectra in the room temperature for TiO$_2$ (blue line) and TiO$_2$ with titania vacancies (orange line).

The general physicochemical and photocatalytic characteristics of the obtained defective d-TiO$_2$-Pt/Cu and Fe$_3$O$_4$@SiO$_2$/d-TiO$_2$-Pt/Cu samples, i.e., BET surface area, pore volume, calculated bandgap (Eg) and phenol degradation efficiency in UV-Vis and Vis light are presented in Tables 1 and 2.

Table 1. Physicochemical and photocatalytic characteristics of the obtained defective d-TiO$_2$-Pt/Cu samples.

Sample	BET (m$^2 \cdot$g^{-1})	V Pores (cm$^3 \cdot$g^{-1})	Eg (eV)	Rate Constant k (min^{-1})·10^{-2} UV-Vis	Phenol Removal (%) UV-Vis	Rate Constant k (min^{-1})·10^{-2} Vis	Phenol Removal (%) Vis
TiO$_2$	169	0.0836	3.2	0.98	46	0.03	3
d-TiO$_2$_20	172	0.0847	2.7	0.79	40	0.17	13
d-TiO$_2$_75	167	0.0826	2.9	0.52	29	0.03	4
TiO$_2$-Pt0.05	166	0.0819	3.2	2.30	76	0.22	14
d-TiO$_2$_20-Pt0.05	148	0.0727	2.85	1.10	48	0.33	20
d-TiO$_2$_75-Pt0.05	164	0.0808	2.9	1.02	46	0.26	15
d-TiO$_2$_20-Pt0.1	152	0.0747	2.7	1.42	57	0.41	22
d-TiO$_2$_20-Cu0.1	101	0.0499	2.9	0.51	30	0.12	7
d-TiO$_2$_20-Pt0.1/Cu0.1	152	0.0744	2.75	1.47	59	0.27	17

Table 2. Physicochemical and photocatalytic characteristics of the obtained magnetic Fe$_3$O$_4$@SiO$_2$/d-TiO$_2$-Pt/Cu samples.

Sample	BET (m$^2 \cdot$g^{-1})	V Pores (cm$^3 \cdot$g^{-1})	Rate Constant k (min^{-1})·10^{-2} UV-Vis	Phenol Removal (%) UV-Vis	Rate Constant k (min^{-1})·10^{-2} Vis	Phenol Removal (%) Vis
Fe$_3$O$_4$@SiO$_2$/d-TiO$_2$_20	117	0.0578	0.41	20	0.13	8
Fe$_3$O$_4$@SiO$_2$/d-TiO$_2$_20-Pt0.05	115	0.0568	0.15	11	0.08	4
Fe$_3$O$_4$@SiO$_2$/d-TiO$_2$_20-Pt0.1	122	0.0602	0.41	19	0.19	11
Fe$_3$O$_4$@SiO$_2$/d-TiO$_2$_20-Cu0.1	117	0.0579	0.22	15	0.05	2
Fe$_3$O$_4$@SiO$_2$/d-TiO$_2$_20-Pt0.1/Cu0.1	117	0.0580	0.36	22	0.26	11

The BET surface area of pure TiO_2 obtained from Titanium(IV) butoxide (TBT) hydrolysis in water and defective TiO_2 samples was similar and ranged from 167 to 172 $m^2 \cdot g^{-1}$. The specific surface area of the metal-modified d-TiO_2 samples fluctuated from 166 to 101 $m^2 \cdot g^{-1}$ and depended on the type and amount of metallic species deposited on d-TiO_2 surface. The samples modified with Pt NPs revealed a higher BET surface area of about 148 $m^2 \cdot g^{-1}$ compared to d-TiO_2 modified with copper oxide (101 $m^2 \cdot g^{-1}$), and bimetallic Pt/Cu NPs (152 $m^2 \cdot g^{-1}$). The relations between photoactivity in UV-Vis and Vis light range versus BET surface area are also shown in Tables 1 and 2. The obtained results indicated that not so much the surface area but rather the presence of Ti defects and modification with metal nanoparticles caused the enhanced photoactivity of the obtained photocatalysts. Moreover, as shown in Table 2, the addition of surface-modifying metal nanoparticles, as well as further deposition of d-TiO_2-Pt/Cu on magnetic matrice, did not affect the magnitude order of the BET surface area, which remained in the range of 101 to 172 $m^2 \cdot g^{-1}$ for d-TiO_2_20-Cu0.1 and d-TiO_2_20, respectively.

The energy bandgaps for all samples were calculated from the plot of (Kubelka–Munk·E)$^{0.5}$ versus E, where E is energy equal to hv, and summarized in Table 1. The samples consist of defective TiO_2 exhibited narrower bandgap of 2.7–2.9 eV compared to TiO_2 and TiO_2-Pt0.05 photocatalysts. Moreover, for all metal-modified defective photocatalysts, the bandgap value, calculated from Kubelka–Munk, transformation did not change, compared to d-TiO_2 matrice, indicating surface modification than doping [11].

The XRD patterns for selected d-TiO_2-Pt/Cu and Fe_3O_4@SiO_2/d-TiO_2-Pt/Cu samples are presented in Figures 1 and 2, with a detailed phase composition and crystalline sizes for all photocatalysts being listed in Tables 3 and 4. Peaks marked "A", "R", and "B" corresponds to anatase, rutile, and brookite phases, respectively. Both crystalline structures (anatase and brookite) appeared for pure TiO_2 prepared by the sol–gel method. For Pt-modified TiO_2 anatase was the major phase, whereas brookite existed as the minor phase. The average crystallite size of anatase was 5–6 nm. The preparation of d-TiO_2 photocatalysts proceeded in the oxidative environment. The introduction into the crystal structure of various types of defects promotes the transformation of anatase to rutile at lower temperatures. Therefore, for the samples obtained in the presence of HIO_3 as the oxidizing agent, after the annealing process the percentage of anatase (the most intense peak at 25° 2θ, with the (101) plane diffraction, ICDD's card No. 7206075) was decreased in favor of (110) rutile, with the peak at 31° 2θ (ICDD's card No. 9004141), even below the anatase to rutile phase transformation temperature [35–37]. For the samples d-TiO_2_75 and d-TiO_2_75-Pt0.05, the dominant phase was rutile with a crystallite size of about 6 nm. Further, the surface modification with plasmonic platinum and semi-noble copper did not cause changes in anatase crystallite size, remaining about 5–6 nm size. The percentage of the brookite phase increased to 8.5% and 13% for d-TiO_2 _20-Pt0.1/Cu0.1, and d-TiO_2_20-Pt0.1 samples, respectively. It resulted from the additional thermal treatment after metal nanoparticles deposition on the photocatalyst surface. Moreover, Pt and Cu modification of TiO_2 did not cause the shift of the peaks in the XRD pattern. The presence of platinum and copper deposited on TiO_2 was not approved (no peaks for platinum or copper) due to low content (0.05–0.1 mol%) and nanometric size.

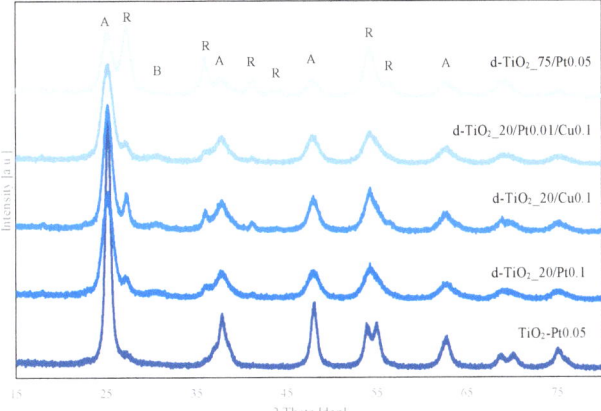

Figure 2. X-ray diffractometry (XRD) patterns for selected defective d-TiO$_2$-Pt/Cu photocatalysts (A—anatase, B—brookite, and R—rutile).

Table 3. Crystalline phases characteristic for obtained defective d-TiO_2-Pt/Cu samples.

Sample	Crystalline Size and Phase Content					
	Anatase		Rutile		Brookite	
	Size (nm)	Phase Content (wt.%)	Size (nm)	Phase Content (wt.%)	Size (nm)	Phase Content (wt.%)
TiO_2	5.97 ± 0.04	95.5 ± 1	-	-	6.1 ± 0.3	4.5 ± 1
d-TiO_2_20	5.14 ± 0.03	96 ± 1.0	-	-	4.0 ± 0.6	3.5 ± 0.5
d-TiO_2_75	5.67 ± 0.05	21 ± 3.5	6.57 ± 0.09	80 ± 2	-	-
TiO_2-Pt0.05	5.71 ± 0.06	91 ± 0.5	-	-	5.7 ± 0.6	9 ± 1
d-TiO_2_20-Pt0.05	5.66 ± 0.03	85 ± 8	9.8 ± 0.7	9 ± 1.0	4.9 ± 0.3	6 ± 0.5
d-TiO_2_75-Pt0.05	5.49 ± 0.04	48 ± 2	7.53 ± 0.12	52 ± 2	-	-
d-TiO_2_20-Pt0.1	5.58 ± 0.03	81 ± 8	7.6 ± 0.8	6 ± 1	4.8 ± 0.3	13 ± 1
d-TiO_2_20-Cu0.1	6.62 ± 0.04	72 ± 11	11.7 ± 0.4	9 ± 2	1.52 ± 0.08	8.5 ± 1
d-TiO_2_20-Pt0.1/Cu0.1	5.52 ± 0.03	83 ± 12	9.5 ± 0.6	8 ± 2	5.1 ± 0.3	8.5 ± 1

Table 4. Crystalline phases characteristic for the obtained magnetic Fe_3O_4@SiO_2/d-TiO_2_20-Pt/Cu samples.

Sample	Crystalline Size and Phase Content					
	Anatase		Rutile		Magnetite	
	Size (nm)	Phase Content (wt.%)	Size (nm)	Phase Content (wt.%)	Size (nm)	Phase Content (wt.%)
Fe_3O_4@SiO_2/d-TiO_2_20	5.19 ± 0.05	71 ± 1.5	8.6 ± 0.5	8 ± 0.5	46.1 ± 1.1	21 ± 0.5
Fe_3O_4@SiO_2/d-TiO_2_20-Pt0.05	5.60 ± 0.05	71 ± 1.5	8.9 ± 0.5	7 ± 0.5	45.7 ± 1.4	21 ± 0.5
Fe_3O_4@SiO_2/d-TiO_2_20-Pt0.1	5.49 ± 0.05	68 ± 2	9.1 ± 0.6	7 ± 1	47.2 ± 4.0	24 ± 0.5
Fe_3O_4@SiO_2/d-TiO_2_20-Cu0.1	7.81 ± 0.17	57 ± 2	13.3 ± 1.4	5 ± 1	37.1 ± 1.8	28 ± 1.5
Fe_3O_4@SiO_2/d-TiO_2_20-Pt0.1/Cu0.1	5.48 ± 0.05	69 ± 1	7.9 ± 0.4	8 ± 0.5	42.6 ± 3.3	22 ± 1

The XRD analysis of $Fe_3O_4@SiO_2/d-TiO_2-Pt/Cu$ confirmed the formation of a magnetic composite, and, as observed in Figure 3 and Table 4, there was no significant difference between the diffraction patterns of the obtained magnetic photocatalysts modified with Pt/Cu NPs. The presence of pure magnetite, with diffraction peaks at 30.2°, 35.6°, 43.3°, 57.3°, and 62.9° 2θ corresponding to (220), (311), (400), (511), and (440) cubic inverse spinel planes (ICDD's card No. 9005813) was confirmed for all $Fe_3O_4@SiO_2/d-TiO_2-Pt/Cu$ magnetic photocatalysts. The decrease in Fe_3O_4 peaks intensity was caused by the formation of tight non-magnetic shell on the core surface, which was previously described by Zielińska-Jurek et al. [26]. The broad peak at 15–25° 2θ corresponds to amorphous silica [38,39]. The content of the magnetite crystalline phase varied from 21% to 28% for $Fe_3O_4@SiO_2/d-TiO_2\_20-Pt0.05$ and $Fe_3O_4@SiO_2/d-TiO_2\_20-Cu0.1$, respectively. At the same time, TiO_2 crystallite size and anatase to rutile phase content ratio remained unchanged for $Fe_3O_4@SiO_2/d-TiO_2\_20$ and $Fe_3O_4@SiO_2/d-TiO_2\_20-Pt/Cu$ samples. No other crystalline phases were identified in the XRD patterns, which indicated the crystal purity of the obtained composites.

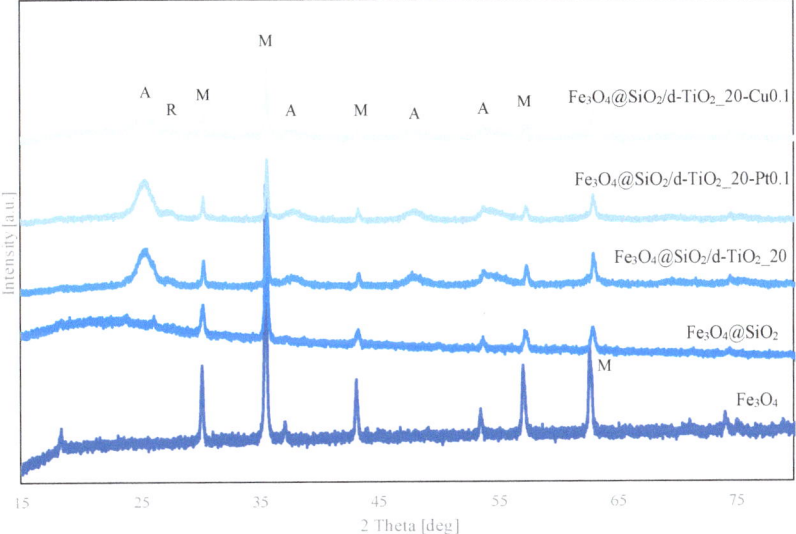

Figure 3. XRD patterns of magnetic photocatalysts, compared with Fe_3O_4 and $Fe_3O_4@SiO_2$ (A—anatase, R—rutile, and M—magnetite).

The photoabsorption properties of metal-modified defective d-TiO_2 samples were studied by diffuse reflectance spectroscopy, and exemplary data are shown in Figure 4. Comparing to pure TiO_2 photocatalyst, introducing platinum as a surface modifier caused an increase of absorption in the visible light region, however, without shifting a maximum, as presented for sample TiO_2–Pt0.05. Modification of defective d-TiO_2 with Pt and Cu was associated with a further increase of Vis light absorbance and proportional to the amount of the deposited metal. Moreover, the deposition of Pt caused a more significant absorbance increment than the same modification with Cu species.

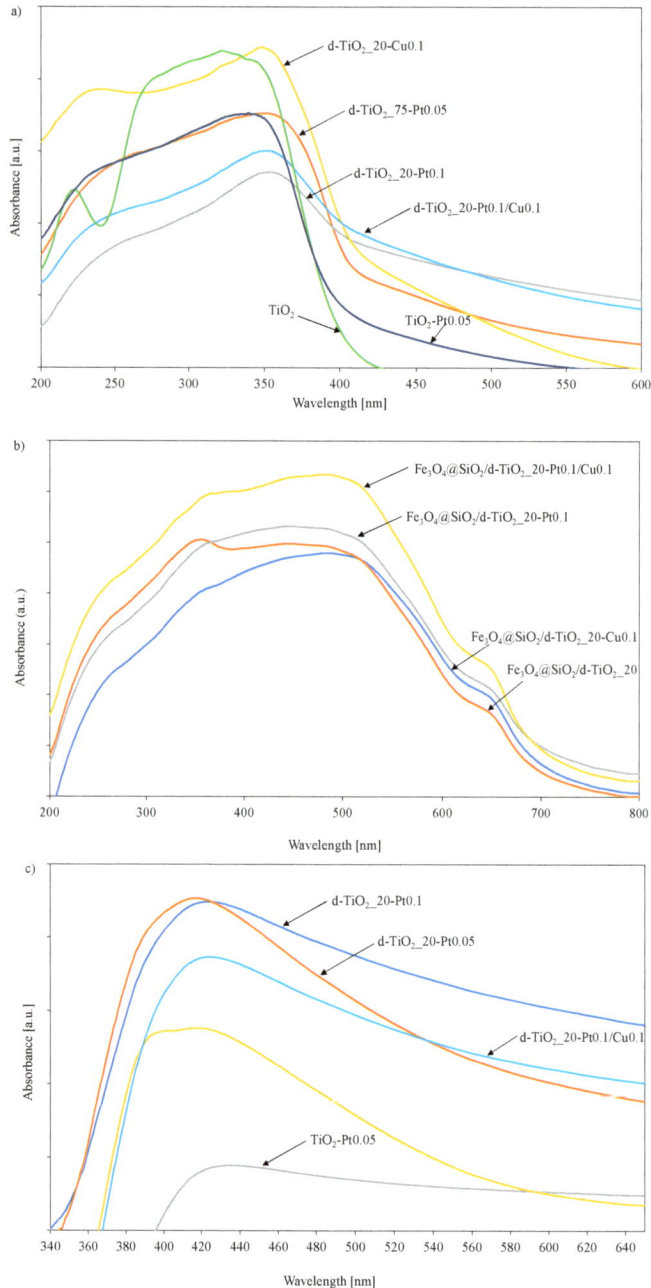

Figure 4. DR/UV-Vis spectra for selected d-TiO$_2$-Pt/Cu (**a**) and Fe$_3$O$_4$@SiO$_2$/d-TiO2-Pt/Cu (**b**) photocatalysts together with Pt and Cu plasmon determination with bare TiO$_2$ as a reference (**c**).

Defective d-TiO$_2$-Pt/Cu deposited on Fe$_3$O$_4$@SiO$_2$ core were characterized by extended light absorption ranged to 700 nm. It could be observed that the described absorption properties in the Vis light for metal-modified TiO$_2$ and absorption properties of final composites have been preserved.

The presence of Localized Surface Plasmon Resonance (LSPR) peaks for Pt and Cu were confirmed based on DR-UV/Vis spectra measurements with pure TiO_2 as a reference (see in Figure 4c). Platinum surface plasmon resonance was observed at the wavelength of about 410–420 nm [33,40]. Electron transfer between Cu(II) and valence band of titanium(IV) oxide could be confirmed by absorption increment from 400 to 450 nm. The typical LSPR signal for zero valent copper at 500–580 nm was not observed, suggesting that Cu is mainly present in its oxidized forms [41,42].

To confirm the presence of noble metal and semi-noble metal NPs on defective TiO_2 surface, the XPS analyses for the selected photocatalysts and deconvolution of Pt 4f and Cu 2p were performed, and the results are presented in Figure 5. Platinum species deposited on the titania surface were designated by deconvolution of Pt 4f peak into two components: Pt $4f_{7/2}$ and Pt $4f_{5/2}$. According to the literature, Pt $4f_{7/2}$ peak, with binding energies in the range of 74.2 to 75.0 eV, refers to the Pt^0, while Pt $4f_{5/2}$ peak, appearing at 77.5–77.9 eV is assigned to Pt^{4+} [11]. The main peaks for Cu 2p appeared as Cu $2p_{3/2}$ and Cu $2p_{1/2}$ at 934 eV and 952 eV. Both of those peaks are commonly attributed to Cu^+ and Cu^{2+} ions [13,43,44]. Obtained data indicated that both Pt and Cu species were successfully deposited on the titania surface.

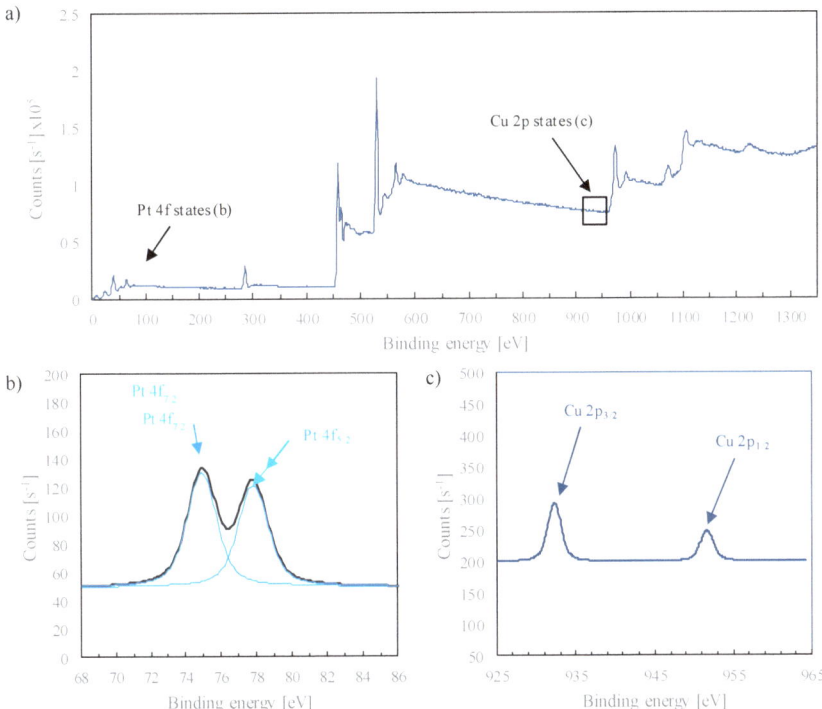

Figure 5. XPS spectra for d-TiO_2_20-Pt0.1/Cu0.1 sample (**a**) with the deconvolution for Pt 4f (**b**) and Cu 2p (**c**).

Moreover, the presence of Pt NPs at the surface of the magnetic nanocomposites was also confirmed by microscopy analysis. As presented in Figure 6, the formation of SiO_2/TiO_2 shell, with a thickness of about 20 nm, tightly covering magnetite nanoparticles was observed. Platinum nanoparticles with a diameter of about 10–20 nm were evenly distributed on the d-TiO_2 layer.

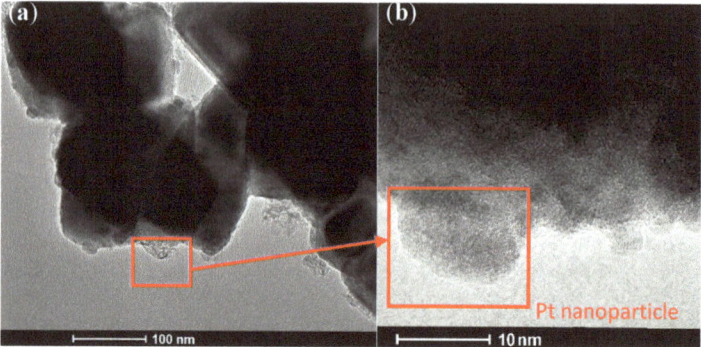

Figure 6. Transmission electron microscopy (TEM) analysis for Fe_3O_4@SiO_2/d-TiO_2_20/Pt0.1 sample (**a**) magnification on Pt nanoparticle (**b**).

2.2. Photocatalytic Activity of d-TiO_2-Pt/Cu and Fe_3O_4@SiO_2/d-TiO_2-Pt/Cu Photocatalysts

The effect of Pt and Cu presence on the properties of defective d-TiO_2 photocatalysts was evaluated in reaction of phenol degradation under UV-Vis and Vis light irradiation. The results, presented as the efficiency of phenol degradation as well as phenol degradation rate constant k, are given in Figures 7 and 8. Additionally, the effect of the electron (e^-), hole (h^+), hydroxyl radical ($^\bullet OH$), and superoxide radical ($^\bullet O_2^-$) scavengers were investigated and presented in Figure 9.

Among analyzed metal-modified photocatalysts, TiO_2–Pt0.05 revealed the highest phenol degradation in UV-Vis light. After 60 min of irradiation, about 76% of phenol was degraded. After introducing plasmonic platinum and semi-noble copper species as a surface modifiers, UV-Vis photoactivity of defective d-TiO_2 samples increased to 59%. The degradation rate constant k increased to 1.47×10^{-2} min^{-1} compared to d-TiO_2_20 (0.79×10^{-2} min^{-1}), and d-TiO_2_75 (0.52×10^{-2} min^{-1}) photocatalysts. Nonetheless, the most significant changes were observed during the photocatalytic process in visible light ($\lambda > 420$ nm). Modifying with 0.05 mol% of Pt, the surface of almost inactive in Vis light d-TiO_2_75 resulted in three-times higher photocatalytic activity under visible light. Therefore, a highly positive effect of metal surface modification of defective d-TiO_2 photocatalyst surface was noticed. It resulted from better charge carriers' separation and decreasing the electron-hole recombination rate. Moreover, the narrower bandgap of the defective d-TiO_2 (in comparison with pure TiO_2) and modification with Pt possessing surface plasmon resonance properties, could also enhance visible light absorption and consequently led to photocatalytic activity increase.

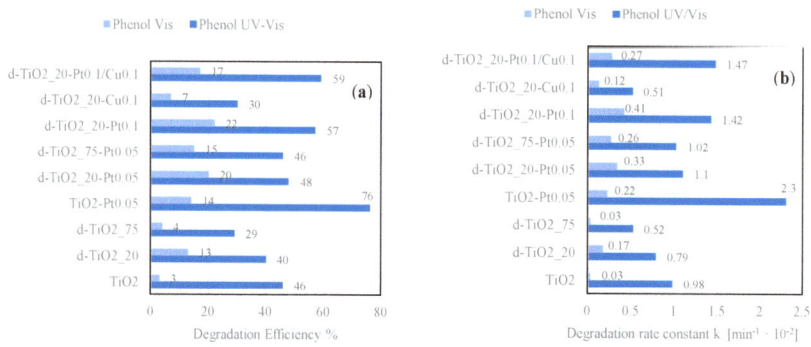

Figure 7. Efficiency of phenol degradation in UV-Vis and Vis light for d-TiO_2-Pt/Cu photocatalysts, presented as % of degradation (**a**) and rate constant k (**b**).

Pure magnetite, coated with inert silica, did not affect the photocatalytic process. Furthermore, the $Fe_3O_4@SiO_2/d\text{-}TiO_2$ composite modified with Pt NPs, and bimetallic Pt/Cu NPs revealed the highest photocatalytic activity in Vis light range. The phenol degradation rate constant in Vis light was 2-times higher for $Fe_3O_4@SiO_2/d\text{-}TiO_2\text{-}Pt/Cu$ compared to $Fe_3O_4@SiO_2/d\text{-}TiO_2$ sample. However, the obtained magnetic photocatalysts had similar photocatalytic activity in UV-Vis light, almost regardless of the surface modification of $d\text{-}TiO_2$ with noble metals. It probably resulted from larger Pt particles (~20 nm) deposition at the surface of $Fe_3O_4@SiO_2/d\text{-}TiO_2$ composite than for TiO_2–Pt0.05 with particles size of about 2–3 nm. Previously, we have reported that the size of noble metal nanoparticles, especially platinum, deposited on the TiO_2 surface strictly depends on the semiconductor surface area, as well as its crystal lattice defects [33,45]. Fine metal particles are produced on the TiO_2 surface with a developed specific surface area with a high density of oxygen traps and nucleation sites, and the highest photocatalytic activity is noticed for Pt-modified photocatalyst, where the size of Pt is below 3 nm [33]. In the present study, Pt nanoparticles' average diameter was about 20 nm as a result of the deposition of Pt ions and their reduction on formed particles' defects. Therefore, the lower metal/semiconductor interface resulted in a decrease in photocatalytic activity under UV-Vis light irradiation.

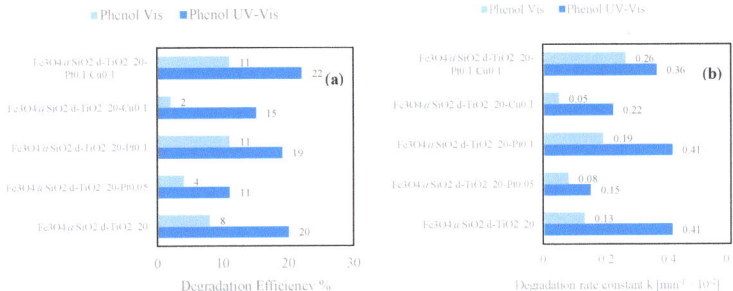

Figure 8. Efficiency of phenol degradation in UV-Vis and Vis light for magnetic $Fe_3O_4@SiO_2/d\text{-}TiO_2\text{-}Pt/Cu$ photocatalysts, presented as % of degradation (**a**) and rate constant k (**b**).

For the final stability and reusability test, the most active defective photocatalyst was selected. For sample d-TiO2_20/Pt0.1/Cu0.1, three 1-h-long subsequent cycles of phenol degradation under UV-Vis light were performed. The obtained results are presented in Figure 9.

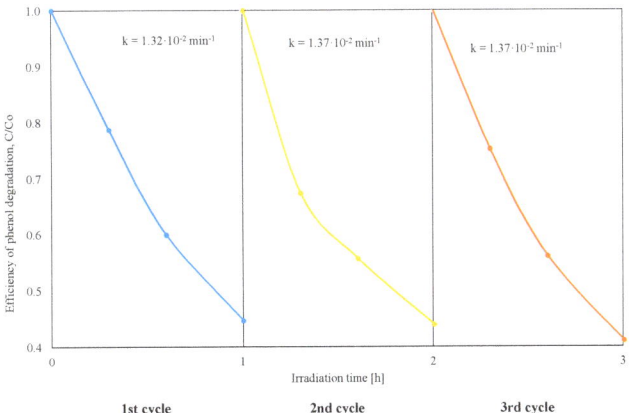

Figure 9. Efficiency of UV-Vis phenol degradation in the presence of defective d-TiO2_20/Pt0.1/Cu0.1 photocatalyst measured in the three subsequent cycles.

There was no significant change in phenol degradation rate constant after the second and third cycles. Thus, the analyzed photocatalyst revealed good stability and reusability.

Furthermore, the reactive species were investigated to understand the photocatalytic reaction mechanism. Benzoquinone (BQ), silver nitrate (SN), ammonium oxalate (AO), and tert-butanol (t-BuOH) were used as superoxide radical anions ($\cdot O_2^-$), electrons (e^-), holes (h^+), and hydroxyl radicals ($\cdot OH$) scavengers, respectively. Obtained results, presented as phenol degradation rate constant k, in comparison to the photodegradation process without scavengers, are presented in Figure 10. The most significant impact on phenol degradation reaction in the presence of metal-modified d-TiO$_2$ was observed for superoxide radicals. After introducing to the photocatalyst suspension BQ solution, the phenol degradation efficiency was significantly inhibited. A slight decrease was also observed in the presence of SN as an electron trap. On the other hand, the addition of AO and t-BuOH did not decrease the phenol degradation rate.

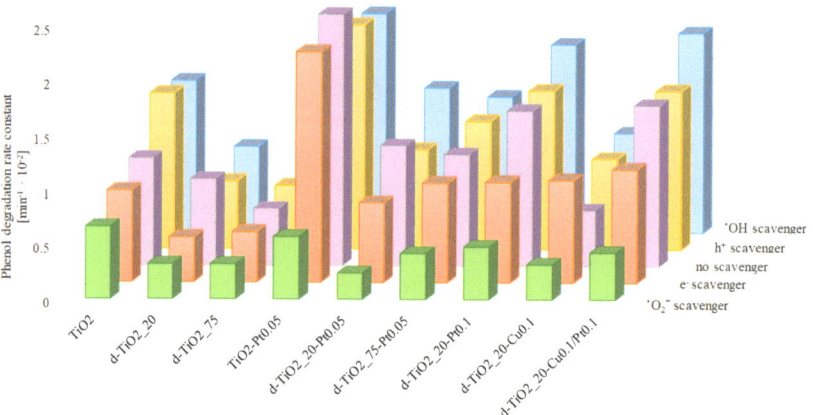

Figure 10. Photocatalytic degradation of phenol for defective Fe$_3$O$_4$@SiO$_2$/d-TiO$_2$-Pt/Cu photocatalysts in the presence of e^-, h^+, $\bullet O_2^-$, and $\bullet OH$ scavengers.

Modification of TiO$_2$ resulted in the shift of the valence band as was revealed from the analysis of Mott–Schottky plot, where the relation between applied potential vs. C_{sc}^{-2} is presented (see in Figure 11). According to the intersection with E axis the flat band potential was estimated. In the case of pure titania it equals to −1.2 V, whereas for d-TiO$_2$_20-Pt0 the value of −1.13 was reached. In order to prepare energy diagram of both materials given in Figure 12, the values of bandgap energy was taken into account. As could be seen, for the modified material the position both the conduction and valence band are shifted. According to Monga et al. [46] the Schottky barrier formed at the metal-TiO$_2$ interface affecting the efficiency of e- transfer. The lowering of the CB band edge is in accordance with the literature indicating that the work function of the metal prone decrease of the CB location. Then, the Schotky barrier is decreased at the metal/semiconductor heterojunction. As a result, the transfer of the photoexcited electron from metal NPs to titania is facilitated and plays important role in photocatalytic activity improvement. The introduction of titanium defects to the TiO$_2$ crystal structure also resulted in narrowing the bandgap from 3.2 to 2.7 eV.

Based on the presented results, a schematic mechanism of UV-Vis phenol degradation in the presence of metal-modified defective Fe$_3$O$_4$@SiO$_2$/d-TiO$_2$-Pt/Cu photocatalyst was proposed and shown in Figure 12. After irradiation of the photocatalyst surface with UV-Vis light, electrons from the Pt are injected to the conduction band of titania and then utilized in oxygen reduction to form reactive oxygen radicals. The path of phenol degradation led through several intermediates, such as benzoquinone, hydroquinone, catechol, resorcinol, oxalic acid, and finally, to complete mineralization to CO_2 and H_2O [47–49]. An analysis of possible charge carriers' impact revealed that for photoactivity of d-TiO_2-Pt/Cu, they are responsible for mainly generated superoxide radicals. The phenol degradation mechanism proceeded by the generation of reactive oxygen species, e.g., $^{\bullet}O_2^{-}$, which attacked the phenol ring, resulting in benzoquinone and hydroquinone formation confirmed by high-performance liquid chromatography (HPLC) analyses. Moreover, during the photoreaction, the concentration of formed intermediates decreased, which suggests mineralization of recalcitrant chemicals to simple organic compounds.

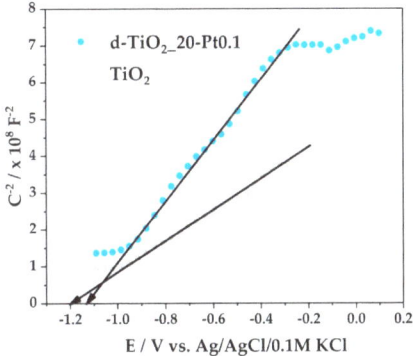

Figure 11. The Mott–Schottky plot for the bare and Pt modified d-TiO_2.

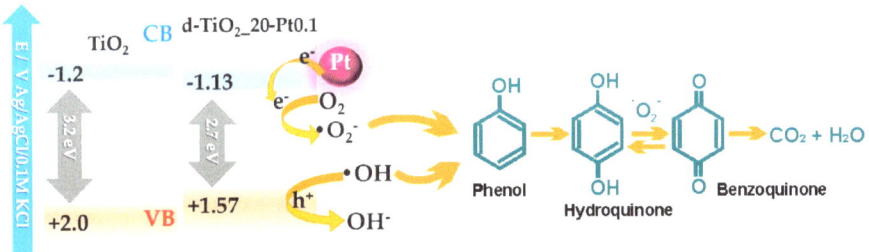

Figure 12. Energy diagram depicting the position of valence and conduction bands of bare and modified titania including the indication of charge transfer within the Schottky junction, and schematic illustration of phenol degradation mechanism over defective Fe_3O_4@SiO_2/d-TiO_2-Pt/Cu photocatalysts.

3. Materials and Methods

Titanium(IV) oxide organic precursor: titanium(IV) butoxide (>99%) was provided by Alfa Aesar (Haverhill, MA, USA). Iodic acid (99.5%), sodium borohydride (99%), chloroplatinate acid hydrate ($H_2PtCl_4 \cdot xH_2O$) (99.9%) and copper nitrate trihydrate ($Cu(NO_3)_2 \cdot 3H_2O$) (99–104%), used for TiO_2 structure and surface modification, were purchased from Sigma (Poznan, Poland). Ferrous ferric oxide (Fe_3O_4, 97%) with a declared particles size of 50 nm was purchased from Aldrich (Poznan, Poland). Tetraethyl orthosilicate (TEOS) was provided by Aldrich and was used as a precursor for the inert interlayer of magnetic nanoparticles. Ammonium hydroxide solution (25%) was purchased from Avantor (Gliwice, Poland). Chemicals for w/o microemulsions preparation, such as cyclohexane and 2-propanol, were purchased from Avantor. Cationic surfactant, hexadecyltrimethylammonium bromide (CTAB), was provided by Sigma Aldrich (Poznan, Poland). Acetonitrile and orthophosphoric acid (85%) for HPLC mobile phase preparation were provided by Merck (Darmstadt, Germany) and VWR (Gdansk, Poland), respectively. Phenol, used as a model organic recalcitrant pollutant in photocatalytic activity measurements, was purchased from VWR. For the titania paste formation polyethylene glycol (PEG) from Sigma Aldrich (Poznan, Poland) was used, while Na_2SO_4 used for electrolyte preparation was purchase from VWR. All reagents were used without further purification.

3.1. Preparation of Defective TiO_2-Pt/Cu Photocatalysts

Defective TiO_2 (marked as d-TiO_2) was obtained by the hydrothermal method assisted with the annealing process. Titanium(IV) butoxide (TBT) and iodic acid (HIO_3) were used as a TiO_2 precursor and oxidizing environment for titanium vacancies formation, respectively. Briefly, the appropriate amount of HIO_3 (presented in Table 5) was dissolved in 80 cm^3 of distilled water. After that, 10 cm^3 of TBT was added dropwise, and the obtained suspension was stirred for 1 h with magnetic stirring at room temperature. In the next step, the suspension was transferred into a Teflon-lined autoclave for thermal treatment at 110 °C for 24 h. The resultant precipitate was centrifuged, dried at 70 °C and then calcined at 300 °C for 3 h.

Table 5. HIO_3 concentrations used for the synthesis of each d-TiO_2 photocatalyst.

Sample	Oxidant Concentration [mol%]	Mass of Added Oxidant [g]
TBT	0	0
d-TiO_2_20	20	1.032
d-TiO_2_75	75	3.869

The obtained defective d-TiO_2 photocatalysts were modified using platinum and copper nanoparticles by the co-precipitation method. In this regard, d-TiO_2 was dispersed in 50 cm^3 of deionized water, and Pt/Cu precursor solutions (0.05 and 0.1 mol% of Pt and 0.1 mol% of Cu with respect to TiO_2) were added. After that, $NaBH_4$ solution was introduced to reduce the metals ions followed by their deposition on the titania surface. The mole ratio of metal ions to $NaBH_4$ was 1:3. After the reduction process, the photocatalyst suspension was mixed for 2 h, and the d-TiO_2-Pt/Cu nanoparticles were separated, washed with deionized water, and dried at 80 °C to dry mass. The final step was calcination at 300 °C for 3 h.

3.2. Preparation of Magnetic Fe_3O_4@SiO_2/d-TiO_2-Pt/Cu Nanocomposites

Previously obtained d-TiO_2-Pt/Cu nanoparticles were deposited on a magnetic substrate as a thin photocatalytic active shell. Magnetite (Fe_3O_4) was selected as a core of the designed composite due to its excellent magnetic properties (high Ms value and low Hc), which enable us to separate obtained photocatalyst in the external magnetic field. Silica was used as an interlayer to isolate Fe_3O_4 from TiO_2 and suppress possible electron transfer between them. The magnetic photocatalysts were obtained in the w/o microemulsion system based on changes in the particles surface charge as a function of pH, described in the previous study [26].

Firstly, commercially available Fe_3O_4 nanoparticles with nominate particles diameter of 50 nm were dispersed in water at pH 10. The prepared suspension was then introduced to cyclohexane/isopropanol (100:6 volume ratio) mixture in the presence of cationic surfactant and cetyltrimethylammonium bromide (CTAB) creating stable w/o microemulsion system with water nanodroplets dispersed in the continuous oil phase. The molar ratio of water to surfactant was set at 30. After the microemulsion stabilization, the corresponding amount of tetraethyl orthosilicate (TEOS) was added, resulting in the formation of SiO_2 interlayer on Fe_3O_4 core, after ammonia solution introduced into the microemulsion system. The molar ratio of TEOS to Fe_3O_4 was 8:1, and NH_4OH to TEOS was 16:1. The microemulsion was destabilized using acetone and obtained nanocomposite Fe_3O_4@SiO_2 was separated, washed with ethanol and water, dried at 70 °C to dry mass, and calcined at 400 °C for 2 h.

In the second step, previously obtained Fe_3O_4@SiO_2 particles were combined with d-TiO_2-Pt/Cu in order to create photocatalytic active nanomaterial. The reversed-phased microemulsion system at pH 10 was used, and Fe_3O_4 to the TiO_2 molar ratio was equaled to 1:4 [27]. The junction between magnetic/SiO_2 and photocatalytic layers was promoted by their opposite surface charges, provided by the presence of CTAB at the basic conditions. The as-obtained $Fe_3O_4@SiO_2$/d-TiO_2-Pt/Cu samples, after their separation and purification using water and ethanol, were dried at 70 °C to dry mass and calcined at 300 °C for 2 h.

3.3. Characterization of the Obtained Magnetic Photocatalysts

The XRD analyses were performed using the Rigaku Intelligent X-ray diffraction system SmartLab equipped with a sealed tube X-ray generator (a copper target; operated at 40 kV and 30 mA). Data was collected in the 2θ range of 5–80° with a scan speed and scan step of $1° \cdot min^{-1}$ and 0.01°, respectively. The analyses were based on the International Centre for Diffraction Data (ICDD) databased. The crystallite size of the photocatalysts in the vertical direction to the corresponding lattice plane was determined using Scherrer's equation with Scherrer's constant equal to 0.891. Quantitative analysis, including phase composition with standard deviation, was calculated using the Reference Intensity Ratio (RIR) method from the most intensive independent peak of each phase.

Nitrogen adsorption-desorption isotherms (BET method for the specific surface area) were recorded using the Micromeritics Gemini V (model 2365) (Norcross, GA, USA) instrument at 77 K (liquid nitrogen temperature).

Light absorption properties were measured using diffuse reflectance (DR) spectroscopy in the range of 200–800 nm. The bandgap energy of obtained samples was calculated from $(F(R) \cdot E)^{0.5}$ against E graph, where E is photon energy, and F(R) is Kubelka–Munk function, proportional to the radiation's absorption. The measurements were carried out using ThermoScientific Evolution 220 Spectrophotometer (Waltham, MA, USA) equipped with a PIN-757 integrating sphere. As a reference, $BaSO_4$ was used.

X-ray photoelectron spectroscopy (XPS) measurements were conducted using Escalab 250Xi multi-spectrometer (Thermofisher Scientific) using Mg K X-rays.

The morphology and distribution size for $Fe_3O_4@SiO_2$/d-TiO_2-Pt as a reference magnetic nanocomposite sample was further analyzed using HR-TEM imagining (Tecnai F20 X-Twin, FEI Europe) together with elements identification in nanometric scale by EDS mapping.

Electron paramagnetic resonance (EPR) spectroscopy was used for intrinsic defects formation confirmation. Measurements were conducted using RADIOPAN SE/X-2547 spectrometer (Poznań, Poland), operating at room temperature, with frequency in range 8.910984–8.917817 GHz.

3.4. Photocatalytic Activity Analysis

Photocatalytic activity of the obtained samples was evaluated in phenol degradation reaction, both in UV-Vis and Vis light irradiation, using 300 W Xenon lamp (LOT Oriel, Darmstadt, Germany). For the visible light measurements, a cut-off 420 nm filter (Optel, Opole, Poland) was used to obtain a settled irradiation interval. A 0.05 g (2 $g \cdot dm^{-3}$) of a photocatalyst, together with a 20 $mg \cdot dm^{-3}$ phenol solution, was added to a 25 cm^3 quartz photoreactor with an exposure layer thickness of 3 cm and obtained suspension was stirred in darkness for 30 min to provide adsorption-desorption equilibrium. After that, photocatalyst suspension was irradiated under continuous stirring and a power flux (irradiation intensity) of 30 $mW \cdot cm^{-2}$ for 60 min. The constant temperature of the aqueous phase was kept at 20 °C using a water bath. Every 20 min of irradiation, 1.0 cm^3 of suspension was collected and filtered through syringe filters (pore size = 0.2 µm) for the removal of photocatalysts particles. Phenol concentration, as well as a formation of degradation intermediates, were analyzed using reversed-phase high-performance liquid chromatography (HPLC) system, equipped with C18 chromatography column with bound residual silane groups (Phenomenex, model 00F-4435-E0) and a UV-Vis detector with a DAD photodiodes array (model SPD-M20A, Shimadzu). The tests were carried out at 45 °C and under isocratic flow conditions of 0.3 $mL \cdot min^{-1}$ and volume composition of the mobile phase of 70% acetonitrile, 29.5% water and 0.5% orthophosphoric acid. Qualitative and quantitative analysis was performed based on previously made measurements of relevant substance standards and using the method of an external calibration curve.

Phenol removal percentage was calculated from the equation:

$$D\% = \frac{C_o - C_n}{C_o} \cdot 100 \quad (1)$$

where: C_o—phenol initial concentration [mg·dm^{-3}], C_n—phenol concentration during photodegradation [mg·dm^{-3}].

Rate constant k was determined from ln(C_o/C_n) against t plot where C_o and C_n are phenol concentrations [mg·dm^{-3}] and t is degradation time [min]. Rate constant k is equal to directional coefficient "a" of the plot.

In order to evaluate the stability of the obtained photocatalysts, three 1-h-long subsequent cycles of phenol degradation under UV-Vis light using the most active defective d-TiO$_2$_20/Pt0.1/Cu0.1 sample were performed. After each cycle, photocatalyst was separated from the suspension and use in the next cycle without additional treatment.

The effect of charge carrier scavengers was examined by addition into phenol solution 1 cm^3 of 500 mg·dm^{-3} of tert-butyl alcohol (t-BuOH), benzoquinone (BQ), ammonium oxalate (AO), and silver nitrate (SN), respectively.

3.5. Electrochemical Measurements

In order to prepare Mott–Schottky plot the fabricated titania powders were used to form the paste, deposited using doctor-blade technique onto the Pt support. The paste consist of 0.2 g of photocatalyst in 0.1 g of polyethylene glycol (PEG) and 1 cm^3 of deionized water. Finally the calcination was carried out at 400 °C for 5 h with a heating rate 1 °C·min-^{1} ensuring removal of the organic binder. The fabricated electrode material stayed as working electrode tested in three electrode arrangement where Ag/AgCl/0.1M KCl and Pt mesh were used as reference and counter electrode, respectively. The deaerated 0.5 M Na$_2$SO$_4$ was applied as electrolyte. The electrochemical spectroscopy (EIS) impedance data was recorded from the anodic towards cathodic direction. Prior the tests, the investigated samples were not subjected to any preliminary treatment or measurement and their potential was held to reach a steady-state conditions. EIS data were recorded for the single frequency of 1000 Hz in the potential range from +0.1 to −1.2 V vs. Ag/AgCl/0.1 M KCl using a 10 mV amplitude of the AC signal. The capacitance of space charge layer was further calculated from the imaginary part of the measured impedance following the equation [50]:

$$C_{SC} = \frac{-1}{2\pi f Z_{im}} \quad (2)$$

where f stands for the frequency of the AC signal and Z_{im} for the imaginary part of impedance.

4. Conclusions

Surface modification of defective d-TiO$_2$ photocatalyst with platinum and copper nanoparticles resulted in a significant increase in its photocatalytic activity, both in UV-Vis and Vis range. The EPR analysis confirmed the presence of Ti defects in the structure of TiO$_2$ samples. The highest activity in Vis light was noticed for d-TiO$_2$ modified with Pt NPs. It resulted from surface plasmon resonance properties of Pt and narrowing the bandgap of the defective d-TiO$_2$. Among magnetic photocatalysts, the highest activity in Vis light was observed for Pt-modified and Pt/Cu-modified defective d-TiO$_2$ deposited on Fe$_3$O$_4$@SiO$_2$ magnetic core. Analysis of phenol degradation mechanism revealed that superoxide radicals are mainly responsible for phenol oxidation and mineralization. However, the photocatalytic activity in reaction of phenol degradation in UV-Vis light in the presence of Pt-modified Fe$_3$O$_4$@SiO$_2$/d-TiO$_2$ with the Pt particle size of about 20 nm was comparable with the activity of $Fe_3O_4@SiO_2/d$-TiO_2. It resulted from the deposition of Pt NPs in the place of titanium vacancies, and as a consequence formation of larger metal particles due to the seed-mediated growth mechanism on the TiO_2. In this regard, a lower metal/semiconductor interface resulted in a decrease in photocatalytic activity in the UV-Vis spectrum range. Furthermore, the creation of a core-shell magnetic $Fe_3O_4@SiO_2/d$-TiO_2-Pt/Cu nanostructures allowed an effective separation of the obtained magnetic photocatalysts.

Author Contributions: Conceptualization, A.Z.-J.; Formal analysis, Z.B., S.D., J.R., K.S., and A.S.; Funding acquisition, A.Z.-J.; Investigation, Z.B.; Methodology, A.Z.-J. and Z.B.; Project administration, A.Z.-J.; Writing—original draft, A.Z.-J. and Z.B.; Writing—review and editing, A.Z.-J. and Z.B. All authors have read and agreed to the published version of the manuscript.

Funding: This research was funded by Polish National Science Centre (Grant No. NCN 2016/23/D/ST5/01021).

Acknowledgments: This research was supported by Polish National Science Centre, grant no. NCN 2016/23/D/ST5/01021.

Conflicts of Interest: The authors declare no conflict of interest.

References

1. Al-Mamun, M.R.; Kader, S.; Islam, M.S.; Khan, M.Z.H. Photocatalytic activity improvement and application of UV-TiO_2 photocatalysis in textile wastewater treatment: A review. *J. Environ. Chem. Eng.* **2019**, *7*, 103248. [CrossRef]
2. Zhu, D.; Zhou, Q. Action and mechanism of semiconductor photocatalysis on degradation of organic pollutants in water treatment: A review. *Environ. Nanotechnol. Monit. Manag.* **2019**, *12*, 100255. [CrossRef]
3. Ahmad, K.; Ghatak, H.R.; Ahuja, S.M. Photocatalytic Technology: A review of environmental protection and renewable energy application for sustainable development. *Environ. Technol. Innov.* **2020**, *19*, 100893. [CrossRef]
4. Wang, K.; Janczarek, M.; Wei, Z.; Raja-Mogan, T.; Endo-Kimura, M.; Khedr, T.M.; Ohtani, B.; Kowalska, E. Morphology- and Crystalline Composition-Governed Activity of Titania-Based Photocatalysts: Overview and Perspective. *Catalysts* **2019**, *9*, 1054. [CrossRef]
5. Koe, W.S.; Lee, J.W.; Chong, W.C.; Pang, Y.L.; Sim, L.C. An overview of photocatalytic degradation: Photocatalysts, mechanisms, and development of photocatalytic membrane. *Environ. Sci. Pollut. Res.* **2020**, *27*, 2522–2565. [CrossRef] [PubMed]
6. Loeb, S.K.; Alvarez, P.J.J.; Brame, J.A.; Cates, E.L.; Choi, W.; Crittenden, J.; Dionysiou, D.D.; Li, Q.; Li-Puma, G.; Quan, X.; et al. The Technology Horizon for Photocatalytic Water Treatment: Sunrise or Sunset? *Environ. Sci. Technol.* **2019**, *53*, 2937–2947. [CrossRef] [PubMed]
7. Serpone, N. Is the Band Gap of Pristine TiO_2 Narrowed by Anion- and Cation-Doping of Titanium Dioxide in Second-Generation Photocatalysts? *J. Phys. Chem. B* **2006**, *110*, 24287–24293. [CrossRef] [PubMed]
8. Dozzi, M.V.; Selli, E. Doping TiO_2 with p-block elements: Effects on photocatalytic activity. *J. Photochem. Photobiol. C Photochem. Rev.* **2013**, *14*, 13–28. [CrossRef]
9. Diaz-Angulo, J.; Gomez-Bonilla, I.; Jimenez-Tohapanta, C.; Mueses, M.; Pinzon, M.; Machuca-Martinez, F. Visible-light activation of TiO_2 by dye-sensitization for degradation of pharmaceutical compounds. *Photochem. Photobiol. Sci.* **2019**, *18*, 897–904. [CrossRef]
10. Endo-Kimura, M.; Janczarek, M.; Bielan, Z.; Zhang, D.; Wang, K.; Markowska-Szczupak, A.; Kowalska, E. Photocatalytic and Antimicrobial Properties of Ag2O/TiO2 Heterojunction. *ChemEngineering* **2019**, *3*, 3. [CrossRef]
11. Wysocka, I.; Kowalska, E.; Ryl, J.; Nowaczyk, G.; Zielińska-Jurek, A. Morphology, Photocatalytic and Antimicrobial Properties of TiO_2 Modified with Mono- and Bimetallic Copper, Platinum and Silver Nanoparticles. *Nanomaterials* **2019**, *9*, 1129. [CrossRef]
12. Klein, M.; Grabowska, E.; Zaleska, A. Noble metal modified TiO_2 for photocatalytic air purification. *Physicochem. Probl. Miner. Process.* **2015**, *51*, 49–57.
13. Janczarek, M.; Wei, Z.; Endo, M.; Ohtani, B.; Kowalska, E. Silver- and copper-modified decahedral anatase titania particles as visible light-responsive plasmonic photocatalyst. *J. Photonics Energy* **2016**, *7*, 12008. [CrossRef]

14. Wei, Z.; Janczarek, M.; Endo, M.; Wang, K.; Balcytis, A.; Nitta, A.; Mendez-Medrano, M.G.; Colbeau-Justin, C.; Juodkazis, S.; Ohtani, B.; et al. Noble Metal-Modified Faceted Anatase Titania Photocatalysts: Octahedron versus Decahedron. *Appl. Catal. B Environ.* **2018**, *237*, 574–587. [CrossRef] [PubMed]
15. Wu, Q.; Huang, F.; Zhao, M.; Xu, J.; Zhou, J.; Wang, Y. Ultra-small yellow defective TiO_2 nanoparticles for co-catalyst free photocatalytic hydrogen production. *Nano Energy* **2016**, *24*, 63–71. [CrossRef]
16. Liriano-Jorge, C.F.; Sohmen, U.; Özkan, A.; Gulyas, H.; Otterpohl, R. TiO_2 Photocatalyst Nanoparticle Separation: Flocculation in Different Matrices and Use of Powdered Activated Carbon as a Precoat in Low-Cost Fabric Filtration. *Adv. Mater. Sci. Eng.* **2014**, *2014*, 1–12. [CrossRef]
17. Lee, S.-A.; Choo, K.-H.; Lee, C.-H.; Lee, H.-I.; Hyeon, T.; Choi, W.; Kwon, H.-H. Use of Ultrafiltration Membranes for the Separation of TiO_2 Photocatalysts in Drinking Water Treatment. *Ind. Eng. Chem. Res.* **2001**, *40*, 1712–1719. [CrossRef]
18. Ray, S.; Lalman, J.A. Fabrication and characterization of an immobilized titanium dioxide (TiO_2) nanofiber photocatalyst. *Mater. Today Proc.* **2016**, *3*, 1582–1591. [CrossRef]
19. Zielińska-Jurek, A.; Klein, M.; Hupka, J. Enhanced visible light photocatalytic activity of Pt/I-TiO2in a slurry system and supported on glass packing. *Sep. Purif. Technol.* **2017**, *189*, 246–252. [CrossRef]
20. Wei, J.H.; Leng, C.J.; Zhang, X.Z.; Li, W.H.; Liu, Z.Y.; Shi, J. Synthesis and magnetorheological effect of Fe_3O_4-TiO_2 nanocomposite. *J. Phys. Conf. Ser.* **2009**, *149*, 25–29. [CrossRef]
21. Zhang, L.; Wu, Z.; Chen, L.; Zhang, L.; Li, X.; Xu, H.; Wang, H.; Zhu, G. Preparation of magnetic Fe_3O_4/TiO_2/Ag composite microspheres with enhanced photocatalytic activity. *Solid State Sci.* **2016**, *52*, 42–48. [CrossRef]
22. Abbas, M.; Rao, B.P.; Reddy, V.; Kim, C. Fe_3O_4/TiO_2 core/shell nanocubes: Single-batch surfactantless synthesis, characterization and efficient catalysts for methylene blue degradation. *Ceram. Int.* **2014**, *40*, 11177–11186. [CrossRef]
23. Sathishkumar, P.; Viswanathan, R.V.; Anandan, S.; Ashokkumar, M. $CoFe_2O_4$/TiO_2 nanocatalysts for the photocatalytic degradation of Reactive Red 120 in aqueous solutions in the presence and absence of electron acceptors. *Chem. Eng. J.* **2013**, *220*, 302–310. [CrossRef]
24. Jia, Y.; Liu, J.; Cha, S.; Choi, S.; Chang, Y.C.; Liu, C. Magnetically separable Au-TiO_2/nanocube $ZnFe_2O_4$ composite for chlortetracycline removal in wastewater under visible light. *J. Ind. Eng. Chem.* **2017**, *47*, 303–314. [CrossRef]
25. Fu, W.; Yang, H.; Li, M.; Chang, L.; Yu, Q.; Xu, J.; Zou, G. Preparation and photocatalytic characteristics of core-shell structure TiO_2/$BaFe_{12}O_{19}$ nanoparticles. *Mater. Lett.* **2006**, *60*, 2723–2727. [CrossRef]
26. Zielińska-Jurek, A.; Bielan, Z.; Dudziak, S.; Wolak, I.; Sobczak, Z.; Klimczuk, T.; Nowaczyk, G.; Hupka, J. Design and Application of Magnetic Photocatalysts for Water Treatment. The Effect of Particle Charge on Surface Functionality. *Catalysts* **2017**, *7*, 360. [CrossRef]
27. Zielińska-Jurek, A.; Bielan, Z.; Wysocka, I.; Strychalska, J.; Janczarek, M.; Klimczuk, T. Magnetic semiconductor photocatalysts for the degradation of recalcitrant chemicals from flow back water. *J. Environ. Manage.* **2017**, *195*, 157–165. [CrossRef]
28. Wysocka, I.; Kowalska, E.; Trzciński, K.; Łapiński, M.; Nowaczyk, G.; Zielińska-Jurek, A. UV-Vis-Induced Degradation of Phenol over Magnetic Photocatalysts Modified with Pt, Pd, Cu and Au Nanoparticles. *Nanomaterials* **2018**, *8*, 28. [CrossRef]
29. Mrotek, E.; Dudziak, S.; Malinowska, I.; Pelczarski, D.; Ryżyńska, Z.; Zielińska-Jurek, A. Improved degradation of etodolac in the presence of core-shell $ZnFe_2O_4$/SiO_2/TiO_2 magnetic photocatalyst. *Sci. Total Environ.* **2020**, *724*, 138167. [CrossRef]
30. Gad-Allah, T.A.; Fujimura, K.; Kato, S.; Satokawa, S.; Kojima, T. Preparation and characterization of magnetically separable photocatalyst (TiO_2/SiO_2/Fe_3O_4): Effect of carbon coating and calcination temperature. *J. Hazard. Mater.* **2008**, *154*, 572–577. [CrossRef]
31. Fan, Y.; Ma, C.; Li, W.; Yin, Y. Synthesis and properties of Fe_3O_4/SiO_2/TiO_2 nanocomposites by hydrothermal synthetic method. *Mater. Sci. Semicond. Process.* **2012**, *15*, 582–585. [CrossRef]
32. Shi, F.; Li, Y.; Zhang, Q.; Wang, H. Synthesis of Fe_3O_4/C/TiO_2 magnetic photocatalyst via vapor phase hydrolysis. *Int. J. Photoenergy* **2012**, *2012*, 1–8. [CrossRef]
33. Zielińska-Jurek, A.; Wei, Z.; Janczarek, M.; Wysocka, I.; Kowalska, E. Size-Controlled Synthesis of Pt Particles on TiO_2 Surface: Physicochemical Characteristic and Photocatalytic Activity. *Catalysts* **2019**, *9*, 940. [CrossRef]

34. Bielan, Z.; Kowalska, E.; Dudziak, S.; Wang, K.; Ohtani, B.; Zielińska-Jurek, A. Mono- and bimetallic (Pt/Cu) titanium(IV) oxide core-shell photocatalysts with UV/Vis light activity and magnetic separability. *Catal. Today* **2020**, in press. [CrossRef]
35. Gamboa, J.A.; Pasquevich, D.M. Effect of Chlorine Atmosphere on the Anatase-Rutile Transformation. *J. Am. Chem. Soc.* **1992**, *75*, 2934–2938. [CrossRef]
36. Byrne, C.; Fagan, R.; Hinder, S.; McCormack, D.E.; Pillai, S.C. New Approach of Modifying the Anatase to Rutile Transition Temperature in TiO_2 Photocatalysts. *RSC Adv.* **2016**, *6*, 95232–95238. [CrossRef]
37. Ricci, P.C.; Carbonaro, C.M.; Stagi, L.; Salis, M.; Casu, A.; Enzo, S.; Delogu, F. Anatase-To-Rutile Phase Transition In Nanoparticles Irradiated By Visible Light. *J. Phys. Chem. C* **2013**, *117*, 785–7857. [CrossRef]
38. Liu, H.; Jia, Z.; Ji, S.; Zheng, Y.; Li, M.; Yang, H. Synthesis of $TiO_2/SiO_2@Fe_3O_4$ magnetic microspheres and their properties of photocatalytic degradation dyestuff. *Catal. Today* **2011**, *175*, 293–298. [CrossRef]
39. Chi, Y.; Yuan, Q.; Li, Y.; Zhao, L.; Li, N.; Li, X.; Yan, W. Magnetically separable $Fe_3O_4@SiO_2@TiO_2$-Ag microspheres with well-designed nanostructure and enhanced photocatalytic activity. *J. Hazard. Mater.* **2013**, *262*, 404–411. [CrossRef] [PubMed]
40. Zielińska-Jurek, A.; Hupka, J. Preparation and characterization of Pt/Pd-modified titanium dioxide nanoparticles for visible light irradiation. *Catal. Today* **2013**, *230*, 181–187. [CrossRef]
41. Bielan, Z.; Kowalska, E.; Dudziak, S.; Wang, K.; Ohtani, B.; Zielinska-Jurek, A. Mono- and bimetallic (Pt/Cu) titanium(IV) oxide photocatalysts. Physicochemical and photocatalytic data for magnetic nanocomposites' shell. *Data Brief* **2020**, *31*, 105814. [CrossRef]
42. Chan, G.H.; Zhao, J.; Hicks, E.M.; Schatz, G.C.; Van Duyne, R.P. Plasmonic Properties of Copper Nanoparticles Fabricated by Nanosphere Lithography. *Nano Lett.* **2007**, *7*, 1947–1952. [CrossRef]
43. Ghodselahi, T.; Vesaghi, M.A.; Shafiekhani, A.; Baghizadeh, A.; Lameii, M. XPS study of the $Cu@Cu_2O$ core-shell nanoparticles. *Appl. Surf. Sci.* **2008**, *225*, 2730–2734. [CrossRef]
44. Li, B.; Luo, X.; Zhu, Y.; Wang, X. Immobilization of Cu(II) in KIT-6 Supported Co3O4 and Catalytic Performance for Epoxidation of Styrene. *Appl. Surf. Sci.* **2015**, *359*, 609–620. [CrossRef]
45. Zielińska-Jurek, A.; Wei, Z.; Wysocka, I.; Szweda, P.; Kowalska, E. The effect of nanoparticles size on photocatalytic and antimicrobial properties of $Ag-Pt/TiO_2$ photocatalysts. *Appl. Surf. Sci.* **2015**, *353*, 317–325. [CrossRef]
46. Monga, A.; Rather, R.A.; Pal, B. Enhanced co-catalytic effect of Cu-Ag bimetallic core-shell nanocomposites imparted to TiO2 under visible light illumination. *Sol. Energy Mater Sol. Cells* **2017**, *172*, 285–292. [CrossRef]
47. Devi, L.G.; Kavitha, R. A review on plasmonic metal—TiO_2 composite for generation, trapping, storing and dynamic vectorial transfer of photogenerated electrons across the Schottky junction in a photocatalytic system. *Appl. Surf. Sci.* **2016**, *360*, 601–622. [CrossRef]
48. Dang, T.T.T.; Le, S.T.T.; Channei, D.; Khanitchaidecha, W.; Nakaruk, A. Photodegradation mechanisms of phenol in the photocatalytic process. *Res. Chem. Intermed.* **2016**, *42*, 5961–5974. [CrossRef]
49. Esplugas, S.; Gimenez, J.; Contreras, S.; Pascual, E.; Rodriguez, M. Comparison of different advanced oxidation processes for phenol degradation. *Water Res.* **2002**, *36*, 1034–1042. [CrossRef]
50. Beranek, R. (Photo)electrochemical methods for the determination of the band edge positions of TiO_2-based nanomaterials. *Adv. Phys. Chem.* **2011**, *2011*, 80–83. [CrossRef]

© 2020 by the authors. Licensee MDPI, Basel, Switzerland. This article is an open access article distributed under the terms and conditions of the Creative Commons Attribution (CC BY) license (http://creativecommons.org/licenses/by/4.0/).

Article

Synthesis of Au–Ag Alloy Nanoparticle-Incorporated AgBr Crystals

Shin-ichi Naya [1], Musashi Fujishima [2] and Hiroaki Tada [1,2,*]

1 Environmental Research Laboratory, Kindai University, 3-4-1, Kowakae, Higashi-Osaka, Osaka 577-8502, Japan
2 Graduate School of Science and Engineering, Kindai University, 3-4-1, Kowakae, Higashi-Osaka, Osaka 577-8502, Japan
* Correspondence: h-tada@apch.kindai.ac.jp; Tel.: +81-6-6721-2332; Fax: +81-6-6727-2024

Received: 1 August 2019; Accepted: 30 August 2019; Published: 3 September 2019

Abstract: Nanoscale composites consisting of silver and silver halide (Ag–AgX, X = Cl, Br, I) have attracted much attention as a novel type of visible-light photocatalyst (the so-called plasmonic photocatalysts), for solar-to-chemical transformations. Support-free Au–Ag alloy nanoparticle-incorporated AgBr crystals (Au–Ag@AgBr) were synthesized by a photochemical method. At the initial step, Au ion-doped AgBr particles were prepared by adding an aqueous solution of $AgNO_3$ to a mixed aqueous solution of KBr and $HAuBr_4$. At the next step, UV-light illumination (λ = 365 nm) of a methanol suspension of the resulting solids yielded Au–Ag alloy nanoparticles with a mean size of approximately 5 nm in the micrometer-sized AgBr crystals. The mole percent of Au to all the Ag in Au–Ag@AgBr was controlled below < 0.16 mol% by the $HAuBr_4$ concentration in the first step. Finite-difference time-domain calculations indicated that the local electric field enhancement factor for the alloy nanoparticle drastically decreases with an increase in the Au content. Also, the peak of the localized surface plasmon resonance shifts towards longer wavelengths with increasing Au content. Au–Ag@AgBr is a highly promising plasmonic photocatalyst for sunlight-driven chemical transformations due to the compatibility of the high local electric field enhancement and sunlight harvesting efficiency.

Keywords: silver halide; silver–gold alloy nanoparticle; local electric field enhancement; plasmonic photocatalyst

1. Introduction

In view of energy and environmental issues, solar energy utilization for the production of useful chemicals and for the decomposition of harmful environmental pollutants has become increasingly important. Nanoparticles (NPs) made from Ag and Au have high absorption due to the localized surface plasmon resonance (LSPR). The LSPR excitation-driven photocatalysts represented by Au NP-loaded TiO_2 (Au/TiO_2) have emerged as a new type of visible-light photocatalysts [1,2]. Among the plasmonic photocatalysts, silver–silver halides (Ag–AgX, X = Cl, Br, I) also exhibit visible-light activity for important chemical reactions including hydrogen evolution [3,4] and CO_2 reduction [5–7]. Recently, Ag NP-incorporated AgBr crystals on TiO_2 have been shown to work as a plasmonic photocatalyst via the local electric field enhancement (LEFE) mechanism [8]. In this case, the key to improving the solar-to-chemical conversion efficiency is the compatibility of high LEFE and sunlight harvesting efficiency (LHE) [9]. Figure 1a shows the absorption spectra of spherical Au–Ag alloy NPs in water calculated as a function of the Au mole fraction (x) by the finite-difference time-domain (FDTD) method. Most LSPR absorption of Ag NPs is located at a wavelength region below 400 nm, while the absorption spectrum for Au NPs matches well with the solar spectrum. In the Au_x–Ag_{1-x} alloy system, the LSPR peak redshifts from 390 nm at x = 0 to 530 nm at x = 1. The light absorption or the generation rate of the photocharge carriers is proportional to the electric field squared $|E|^2$ [10]. To indicate the plasmonic enhancement, the maximum local electric field enhancement factor (EF_{max}) is defined by Equation (1):

$$EF_{max} = (E_{max}/E_0)^2 \quad (1)$$

where E_0 is the amplitude of incident electric field.

Figure 1b shows the EF_{max} calculated for various x values. Ag NPs ($x = 0$) possess much more intense EF_{max} than Au NPs ($x = 1$), and in the alloy system, the EF_{max} drastically decreases with an increase in x. Thus, precise control of the alloy composition would enhance the photocatalytic activity through the fulfilment of the optical requirements. Although the control of the Au–Ag alloy composition is generally difficult because of the large difference in the reduction potentials of Ag$^+$ and Au^{3+} ions [11], Au–Ag alloy NPs have been synthesized by co-reduction of HAuCl$_4$ and AgNO$_3$ with NaBH$_4$ [12,13], citric acid [14], starch [15] or wolfberry fruit extract [16], and by γ-ray irradiation [17]. We have recently reported a photochemical method for preparing Au–Ag alloy NP-incorporated AgBr crystals on TiO$_2$ (Au–Ag@AgBr/TiO$_2$) [18].

In this study, a method has been developed to synthesize support-free Au–Ag@AgBr crystals with varying alloy compositions. The characterization of the samples and FDTD calculations for the model system indicate that Au–Ag@AgBr crystals are a promising material for plasmonic photocatalysis.

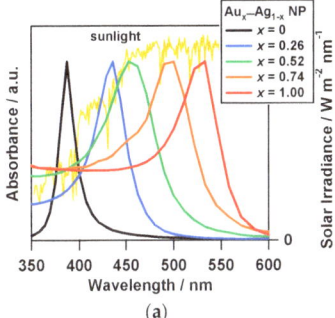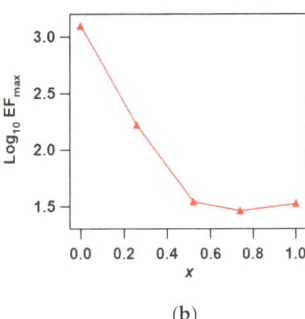

Figure 1. (a) Finite-difference time-domain (FDTD)-calculated absorption spectra of Au$_x$–Ag$_{1-x}$ alloy nanoparticles (x, mole fraction of Au) with a diameter of 5 nm in water. The absorption intensity is normalized with respect to the peak. (b) Maximum local electric field enhancement factor (EF_{max}) as a function of x. This data was cited from reference [18].

2. Results and Discussion

2.1. Synthesis of Au–Ag@AgBr

The Cl$^-$ ligands of HAuCl$_4$ were substituted to Br$^-$ ligands by the addition of KBr aqueous solution. The resulting HAuBr$_4$ is further reduced to HAuBr$_2$ by water in the presence of AgNO$_3$ (Equation (2)) [18] due to the negative standard Gibbs energy of the reaction ($\Delta_r G_0 = -57.6$ kJ mol^{-1}).

$$AuBr_4^- + 2Ag^+ + H_2O \rightarrow AuBr_2^- + 2AgBr + 2H^+ + 1/2O_2, \quad (2)$$

Then, the slow addition of AgNO$_3$ aqueous solution to KBr aqueous solution containing HAuBr$_4$ at varying concentrations (C_{sol}) yields Au$^+$ ion-doped AgBr particles (AgBr:Au$^+$). The mole percent of Au to Ag in AgBr:Au$^+$ ($y = ((Au \text{ mole}/Ag \text{ mole}) \times 100)$) was determined by inductively coupled plasma spectroscopy. Figure 2a shows the relationship between y and C_{sol}. The y value monotonically increases with an increase in C_{sol}, and thus the Au-doping amount can be precisely controlled by the Au-complex concentration. X-ray diffraction (XRD) measurements were carried out for AgBr:Au$^+$ with varying Au$^+$ ion-doping amounts. As shown in Figure 2b, diffraction peaks at $2\theta = 26.8°$, $31.0°$, $44.4°$,

55.1°, 64.5°, 73.2° were indexed as the diffraction from the (111), (200), (220), (222), (400), and (331) crystal planes of AgBr, respectively, and are observed in every sample.

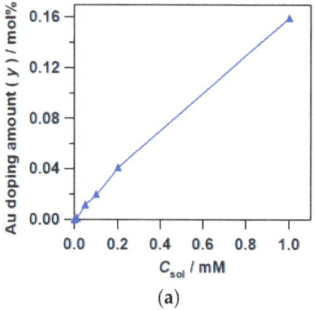 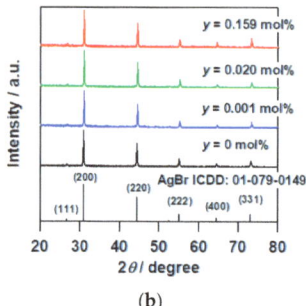

(a) (b)

Figure 2. (a) Plots of Au doping amount (y) as a function of the added HAuCl$_4$ concentration (C_{sol}). (b) X-ray diffraction (XRD) patterns for AgBr:Au$^+$ and AgBr reference pattern of the international center for diffraction data (ICDD).

Next, a methanol suspension of AgBr:Au$^+$ was illuminated by ultraviolet light emitting diode (UV-LED) (λ = 365 nm). Figure 3a shows the scanning electron microscopy (SEM) image of the sample (y = 0.159 mol%) obtained after UV-light irradiation. Micrometer-sized particles are observed, and the other samples with different Au-doping amounts had a similar size and shape. In order to directly confirm the formation of metal NPs in AgBr, the surrounding AgBr was selectively dissolved by an aqueous solution of 7.5 M NaCl and 40 mM octadecyltrimethylammonium chloride. Figure 3b shows a high resolution transmission electron microscopy (HR-TEM) image for a metal NP obtained after dissolving the AgBr matrix of irradiated Au–Ag@AgBr (y = 0.041 mol%). The particle size is approximately 5 nm, and the observed d-spacing is in agreement with the values of Au (111) and Ag (111) planes (Au(111) = 0.236 nm (the international center for diffraction data (ICDD) No. 00-004-0784), Ag(111) = 0.237 nm (ICDD No. 01-071-3752)). Figure 3c shows energy dispersive X-ray spectroscopy (EDX) line-elemental analysis for the metal NP. While the intensity of Au is smaller compared with that of Ag due to the small doping amount of Au (y = 0.041 mol%), Ag and Au co-exist homogeneously in the metal NP in contrast to the core-shell structure [19]. Evidently, the Au–Ag alloy NP-incorporated AgBr crystals are formed by this solid-phase photochemical reaction. No change in the XRD patterns of AgBr:Au$^+$ was observed before and after irradiation, most likely due to the small amounts of the metal NPs generated.

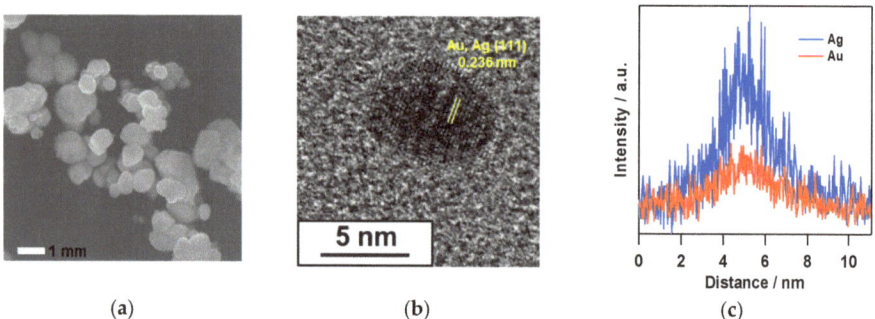

(a) (b) (c)

Figure 3. (a) Scanning electron microscopy (SEM) image of Au–Ag@AgBr (y = 0.159 mol%). High resolution transmission electron microscopy (HR-TEM) image (b) and energy dispersive X-ray spectroscopy (EDX) analysis (c) of metal nanoparticles obtained from Au–Ag@AgBr (y = 0.041 mol%) by dissolving AgBr.

2.2. Optical Properties of Au–Ag@AgBr

Figure 4a,b compares the UV-Vis absorption spectra of AgBr:Au$^+$ with varying y before (Figure 4a) and after (Figure 4b) irradiation by UV-LED. Before irradiation, all samples possess strong absorption below 470 nm due to the interband transition of AgBr with a very weak LSPR absorption. After irradiation, the LSPR greatly intensifies around 500–600 nm, which also indicates the growth of Ag and/or Au NPs in AgBr by irradiation.

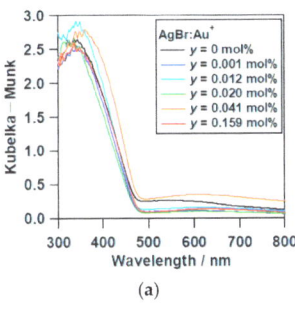
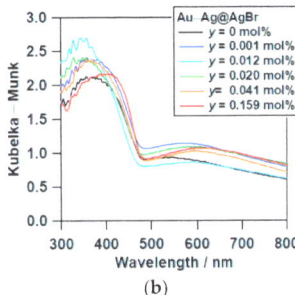

Figure 4. (**a**) UV-visible absorption spectra for AgBr:Au$^+$ with varying y before (**a**) and after (**b**) irradiation with UV light.

Figure 5a shows the UV-Vis absorption spectra of the Au–Ag colloids obtained by dissolving the AgBr matrix of Au–Ag@AgBr. At $y = 0$, the LSPR peak is located at 414 nm, and the increase in y causes the redshift in the LSPR peak and its significant broadening. A similar trend was observed for the Au–Ag colloids obtained from Au–Ag@AgBr/mesoporous-TiO$_2$ in the previous report, although the alloy composition could not be determined [18]. Figure 5b shows the LSPR peak position (λ_{max}) for the Au–Ag colloids as a function of y. In each case, the λ_{max} monotonically increases with an increase in y. Thus, Au–Ag alloy NP-incorporated AgBr crystals can be formed by the solid-phase photochemical reaction, and the alloy composition can be simply controlled by the Au doping amount.

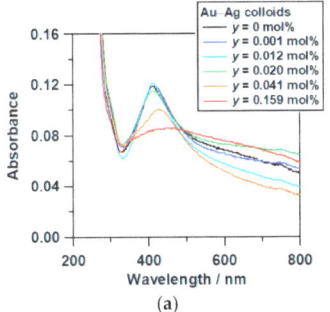
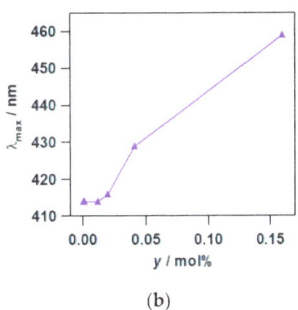

Figure 5. (**a**) UV-visible absorption spectra for Au–Ag aqueous colloids prepared from Au–Ag@AgBr by dissolving AgBr. (**b**) Plots of λ_{max} of localized surface plasmon resonance (LSPR) as a function of Au-doping amount (y).

2.3. Mechanism on the Formation of Au–Ag@AgBr

We propose a reaction mechanism on the solid-state photochemical formation of Au–Ag alloy NP-incorporated AgBr crystals by taking the results of the previous density functional theory calculations [18] (Scheme 1). The Au$^+$ ions doped into AgBr substitute the interstitial Ag ions (Ag$_i^+$) to occupy the interstitial sites (Au$_i^+$). The rise in C_{sol} increases the mole ratio of the Au$_i^+$ ions (m) to Ag$_i^+$ ions (n). Irradiation of AgBr by photons with more energy (hv) than the band gap excites the electrons from the valence band (VB) to the conduction band (CB) (Equation (3)). Methanol works as a sacrificial electron donor for the VB-holes (h^+_{VB}) (Equation (4)). On the other hand, the CB-electrons (e^-_{CB}) can be trapped by the levels of the interstitial ions Ag_i^+ or Au_i^+ (M_i^+) (Equation (5)). Due to the large electronegativity, the resulting metal atom (M) collects e^-_{CB} to generate a metal anion (Equation (6)) [20]. Ag^+ ions can easily migrate in an AgBr lattice with a low activation energy (~4 kJ mol^{-1}) [21]. Thus, the Ag_i^+ and Au_i^+ ions can migrate to the Ag and Au metal anions to generate a diatomic metal cluster (Equation (7)). Through repetition of these processes, Au_x–Ag_{1-x} alloy NPs grow in AgBr crystals, and the mole ratio of Au ($x = m/(m + n)$) can be changed by C_{sol} or the amount of doped Au^+ ions (y) (Equation (8)).

$$AgBr + h\nu \rightarrow e^-_{CB} + h^+_{VB}, \qquad (3)$$

$$h^+_{VB} + CH_3OH \rightarrow OP, \qquad (4)$$

where OP denotes the oxidized product.

$$M_i^+ + e^-_{CB} \rightarrow M, \qquad (5)$$

where the Au or Ag metal atom is abbreviated as M.

$$M + e^-_{CB} \rightarrow M^-, \qquad (6)$$

$$M^- + M_i^+ \rightarrow M^-M, \qquad (7)$$

$$M - M + M_i^+ + e^-_{CB} \rightarrow \cdots \rightarrow Au_m\text{–}Ag_n, \qquad (8)$$

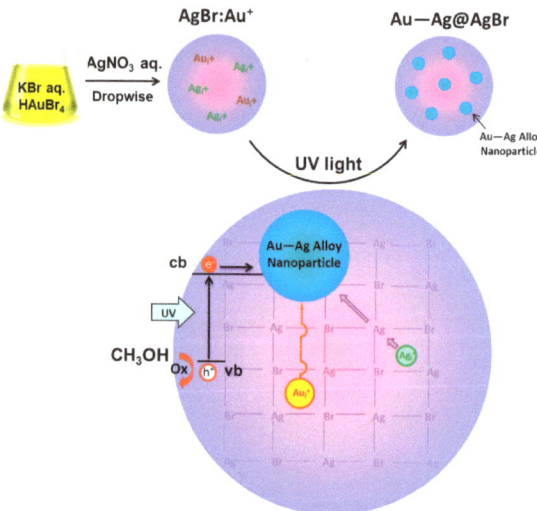

Scheme 1. Proposed mechanism for the synthesis of Au–Ag alloy nanoparticle-incorporated AgBr crystals.

3. Experimental Section

3.1. Catalyst Preparation and Characterization

An aqueous solution of KBr (0.1 M, 80 mL) containing HAuBr$_4$ (0–1 mM) was prepared by dissolution of HAuCl$_4$ and KBr in distilled water. AgNO$_3$ aqueous solution (0.1 M, 40 mL) was added dropwise slowly to the solution at 0.5 mL min^{-1} using a Perista pump. The resulting suspension was stirred at room temperature for 1 h. The particles collected by centrifugation were washed with distilled water three times and dried in vacuo to obtain AgBr:Au$^+$. The amounts of Ag and Au in the reaction solutions before and after the formation of AgBr particles were determined by inductively coupled plasma spectroscopy (ICPS-7000, Shimadzu). From the difference in each amount, the mole numbers of Ag and Au contained in AgBr:Au$^+$ were calculated. The particles (100 mg) were re-dispersed into methanol (20 mL), and illuminated by UV-light (λ = 365 nm, the light intensity integrated from 310 to 420 nm ($I_{310-420}$) = 4.0 mWcm^{-2}) at room temperature for 15 min. The particles collected by centrifugation were washed with distilled water three times and dried in vacuo to obtain Au–Ag@AgBr. The sample morphology was characterized by scanning electron microscopy (SEM, Hitachi S-800) at an acceleration voltage of 10 kV. Further, the samples were observed by transmission electron microscopy (TEM) at an applied voltage of 200 kV (JEM-2100F, JEOL). X-ray diffraction (XRD) was measured by a Mini Flex X-ray diffractometer (Rigaku) operating at 40 kV and 100 mA. The scans were collected in the range from 20° to 90° (2θ) by the use of Cu Kα radiation (λ = 1.545 Å). Diffuse reflectance UV-Vis-NIR spectra of the samples were recorded on a UV-2600 spectrometer (Shimadzu) with an integrating sphere unit (Shimadzu, ISR-2600Plus) by using a quartz cell at room temperature. The reflectance (R_∞) was recorded with respect to a reference of BaSO$_4$, and the Kubelka–Munk function [$F(R_\infty)$] expressing the relative absorption coefficient was calculated by the equation $F(R_\infty) = (1 - R_\infty)^2/2R_\infty$.

3.2. FDTD Calculaitons

According to the method previously reported [17], the local electric field of Au$_x$–Ag$_{1-x}$ alloy NPs was analyzed by the three-dimensional (3D) finite-difference time-domain (FDTD) method using FDTD Solutions (Lumerical Solutions, Inc. Vancouver, BC, Canada). For the models constructed by fixing the metal particle size at 5 nm, the calculations were performed in a water medium using the optical constants (refractive index and extinction coefficient) previously reported [22].

4. Conclusions

The optical property and local electric field enhancement of Au–Ag alloy NPs strongly depends on the alloy composition. This study has presented a photochemical method for synthesizing support-free Au–Ag alloy NP-incorporeated AgBr crystals (Au–Ag@AgBr) with the alloy composition controlled. We anticipate that the present Au–Ag@AgBr can be a promising plasmonic photocatalyst for efficient solar-to-chemical transformations through fine-tuning of the alloy composition.

Author Contributions: S.N. prepared the catalysts, and conducted characterization. M.F. performed FDTD calculation. H.T. supervised the experimental work and data analysis.

Funding: This research was funded by JSPS KAKENHI a Grant-in-Aid for Scientific Research (C), Grant Numbers 15K05654 and 18K05280, by the Futaba Foundation, and by the MEXT Supported Program for the Strategic Research Foundation at Private Universities.

Acknowledgments: This work was partially supported by JSPS KAKENHI a Grant-in-Aid for Scientific Research (C), Grant Numbers 15K05654 and 18K05280, by the Futaba Foundation, and by the MEXT Supported Program for the Strategic Research Foundation at Private Universities.

Conflicts of Interest: The authors declare no conflict of interest.

References

1. Panayotov, D.A.; Morris, J.R. Surface chemistry of Au/TiO$_2$: Thermally and photolytically activated reactions. *Surf. Sci. Rep.* **2016**, *71*, 77–271. [CrossRef]
2. Tada, H. Size, shape and interface control in gold nanoparticle-based plasmonic photocatalysts for solar-to-chemical transformations. *Dalton Trans.* **2019**, *48*, 6308–6313. [CrossRef] [PubMed]
3. Kakuta, N.; Goto, N.; Ohkita, H.; Mizushima, T. Silver Bromide as a Photocatalyst for Hydrogen Generation from CH$_3$OH/H$_2$O Solution. *J. Phys. Chem. B* **1999**, *103*, 5917–5919. [CrossRef]
4. An, C.; Wang, J.; Wang, S.; Zhang, Q.-H. Plasmonic enhancement of photocatalysis over Ag incorporated AgI hollow nanostructures. *RSC Adv.* **2014**, *4*, 2409–2413. [CrossRef]

5. An, C.; Wang, J.; Qin, C.; Jiang, W.; Wang, S.; Li, Y.; Zhang, Q. Synthesis of Ag@AgBr/AgCl heterostructured nanocashews with enhanced photocatalytic performance via anion exchange. *J. Mater. Chem.* **2012**, *22*, 13153. [CrossRef]
6. Asi, M.A.; Zhu, L.; He, C.; Sharma, V.K.; Shu, D.; Li, S.; Yang, J.; Xiong, Y. Visible-light-harvesting reduction of CO2 to chemical fuels with plasmonic Ag@AgBr/CNT nanocom-posites. *Catal. Today* **2013**, *216*, 268–275. [CrossRef]
7. Marszewski, M.; Cao, S.; Yu, J.; Jaroniec, M. Semiconductor-based photocatalytic CO_2 conversion. *Mater. Horiz.* **2015**, *2*, 261–278. [CrossRef]
8. Hayashido, Y.; Naya, S.-I.; Tada, H. Local Electric Field-Enhanced Plasmonic Photocatalyst: Formation of Ag Cluster-Incorporated AgBr Nanoparticles on TiO_2. *J. Phys. Chem. C* **2016**, *120*, 19663–19669. [CrossRef]
9. Tada, H.; Naya, S.-I.; Fujishima, M. Water splitting by plasmonic photocatalysts with a gold nanoparticle/cadmium sulfide heteroepitaxial junction: A mini review. *Electrochem. Commun.* **2018**, *97*, 22–26. [CrossRef]
10. Atkins, P.; de Paula, J. *Physical Chemistry*, 8th ed.; Oxford University Press: New York, NY, USA, 2006.
11. Gao, C.; Hu, Y.; Wang, M.; Chi, M.; Yin, Y. Fully Alloyed Ag/Au Nanospheres: Combining the Plasmonic Property of Ag with the Stability of Au. *J. Am. Chem. Soc.* **2014**, *136*, 7474–7479. [CrossRef] [PubMed]
12. Hostetler, M.J.; Zhong, C.-J.; Yen, B.K.H.; Anderegg, J.; Gross, S.M.; Evans, N.D.; Porter, M.; Murray, R.W. Stable, Monolayer-Protected Metal Alloy Clusters. *J. Am. Chem. Soc.* **1998**, *120*, 9396–9397. [CrossRef]
13. Zielińska-Jurek, A.; Kowalska, E.; Sobczak, J.W.; Lisowski, W.; Ohtani, B.; Zaleska, A. Preparation and characterization of monometallic (Au) and bimetallic (Ag/Au) modified-titania photocatalysts activated by visible light. *Appl. Catal. B Environ.* **2011**, *101*, 504–514. [CrossRef]
14. Link, S.; Wang, Z.L.; El-Sayed, M.A. Alloy Formation of Gold–Silver Nanoparticles and the Dependence of the Plasmon Absorption on Their Composition. *J. Phys. Chem. B* **1999**, *103*, 3529–3533. [CrossRef]
15. Sun, L.; Lv, P.; Li, H.; Wang, F.; Su, W.; Zhang, L. One-step synthesis of Au–Ag alloy nanoparticles using soluble starch and their photocatalytic performance for 4-nitrophenol degradation. *J. Mater. Sci.* **2018**, *53*, 15895–15906. [CrossRef]
16. Sun, L.; Yin, Y.; Lv, P.; Su, W.; Zhang, L. Green controllable synthesis of Au–Ag alloy nanoparticles using Chinese wolfberry fruit extract and their tunable photocatalytic activity. *RSC Adv.* **2018**, *8*, 3964–3973. [CrossRef]
17. Ray, P.; Clément, M.; Martini, C.; Abdellah, I.; Beaunier, P.; Rodriguez-Lopez, J.L.; Huc, V.; Remita, H.; Lampre, I. Stabilisation of small mono- and bimetallic gold–silver nanoparticles using calix [8] arene derivatives. *New J. Chem.* **2018**, *42*, 14128–14137. [CrossRef]
18. Naya, S.-I.; Hayashido, Y.; Akashi, R.; Kitazono, K.; Soejima, T.; Fujishima, M.; Kobayashi, H.; Tada, H. Solid-Phase Photochemical Growth of Composition-Variable Au–Ag Alloy Nanoparticles in AgBr Crystal. *J. Phys. Chem. C* **2017**, *121*, 20763–20768. [CrossRef]
19. Negishi, R.; Naya, S.-I.; Kobayashi, H.; Tada, H. Gold(Core)-Lead(Shell) Nanoparticle-Loaded Titanium(IV) Oxide Prepared by Underpotential Photodeposition: Plasmonic Water Oxidation. *Angew. Chem. Int. Ed.* **2017**, *56*, 10347–10351. [CrossRef] [PubMed]
20. Tada, H.; Kiyonaga, T.; Naya, S.-I. Rational design and applications of highly efficient reaction systems photocatalyzed by noble metal nanoparticle-loaded titanium(iv) dioxide. *Chem. Soc. Rev.* **2009**, *38*, 1849. [CrossRef] [PubMed]
21. Tani, T. *Silver Nanoparticles from Silver Halide Photography to Plasmonics*; Oxford University Press: New York, NY, USA, 2015.
22. Rodríguez, O.P.; Caro, M.; Rivera, A.; Olivares, J.; Perlado, J.M.; Caro, A. Optical properties of Au-Ag alloys: An ellipsometric study. *Opt. Mater. Express* **2014**, *4*, 403. [CrossRef]

© 2019 by the authors. Licensee MDPI, Basel, Switzerland. This article is an open access article distributed under the terms and conditions of the Creative Commons Attribution (CC BY) license (http://creativecommons.org/licenses/by/4.0/).

Review

Recent Advances in the Design and Photocatalytic Enhanced Performance of Gold Plasmonic Nanostructures Decorated with Non-Titania Based Semiconductor Hetero-Nanoarchitectures

Jose I. Garcia-Peiro [1,2,3], Javier Bonet-Aleta [1,2,3], Carlos J. Bueno-Alejo [4] and Jose L. Hueso [1,2,3,*]

1. Instituto de Nanociencia y Materiales de Aragon (INMA), CSIC-Universidad de Zaragoza, 50009 Zaragoza, Spain; joseignacio.garcia.peiro@gmail.com (J.I.G.-P.); jbaleta@unizar.es (J.B.-A.)
2. Department of Chemical Engineering and Environmental Technology (IQTMA), University of Zaragoza, 50018 Zaragoza, Spain
3. Networking Research Center on Bioengineering Biomaterials and Nanomedicine (CIBER-BBN), 28029 Madrid, Spain
4. Department of Chemistry and Molecular and Cell Biology, University of Leicester, Leicester LE1 7RH, UK; carlosj_bueno@yahoo.es
* Correspondence: jlhueso@unizar.es; Tel.: +34-876555442

Received: 3 November 2020; Accepted: 9 December 2020; Published: 14 December 2020

Abstract: Plasmonic photocatalysts combining metallic nanoparticles and semiconductors have been aimed as versatile alternatives to drive light-assisted catalytic chemical reactions beyond the ultraviolet (UV) regions, and overcome one of the major drawbacks of the most exploited photocatalysts (TiO_2 or ZnO). The strong size and morphology dependence of metallic nanostructures to tune their visible to near-infrared (vis-NIR) light harvesting capabilities has been combined with the design of a wide variety of architectures for the semiconductor supports to promote the selective activity of specific crystallographic facets. The search for efficient heterojunctions has been subjected to numerous studies, especially those involving gold nanostructures and titania semiconductors. In the present review, we paid special attention to the most recent advances in the design of gold-semiconductor hetero-nanostructures including emerging metal oxides such as cerium oxide or copper oxide (CeO_2 or Cu_2O) or metal chalcogenides such as copper sulfide or cadmium sulfides (CuS or CdS). These alternative hybrid materials were thoroughly built in past years to target research fields of strong impact, such as solar energy conversion, water splitting, environmental chemistry, or nanomedicine. Herein, we evaluate the influence of tuning the morphologies of the plasmonic gold nanostructures or the semiconductor interacting structures, and how these variations in geometry, either individual or combined, have a significant influence on the final photocatalytic performance.

Keywords: plasmonics; photocatalysis; heterostructures; semiconductors; NIR; core-shell; Janus-like; yolk-shell; nanorods; chalcogenides

---

1. Introduction

Since the first reported example of heterogeneous photocatalysis in 1911 applied to the degradation of Prussian Blue by ZnO powder and illumination [1], the degree of sophistication and complexity in photocatalyst design has experienced a huge development [2]. In heterogeneous photocatalysis, the process is initiated by the interaction between incident photons and the catalyst. The photon absorption by the catalyst (typically a semiconductor) leads to the promotion of valence band electrons into the conduction band, thereby creating electron-hole pairs. Those carriers can induce the subsequent generation of free radicals (e.g., hydroxyl (OH), superoxide (O_2^-)) to target specific chemical reactions. As a requirement, the energy of the incident photon (hν) must be equal or higher than the energy band gap (E_g) of the catalyst, meaning that the incident electromagnetic wavelength must be energetic enough to overcome the barrier to excite an electron from the highest occupied energy levels to the lowest unoccupied levels. Well-established semiconductor-based photocatalysts, such as TiO_2 or ZnO possess high E_g values (3.05 [3] and 3.3 [4] eV, respectively) and are constrained to the more energetic ranges of the solar spectrum (i.e., UV window representing only up to 5% of the solar radiation) for an effective photoactivation. Abundant efforts have been devoted in the past decades to expand the photocatalytic response of heterogeneous semiconductor photocatalysts towards the visible and infrared ranges in order to maximize the absorption of the solar spectrum [5–9]. One of the most promising and explored strategies has consisted on the combination of semiconductor structures with noble-metal based nanoparticles [10–12]. Recent reviews available in the literature have deepened into the synergistic action of small metal nanoparticles decorating semiconductors and on how the controlled architecture of the latter may have a strong influence on the final photocatalytic outcome [13–20]. Another interesting aspect of metallic nanostructures correlates with their unique optical response that can be modulated upon variations of their specific size and morphology [19–23]. At the interface between the metal surface and other medium (with different dielectric properties), exist a phenomenon known as Localized Surface Plasmon (LSP) that consist on a coherent delocalized electron oscillation leading to the generation of an electromagnetic field both outside and inside the metal. An excitation with radiation of the right wavelength causes a resonance interaction and subsequent collective oscillation of conduction electrons, in the case of metallic materials, due to the restoring force between electrons and nuclei through Coulombic attraction (Figure 1). This phenomenon is called Localized Surface Plasmon Resonance (LSPR), and for metals like Au, Ag, or Cu, the LSPR may take place over a wide range of 400–1300 nm [24] as a function of their size and shape (Figure 1). Thus, metallic nanoparticles emerge as perfect candidates as visible near infrared (NIR) light harvesters to combine and improve the efficiency of semiconductor photocatalysts.

So far, TiO_2 has been set as one of the most explored semiconductors to form hetero-nanostructures in combination with metals to overcome its limited photo-response beyond UV window [25,26] that can overcome its one of the most widely used semiconductors to carry out photocatalytic reactions. Systematic evaluation of Au-TiO_2 hybrid systems exploring the role and influence of shape, specific configuration, heterojunction conformations, and so on, have been developed in the past years. Numerous and varied architecture designs have been successfully reported including core-shell (concentric and eccentric) [27,28], yolk-shell [29], Janus type structures [28], or even multi-component heterostructures [30,31] and their performance successfully tested towards energy and environmental applications [17,18,31–33]. The generation and assembly of these hetero-nanostructures offers multiple advantages but the number of alternative candidates to TiO_2 still remains as an open challenge. The present review intends to overview the most recent advances described in the literature involving the design of hybrid photocatalysts combining plasmonic Au nanoparticles and non-titania based semiconductor coatings organized in a wide variety of nanoarchitectures (vide infra). Herein, we paid special attention to plasmonic hybrids that involved the selection of anisotropic Au nanostructures (mostly nanorods (AuNRs) and nanostars (AuNSs)) and a controlled growth of semiconductors beyond the most typically studied (i.e., TiO_2 [27,34,35] or ZnO [5,36,37]). This approach allows a fine control of sizes for both metal and semiconductor, reduces the probability of recombination of carriers, and maximizes an intimate contact to form efficient heterojunctions [38]. In contrast, other methodologies lack sufficient control on the size and dispersion of metal nanoparticles or the corresponding supports. Furthermore, many times these semiconductor nanoparticle supports require additional tuning or post-treatments to ensure the exposure of preferential facets that do not necessarily prevent numerous bulk recombination events due to their inherent polydispersity [13]. Hence, recent research innovative trends, such as cancer therapy, require more accurate control of the photocatalysts dimension for proper internalization in cells and accurate reproducibility [39]. Therefore, we consider that the efforts made to improve the generation of novel plasmonic photocatalysts with a controlled size and distribution in a core-shell (or analogous nanoarchitecture) and/or semiconductor supports represents a very promising alternative to other metal-semiconductor configurations [13]. Herein, we have made special emphasis on highlighting the latest achievements with AuNRs or AuNSs as plasmonic cores and different oxides (CeO_2 and Cu_2O) or chalcogenide semiconductors (CuS, CdS) grown with core-shell, Janus, or dumbbell-like configurations. We have correlated the influence of the different architectures with their final photocatalytic response.

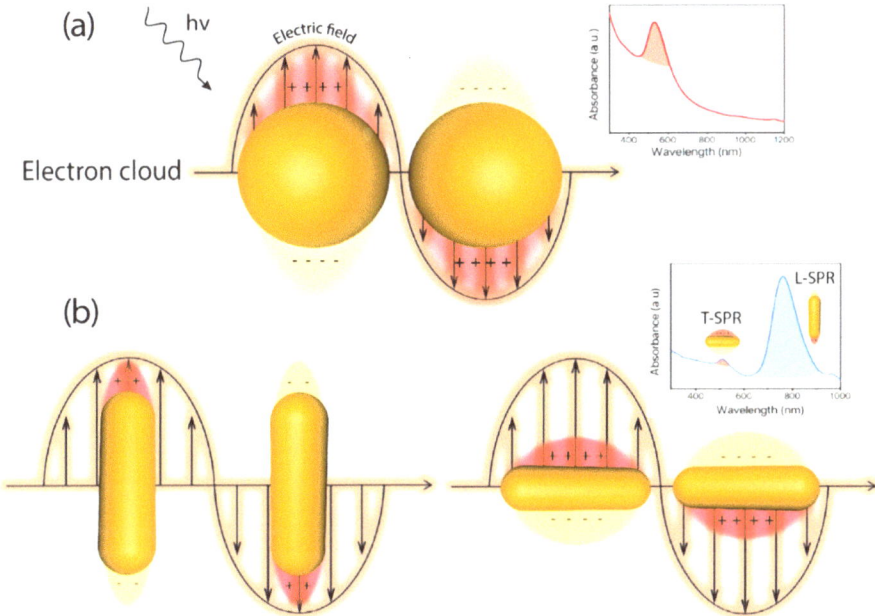

Figure 1. Morphology of the plasmonic nanostructure determines its Localized Surface Plasmon Resonance (LSPR). (**a**) Isotropic Au nanoparticles (AuNPs) possess confined electrons alike in all directions, resulting in a single LSPR band in the visible range; (**b**) the introduction of morphological anisotropy allows LSPR with different absorption maxima. For the particular case of nanorods (AuNR), the longitudinal induced anisotropy entails an LSPR at the near-infrared (NIR) window as the confinement of electrons is different at the AuNR edges or sides.

2. Metal-Semiconductor Hetero-Nanostructures: Different Configurations and Light-Driven Activation Mechanisms

A key feature of plasmonic photocatalysts is related with the photo-induced generation of highly energetic electrons (hot electrons) generated via LSPR [40]. The distribution of these hot electrons can be described by the Fermi equation using an elevated effective temperature [41]. After light absorption, LSPR decay may occur either radiatively, through re-emitted photons, or non-radiatively, for instance through transfer of hot electrons [42], generally through intraband excitations within the conduction band [40,43] thereby causing electrons from occupied energy levels to be excited above the Fermi energy. Hence, after coupling with metals, typical semiconductors used in photocatalysis can capture these hot electrons and generate reactive species by using visible-NIR light. The formation of metal-semiconductor heterojunctions allows hot electrons to be accepted into semiconductor conduction band and carry on the photocatalytic process (Figure 2a). The energetic barrier formed at the metal-semiconductor interface is called Schottky barrier [44]. Hot electrons are injected into semiconductor conduction band when their energy is superior to Schottky barrier energy (E_{SB}) which is lower than the bandgap of semiconductor (E_g) [45]. After the hot electron generation, holes are also generated in the plasmonic structure as illustrated in Figure 2b. For this mechanism to occur there must be a good interaction between the metal and the support, which make critical the synthesis step of these hybrid materials.

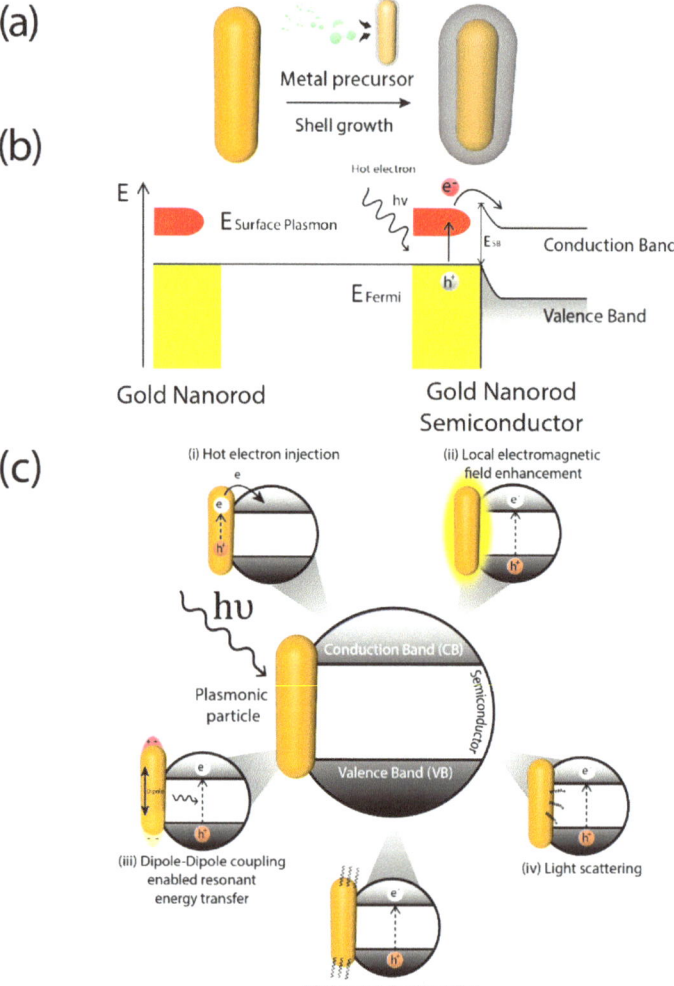

Figure 2. (**a**) Formation of Au-semiconductor heterostructures for the particular case of a core-shell structure. Different wet chemistry approaches are available to create a wide range of morphologies; (**b**) hot electron injection process. When a plasmonic structure as, in example, AuNR is irradiated with light with an energy equal to its LSPR, electrons mainly coming from conduction band are excited. If their energy overcomes the Schottky barrier energy (E_{SB}), they can be injected into the semiconductor conduction band to further perform the photocatalytic process; (**c**) mechanisms involved in plasmonic-semiconductor heterostructures (adapted from [46]). Diverse physical processes may take place between plasmonic nanostructure and semiconductor to promote valence band electrons to conduction band energy levels.

Nevertheless, several mechanisms could be involved independently or, most often, concurrently in photocatalysis using hybrid plasmonic materials [46]. Depending on the interaction of the plasmonic nanoparticle with the support and the electronic characteristic of the latter, plasmonic excitation can improve the photocatalytic properties of materials in several ways: (i) increasing absorption and scattering of light [47]; (ii) enhancing of the localized electric field [48]; (iii) hot charge carriers generation and transfer [49], already mentioned; (iv) dipole induction on non-polar molecules [50]; (v) local heat generation [46,51], depicted in Figure 2c.

Since nature, size, shape, and crystalline structure of the nanoparticle determine the energy of the LSPR [52] and in consequence the wavelength of the light used in photocatalysis, the control of those cited parameters is fundamental for the synthesis of suitable hybrid materials [53–55]. Concretely, the introduction of anisotropy in plasmonic-semiconductor systems adds a superior level of performance. Plasmonic Au cores exhibit a wide range of anisotropic nanostructures (nanoshells, nanorods, hollow spheres, nanoprisms, triangles, cubes, nanostars, urchins, etc.) with different LSPR [8,18,24,31,52,56] (Figure 3a). Typically, plasmon energy is concentrated on the sharpest edges of the anisotropic plasmonic nanostructures of high curvature [57–59] where light harvesting will take place more efficiently [60].

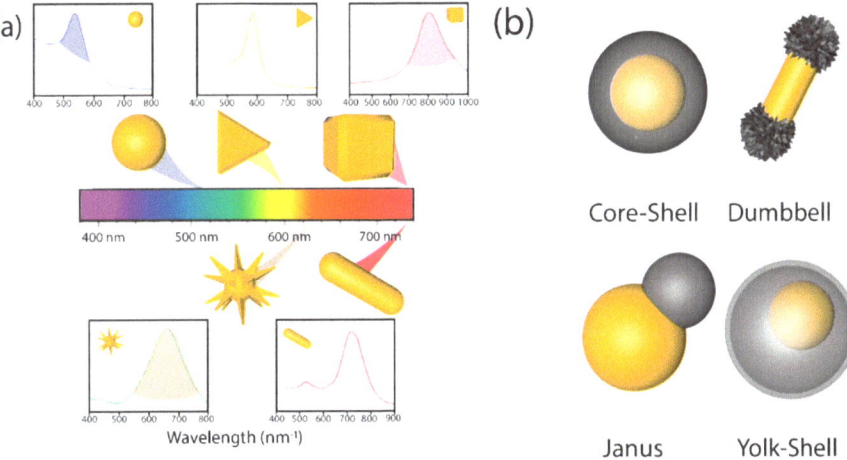

Figure 3. (a) The morphology of Au nanostructure determines LSPR shift. Au nanospheres, prisms, cages, urchin-like or rods exhibit different electron confinement, entailing different UV-vis. spectra; (b) Plasmonic-semiconductor heterostructures with different architectures, including core-shell (plasmonic nanoparticle as core and semiconductor as a surrounding shell), dumbbell (plasmonic AuNR where the semiconductor is selectively deposited on tips), Janus (plasmonic nanoparticle and semiconductor attached with both surfaces available for the substrates), and yolk-shell (plasmonic nanoparticle inside a voided semiconductor structure).

As shown in Figure 3b, it is possible to differentiate between core-shell, dumbbell, Janus, and yolk-shell configurations attending to the semiconductor distribution around the metallic cores [38] with several implications in their photocatalytic activity. Structures with exposed surfaces of both the metal and the semiconductors (i.e., dumbbell and Janus) exhibit a superior photocatalytic performance due to the continuous exposure of the reactants with the generated charge carriers (e^-/h^+) [27,40,60–64]. Different anisotropic heterostructures have been prepared with SiO_2 [65], Pt [66,67], Fe_2O_3 [68], Cu_2O [60], CeO_2 [63], and TiO_2 [27]. Yolk-shell structures (Figure 3b) are characterized by a hollow shell and an inner plasmonic core with several benefits for the photocatalytic process: (i) the presence of a hollow shell ensures higher specific surface area as it possesses an inner and external part. (ii) The small thickness of shells shortens charge diffusion distance, reducing possible bulk recombination processes, and (iii) void space present allows reflection of light inside the hollow shell, causing light scattering, and boosting the number of available photons [69].

3. Evaluation of Au-CeO$_2$ Anisotropic Hetero-Nanoarchitectures

The combination of plasmonic Au nanostructures with cerium oxide as complex hybrid architectures represents an interesting example of synergistic photocatalyst that maximizes the positive properties of each individual counterpart beyond the most explored TiO$_2$ or ZnO hybrids [70–72]. The redox properties of Ce endow this oxide with large capability to transport oxygen via Ce(IV)-Ce(III) pair redox cycles within its framework [21]. As a result, an extensive number of oxygen vacancies are present in ceria-based materials surface [73], favoring a continuous cycle of catalytic reaction-regeneration in oxidation reactions [21,74–77]. Nevertheless, CeO$_2$ exhibits two major drawbacks for photocatalytic purposes: (i) CeO$_2$ possesses a band gap of 3.2 eV burdening its photocatalytic response to the ultraviolet region; (ii) CeO$_2$ possesses low carrier mobility, hindering its transport to surface after the electron-hole pair photogeneration to react with the targeted substrates [78,79].

Regarding the generation of the plasmonic hybrid photocatalyst, (i) CeO$_2$ as an n-type semiconductor possesses high density of states in its conduction band, which confers a good electron-accepting capability [40] and forms a heterojunction and Schottky barrier with Au to allow a proper hot-electron injection [21,22] (Figure 4). Therefore, the synthesis and development of Au-CeO$_2$ photocatalysts to maximize the redox properties of ceria has boosted the research on optimizing the interaction of these materials through an extended variety of imaginative configurations. Thus far, CeO$_2$ has been successfully deposited onto nanorods (both forming dumbbell [22] and core-shell [21,22,80] structures), spheres [21,81], or hollow cages [82].

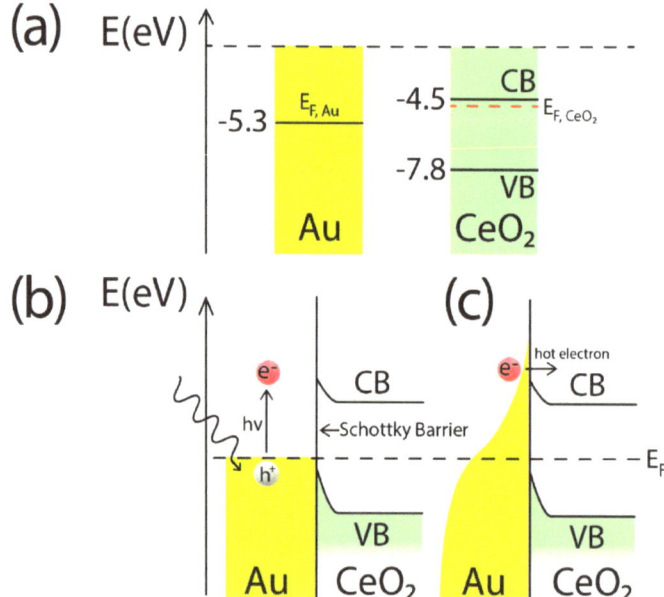

Figure 4. Energy band structure of Au-CeO$_2$ photocatalysts. (**a**) Energy band levels of Au and CeO$_2$ respectively; CeO$_2$ possess a relatively large band gap, needing from UV light to photo-generate charge carriers; (**b**) energy band distribution upon the gold plasmon excitation to (**c**) inject hot electrons into the conduction band of CeO$_2$.

In general, the synthesis of core-shell Au-CeO$_2$ nanostructures has been typically reported via hydrothermal treatment of a chelated Ce precursor [21,22,80–82]. Li et al. [21] reported the first synthesis of Au-CeO$_2$ nanostructures with core-shell configuration. One of the problems facing CeO$_2$ selective deposition on Au nanostructures was the rapid condensation of Ce precursors to form CeO$_2$. By using ethylenediaminetetraacetic acid (EDTA) as chelating agent, fast hydrolysis of Ce (III) ions could be prevented [83]. Thus, tuning Ce (III)/EDTA ratios allowed a fine control of CeO$_2$ deposition rates. Regarding the synthesis of dumbbell and Janus Au-CeO$_2$ systems, two outstanding contributions have been demonstrated in the recent literature by Pan et al. [63] and Jian et al. [22], respectively. It was possible to control the selective growth of CeO$_2$ on AuNR tips by controlling the amount of cetyltrimethylammonium bromide (CTABr) adsorbed onto the Au surface, which acted as a blockade of CeO$_2$ nucleation.

The first reported examples of Au-CeO$_2$ hybrids in photocatalysis were structurally analogous to the conventional counterparts traditionally used in heterogeneous catalysis for CO oxidation reactions. Kominaim et al. [84,85] reported the use of Au nanoparticles dispersed onto nanostructured CeO$_2$ supports to oxidize formic acid under visible light irradiation [84,85]. Nevertheless, Au-CeO$_2$ photocatalysts consisted in the random deposition of Au nanoparticles onto a relatively large CeO$_2$ support that prevented a deep understanding of the photocatalytic mechanism. As mentioned before, Li et al. [21] reported the first core-shell configuration for Au-CeO$_2$ nanoparticles. They tested the use of both spheres and rod-shaped Au plasmonic cores. The Au@CeO$_2$ catalysts exhibited different LSPR absorption maxima at different wavelengths within the visible-NIR region (530, 591, 715 nm, respectively) [21]. These hybrid configurations were employed towards the photooxidation of benzylic alcohol. The catalysts were subjected to calcination post-treatments to remove the excess of the directing surfactant employed to grow the ceria shell (CTABr), which did not possess any influence in the morphology/crystallinity of the sample. The influence of this calcination treatment was thoroughly evaluated upon irradiation with two different excitation sources (Xe lamp ($\lambda > 420$ nm) and laser ($\lambda = 671$ nm)). For Xe lamp, a systematic study of Au@CeO$_{2, \text{LSPR 530nm}}$ was performed. It was found that the Au@CeO$_2$ calcined sample exhibited the highest photocatalytic activity, suggesting that the removal of CTABr enhanced the formation of Schottky barrier between Au and CeO$_2$ and facilitated a better charge carrier transfer in the catalyst. Upon laser irradiation, a similar trend was confirmed for the uncalcined and calcined samples, respectively. Calcined Au@CeO$_{2, \text{LSPR 591nm}}$ showed the highest photocatalytic activity. The authors attributed the enhanced response to the better match between the LSPR absorption maximum of the calcined hybrid (redshifted after calcination to 680 nm) and the incident laser wavelength at 671 nm.

As previously pointed out in Figure 2c (vide supra), not only the hot electron injection pathway can supply electrons in the semiconductor conduction band. Interestingly, the generation of heat caused by one of the decay mechanisms caused by LSPR [86] can also promote the excitation of electrons from the valence band to the conduction band. As control experiments, Li et al. [21] studied the influence of the temperature in the catalytic activity in the absence of light. They found a considerable catalytic activity at the temperature that reached the system under illumination in darkness conditions, suggesting that part of the photocatalytic activity was thermally induced. As a potential mechanism (Figure 5a), the authors suggested a series of redox steps (vide infra): (i) electron injection into Ce(IV) conduction band, which generated Ce(III) species; (ii) O$_2$ adsorbed on previously generated Ce(III) sites forming Ce(IV)-O-O· (iii) that radical can remove α-H of benzyl alcohol yielding Ce(IV)-O-OH; (iv) the radical dehydrogenated benzylic alcohol combines with Ce(IV)-O-OH to produce the final benzaldehyde product and H$_2$O$_2$, and simultaneously, an electron returning to the Au bands to ensure electron-hole pair recombination.

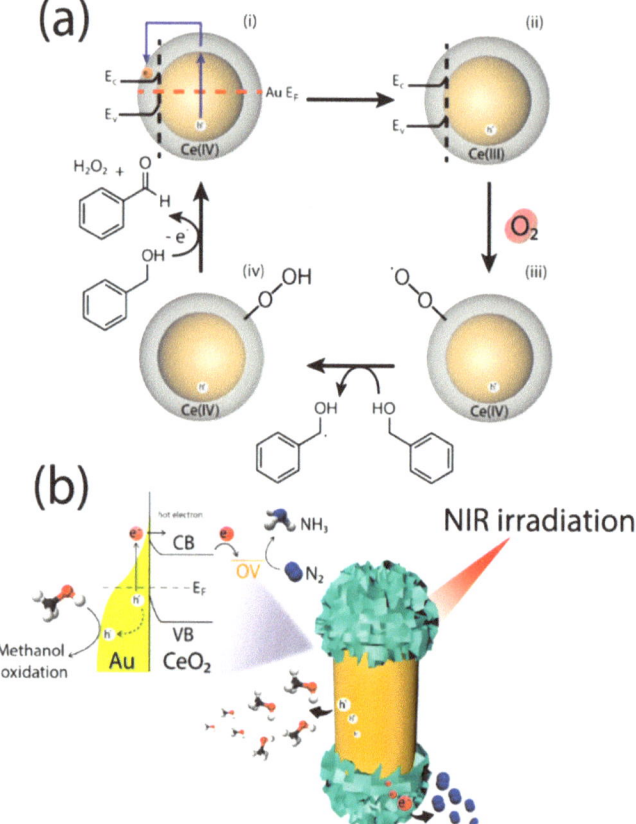

Figure 5. (a) Hot electron injection mechanism for a core-shell Au@CeO$_2$ configuration (adapted from [21]). Hot electrons generate Ce (III) species that can capture and activate O$_2$ which furtherly reacts with the substrate (benzylic alcohol). The generated Ce hydroperoxide (Ce-O-OH) reacts with the radical intermediate, yielding the final aldehyde; (b) photocatalytic mechanism for a dumbbell Au-CeO$_2$ structure (adapted from [22]). The generation and consumption (and thus, photocatalysis) of both hot electrons/holes is enhanced by promoting the accessibility of hole-acceptor molecules to the metallic surface and thanks to the accommodating role of oxygen vacancies (OVs) present in the ceria nanostructures. Core-shell architectures hinder this accessibility, accumulating hot holes and slowing down the catalytic cycle.

Alternatively, Wang et al. [80] demonstrated that the plasmon-induced hot-electron injection under NIR illumination in CeO$_2$ coated AuNRs with Janus configuration accelerated photo Fenton-like reactions using H$_2$O$_2$ as substrate. Again, the hot electron injection in Ce(IV) induced the formation of Ce(III) that acted as active species in Fenton-like reactions [87] with the generation of highly reactive ·OH that subsequently facilitated the degradation of an organic model pollutant. As depicted in Figure 5b, a heterostructure with plasmonic exposed faces possess important photocatalytic advantages. Pan et al. [63] demonstrated that half-encapsulated AuNRs with CeO$_2$ possessed better activity in the catalytic reduction of 4-nitrophenol (4-NP) with NaBH$_4$, in comparison with their core-shell AuNR@CeO$_2$ counterparts. Hot electrons were transferred from the plasmonic AuNR into CeO$_2$, generating holes on Au surface. BH$_4$ can donate electrons to AuNR generating H species that further convert 4-NP into 4-aminophenol. On the other hand, CeO$_2$ is able to withdraw e$^-$ from AuNR that absorbs H^+ from H_2O yielding the same H active species [88]. This phenomenon could not occur in the case of core-shell AuNR@CeO_2 since Au facets remained unexposed to the liquid reaction media and yielded a lower catalytic activity.

Following a similar methodology, Jia et al. [22] combined CeO_2 with AuNRs forming a dumbbell heterostructure (Figure 5b) and evaluated their photocatalytic performance in the N_2 photofixation with NIR irradiation. CeO_2 rich surface in oxygen vacancies (OVs) (Ce(III) sites) chemisorbed N_2 that could be reduced to NH_3 by injected plasmonic hot electrons, breaking triple bond N-N. Due to the exposure of Au facets to the reaction media, as-generated hot holes could be consumed by a hole scavenger (in this particular case, methanol) to close the photocatalytic cycle. Core-shell AuNR@CeO_2 nanostructures prevented the availability of hot holes and the accessibility of CH_3OH to the active sites, which reflected in a huge difference in the NH_3 generation rate (114.3 vs. 18.44 $\mu mol \cdot h^{-1} \cdot g^{-1}$) [22]. Thus, CeO_2 emerges as a promising photocatalytic material in combination with Au plasmonic nanostructures, being the main highlights the regeneration of highly desired OVs due to hot electron transference from Au and the photothermo-induced oxygen mobility [21]. The high oxygen mobility has shown to have influence during the photothermo-catalytic process, as demonstrated by Li's experiments with temperature and in the absence of light [21,89]. CeO_2 as n-type semiconductor [90] and as oxygen ion conductor possess a unique charge separation effect: when a photon is absorbed, an electron is excited from the valence band (orbital O_{2p}) to an empty conduction band (orbital Ce_{4f}), forming a $Ce^{4+}(e^-)/O^{2-}(h^+)$ pair. The as-generated electrons remain localized in Ce (IV) centers favoring the charge separation and consequently, reducing the recombination phenomena (Figure 6) [90,91].

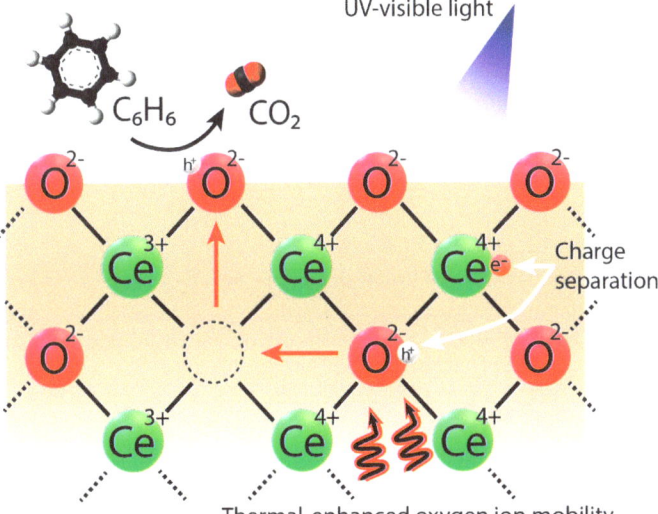

Figure 6. Schematic CeO_2 structure remarking the presence of OV and h^+/e^- separation. Plasmon-photothermal induced local heating in CeO_2 structure also enhances the oxygen ion mobility, facilitating the migration of $O^{2-}(h^+)$ species to the catalyst surface and, thus, enhancing the reaction (adapted from [89]).

Cited reports confirm an effective synergistic effect in the photocatalysis between LSPR and CeO_2 and an enhancement of the photocatalysis with Janus-type heterostructures [22,80]. Hence, synthetized Au-CeO_2 nanostructures interact with visible and NIR light, remarking the promising interest of these materials in environmental, biomedical applications and solar energy harvesting, among others [69,91–98].

4. Evaluation of Au-Cu$_2$O/Au-CuS Anisotropic Heteronanostructures

Cuprous oxide has attracted much attention as a photocatalyst due to its remarkable features as a semiconductor, and its extended variety of active roles in solar and energy applications of paramount interests for the Sustainable Development Goals targeted by most of the worldwide governments' agendas. Copper (I) oxide materials have been widely used in the degradation of organic compounds in solution and their photocatalytic activity has been studied against a large number of potential pollutants, such as organic material [99,100], drugs [101] or nutrients [102]. In addition, studies are also emerging where copper (I) oxide is being used in the generation of molecular hydrogen [32,103]. It is a highly abundant p-type semiconductor on earth with an energy band gap (E_g) of ≈ 2.17 eV that has a low cost and good absorption capabilities in the visible range when compared with other semiconductors [50,104]. However, some drawbacks have been also consistently reported for this oxide such as: (i) rapid recombination of electron-hole pairs; (ii) limited absorption of the semiconductor in the visible range; or (iii) natural tendency to form larger structures or disproportionation (especially in liquid media). The combination with anisotropic plasmonic cores to generate metal-semiconductor heterostructures represents an appealing alternative to overcome these problems [56,105,106]. These hybrid nanostructures offer two remarkable features, the Schottky barrier and the LSPR that improves the photocatalytic response by decreasing the carrier recombination rates and expanding the absorption range to the visible.

The selective growth of copper oxide onto plasmonic Au cores has been a trendy topic widely studied by multiple researchers in recent years [32,56,105–108]. However, the final morphology and selective exposure of specific crystalline facets has also attracted great interest in the field of photocatalysis in terms of both selectivity and reactivity [106,107,109]. The generation of core shell Au-Cu$_2$O hybrid structures with high precision has been reported in the recent literature [107,109,110]. Several parameters such as the nature of the reducing agents, the concentrations of reducing agent or pH values have been systematically evaluated to modulate the morphology, shape, and length of the oxide shells and favor the preferential exposure of selected crystal facets [32,109] (see Figure 7). For instance, Kuo et al. [109] reported the controlled synthesis of different morphologies including cubic structures, truncated cubes, cuboctahedra, or truncated octahedra to octahedra upon tuning the variation of NH$_2$OH. On the other hand, Kuo et al. [106] explored the influence of the different morphologies of AuNPs coated Cu$_2$O shell on the final photocatalytic performance towards the methyl orange (MO) degradation. They showed how core-shell architectures improved MO degradation from 18 to 50 µmol·cm^{-2} in comparison with pristine Cu$_2$O nanostructures and how the influence was higher in the case of the cube-shaped counterparts. This was an example where the plasmonic core helped in the transfer of carriers and promoted a better conductivity in a photo-electronic device. Yuan et al. [107] also developed hybrid AuNRs-copper oxide nanoparticles where they explored variations on the type of reducing agent to tune the preferential octahedral or cubo-octahedral morphologies (see Figure 7a). Both structures were generated by Ostwald ripening process. Smaller crystals were dissolved and redeposited onto larger crystals when different reductants were added. Hydrazine as stronger reductant was reported to promote the copper species to their more stable morphology. They also tested their photocatalytic activity towards the degradation of MO. It was observed how hybrid structures enhanced more activity in comparison with analogous bare semiconductor counterparts. The Au internal core performed in a dual way, LSPR promoted electron injection into the semiconductor and also acted as a charge sinker to increase charge migration and separation. This was a representative example on how the presence of the Au core considerably affects the activity of the system and how the morphology of the external shell (corner-truncated octahedral vs. regular octahedral Cu$_2$O) is also highly relevant in the absorption and degradation of MO. The truncated octahedral core-shell led to 94% MO conversion from an original 10 mg·L^{-1} MO concentration in 80 min [107]. This influence was also reported in Cu$_2$O nanoparticles with different morphologies that also exhibited a much higher photocatalytic activity in rhombohedral shaped structures due to the more selective MO absorption over <110> faces for subsequent further oxidation [111] (see Figure 7a). Recent studies have also reported the use of hollow Au cores to build Au-Cu$_2$O hybrids have also reported a successful tunability of their optical properties depending on their final architecture (Janus vs. core-shell). They reported an effective plasmon-induced energy transfer favored by the hollow Au nature that improved the MO degradation rates [56]. Other recent report by Xu et al. [112] has established a strong influence of the final heterostructure configuration on the capability to generate photocurrent. In this case, the used solid gold seeds incubated with a ligand that shifted the core-shell configuration towards a Janus-like structure.

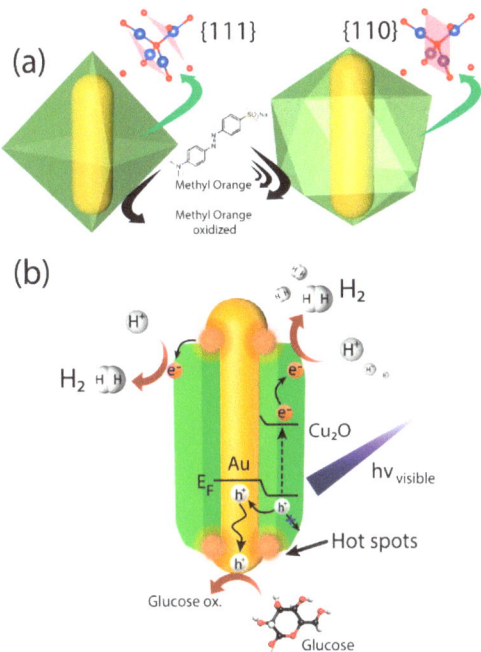

Figure 7. (**a**) Face dependent selectivity of Au@Cu$_2$O photoredox catalysis in MO degradation (adapted from [107]) (**b**) Selective Cu$_2$O deposition for partially/completely coated AuNR (adapted from [32]).

Additional studies have also exploited the synergy of Au- Cu$_2$O hybrid materials. For instance, Yu et al. developed [32] a hot-dog-like hybrid configuration where they selectively deposited copper oxide on the transverse part of large aspect ratio AuNRs, leaving the edges completely exposed [32]. It was possible to change the kind of Cu$_2$O-AuNR architecture from core-shell to tips-exposed hot-dog type structure by reducing the amount of copper precursor added to the synthesis. The preferential epitaxial growth of Cu$_2$O over the transversal part of AuNR promoted a first nucleation over these sides, leaving the edges completely uncoated under low copper concentration conditions. In a sequential manner, they even incorporated very thin TiO$_2$ layers on top of the copper oxide and established a critical evaluation of the photocatalytic activity for each hybrid configuration towards hydrogen production. The charge separation efficiency was significantly higher for the hot-dog configuration than for the bare Cu$_2$O or the fully-covered core-shell counterparts. The hot-dog configuration also exhibited an enhanced photocatalytic response almost 24-fold higher than the Cu$_2$O nanoparticles (80 µmol H$_2$ g^{-1} h^{-1} vs. 3.4 µmol H$_2$ g^{-1} h^{-1}) and four times higher than the core-shell configuration (19.3 µmol H$_2$ g^{-1} h^{-1}). The authors partially attributed the outperformance of the hot-dog configuration to the generation of four hot-spots at the interface of the non-coated AuNR and the semiconductor shells (see Figure 7b) of the hot-dog configuration. They concluded that this interface was more efficient in the transport of carriers due to the direct interaction of holes with glucose (used as sacrifice agent) and to the more efficient generation of hot electrons to induce the reduction step toward H_2 generation (see Figure 7b) [32]. The results were even better in the case of the ternary shell due to the more efficient p-n junction formed with the addition of a titania layer.

Even though ternary heterostructures are not the main focus of this review, it is worth mentioning recent efforts to build Z-schemed TiO_2-Au-Cu_2O photoelectrodes [30] or the addition of a third component of small Au NPs to the Au-Cu_2O hybrids reported by Yu et al. [113]. They designed a more complex core-shell AuNR@Cu_2O (octahedral) architecture additionally decorated with AuNPs (Figure 8). This ternary composite was able to increase the photocatalytic activity of the binary material by incorporating AuNPs, improving the transfer and separation of displaced charges through the semiconductor. Using higher wavelengths where Cu_2O is not able to absorb, light would penetrate through the structure reaching AuNR. Hot electrons could be generated via LSPR effect and injected into the conduction band of the semiconductor. Hot electrons and holes generated in the conduction band of Cu_2O could be promoted from bulk Cu_2O to the surface of the catalyst where the electrons are mainly trapped by Au nanoparticles to form singlet oxygen radical from the O_2 present on the surface. Meanwhile, holes places mainly in the surface of Cu_2O p-type semiconductor are used for the direct MO degradation. An electron transfer "push–pull" synergetic effect was enhanced with the help of both AuNR core and AuNPs on the surface of Cu_2O.

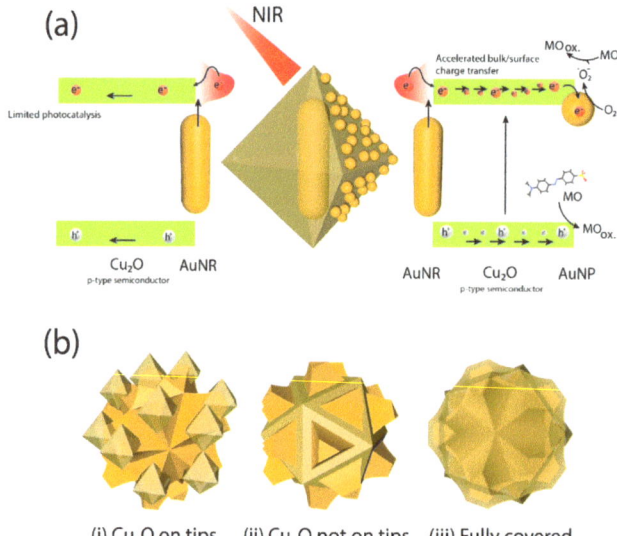

Figure 8. (**a**) Schematic mechanism for methyl orange (MO) photodegradation comparing the binary (left) from the ternary core-shell configuration (right) containing a plasmonic AuNR core, an octahedral Cu_2O shell and an additional layer of small AuNPs for the ternary alternative (adapted from [113]); (**b**) Selective Cu_2O deposition on AuNSs (adapted from [60]).

Sang Woo Han and co-workers conducted a comprehensive and systematic study where they developed hybrid plasmonic structures and explored how the variation of configuration affected the photocatalytic performance for hydrogen production using methanol as hole scavenger [60]. They synthesized plasmonic gold nanostars (Au NSs) and partially deposited copper monoxide (Cu_2O) tuning the stabilizing agent (sodium dodecyl sulfate (SDS), (CTABr), or Polyvinylpyrrolidone (PVP)) (see Figure 8b). The Janus-like structures with <111> vertices completely covered by the semiconductor were achieved by using PVP as a stabilizing agent. A complete core-shell type configuration was achieved with the aid of sodium dodecyl sulfate (SDS) as a stabilizing agent and the structure with the exposed vertices was achieved by combining two surfactants, CTABr and SDS. The nature of each stabilizing agent was crucial for the selective deposition of the semiconductor (Cu_2O) onto the plasmonic Au NSs. Another important parameter to consider is the curvature of the surface of the plasmonic Au cores. In this systematic study, other Au seeds were used to prepare Janus-like architectures. It was observed how more shaped seeds could satisfactorily enhance the proposed architecture. However, it was not possible for less shaped seed structures using PVP as a stabilizing agent. In contrast, samples with less sharp vertex led to the preferential formation of core-shell type structures.

Yu et al. [60] also found that the use of PVP as a capping agent allowed a stronger interaction with low curvature sites from the Au NSs apexes, thereby leading to a preferential nucleation and growth of Cu_2O tips at those positions (see Figure 8b). Homogeneous core-shell heterostructures required the use of SDS instead of PVP. The random assembly and weaker protecting capability of SDS enabled a complete overgrowth of Cu_2O. In vertex-exposed Cu_2O-Au NSs a binary mixture of CTABr/SDS was used. The electrostatic interaction between both surfactants induced the generation of more stable micelles in aqueous solution that preferentially absorbed in the high-curvature sites, leaving the rest of Au nanostructure more accessible for Cu_2O nucleation. To further analyze the influence of the angle and the surfactants in the synthesis of anisotropic structures, AuNRs with different aspect ratios and binary mixtures of two surfactants (CTAB and SDS) were also evaluated by Yu and coworkers. They observed that the formation of Janus-type structures was preferentially induced as the aspect ratio of the AuNRs increased [60]. The authors studied the photocatalytic response of the different AuNSs-copper oxide configurations. A clear correlation between the efficiency of conversion of solar energy and the topology could be established during the hydrogen generation using methanol as sacrificing agent. Better conversions were achieved in partially covered structures compared to fully covered structures. Furthermore, no catalytic response was observed in the absence of the catalyst or in the presence of non-coated plasmonic cores. Furthermore, copper oxide of physical mixtures of both Cu_2O and AuNSs did not render any significant conversion, thereby reinforcing the synergistic role of the plasmonic photocatalyst with a well-defined architecture (especially for the Janus-type configurations).

The comparison of the photocatalytic activities among core shell-type structures with different plasmonic structures (spheres and stars) in their core can be explained considering the excitation capacity of both materials at different wavelengths. Using excitation wavelengths of 700/750 or 800 nm, the sharp surfaces of AuNSs caused a much higher plasmonic excitation than their spherical counterpart. Janus-like architectures present higher catalytic activity than Core-Shell structures. Not covered Au highly increased the final photocatalytic activity of hybrid structure letting the oxidation reaction of hole scavengers to occur for methanol, whereas core-shell architecture dramatically deprives the accessibility of Au surface to hole scavengers.

It is shown how using identical AuNSs cores with a Cu_2O overgrowth forming anisotropic heterostructures, induced completely different photocatalytic outcomes. It was observed a strong influence of Au-Cu_2O plasmon energy transfer. Coated vertex Au nanostar enhanced higher plasmon excitation in comparison to vertex exposed Au NSs configurations and leading the photocatalytic activity of hybrid structures under those experimental conditions [60]. To better understand the interaction mechanism between the plasmonic core and the outer semiconductor layer, the AuNSs were completely coated with a thin insulating SiO_2 layer prior to the subsequent deposition of copper oxide. The hybrids with the SiO_2 interlayer exhibited a considerable decrease in activity, pointing out that the mechanism of action was carried out by plasmon-induced photocatalysis [60].

Au-Cu_2O hybrid structures are being also currently explored for biomedical applications [114–117] such as photodynamic therapy treatment of cancer cells. Xu et al. [118] performed in vitro studies with AuCu-based materials, taking advantage of their ability to generate toxic reactive oxygen species (ROS) and its highly efficient phototherapy. AuNRs were used to selectively deposit copper on one of the sides of the Au structure using hexadecylamine as a protective ligand and sequentially making a coating with Au, which modified the structure inducing a galvanic replacement reaction and decreasing the amount of copper in the sample.

Interestingly, there are additional copper-based heterostructures where the semiconductor deposited over the plasmonic structure of Au is a Cu-based chalcogenide. These semiconductors are not expensive and more stable p-type semiconductors with a well-defined LSPR in the visible region. Furthermore, the amount of free charges in the material can be tuned upon modification of experimental parameters. Generating non-stoichiometric copper deficient structures of CuS induces the formation of transporters in the p-type semiconductor and exhibits subsequent properties such as photothermal catalysis or photocatalysis through the generation of ROS [119]. This type of structure can be made from previously formed copper oxides, giving rise to yolk shell-type structures where the plasmonic core was enclosed in an external copper chalcogenide shell and this can move freely within the voided structure [120]. To carry out the synthesis, it is typically necessary to obtain a copper (I) oxide structure through the conventional protocol, growing copper oxide following an epitaxial growth process. Later, the sample is sulfurized by introducing a sulfide precursor, such as Na_2S in an acidic medium, and it is allowed to evolve until the yolk-shell type structure is formed, maintaining the crystalline faces of the original Cu_2O structure. The catalytic ability of Au-CuS yolk shell structures to form ROS has been used in cancer chemotherapeutic therapies. Zhang et al. used the photothermal and photodynamic properties of this particular hybrid to perform a co-therapy using the resonance energy transfer (RET) mechanism [116].

Wang et al. [121] synthetized hybrid core-shell Au-CuS nanostructures directly without the need to use Cu_2O as a sacrificial agent in the formation of the nanostructure and modified the final properties of the photocatalyst through the non-stoichiometric growth of CuS at different amounts of copper. They carried out a systematic study in which they evaluated the photocatalytic properties at different wavelengths, ranging from UV to Infrared through the visible spectrum and determined the influence of the different parts of the hybrid nanoparticle on the activity of the catalysts. The plasmonic characteristics of the Au nucleus could be measured by changing its dimensions and modifying the thickness of the CuS coating deposited around it. On the other hand, the plasmonic characteristics of the copper sulfide shell could be modified by changing the concentration of free holes, which is closely linked with the crystalline phase of copper sulfide and the Cu:S ratios found.

Nanostructured Au can act as a photosensitizer (see Figure 2), being able to generate more electron-hole pairs in the semiconductor by means of RET mechanisms and thus improve the photocatalytic activity in the visible spectrum. Moreover, charge recombination in the surface of the semiconductor can be partially suppressed with the help of metals and charge migration through the bulk of Cu_2O could be enhance more easily in the presence of Au. The controlling deposition of copper around Au nanostructure enables an overall downsize of the hybrid nanoparticles compared with conventional Cu_2O nanostructures. It also allows a more selective deposition of Cu_2O. All these encountered possibilities appearing with hybrid structures have a positive influence to enhance the photocatalytic response of the hybrid.

5. Evaluation of Au-CdS Anisotropic Heteronanostructures

Other semiconductors commonly used in photocatalysis reactions are metal chalcogenides. However, some of their most remarkable features, such as the absorption of light at certain wavelengths or the separation of charges are far from optimal. For this reason, alternatives have been proposed to improve the photocatalytic properties of these systems when we use a range of wavelengths that encompasses the visible or infrared spectrum while achieving a better and more effective charge separation. An interesting alternative that arises from the need to solve this problem is to create hybrid metal-semiconductor nanostructures that allow the use of the visible-NIR spectrum thanks to the metal's plasmonic properties. Moreover, a wisely metal incorporation in semiconductor to form hybrid structure also improve the charge separation and its photocatalytic activity. Metal chalcogenides have been in the spotlight for the past years due to their versatility and potential number of applications in the energy field (i.e., photovoltaics, imaging). In addition, chalcogenides semiconductors have been also studied for photocatalytic emerging applications in solar energy conversion and renewable energy development [34,122–129].

Great interest has been devoted in recent years to the development of photocatalysts able to exploit the sun's energy to generate combustible, thus trying to reduce the amount of fossil fuels used [130,131]. One of the most interesting alternatives is H_2 generation as fuel using solar energy and semiconductor-based photocatalysts are postulated as one of the most promising alternatives to achieve this objective [132]. Some interesting features for solar-assisted photocatalysis in H_2 generation could be: [133] (i) It is necessary to find a semiconductor that has a narrow band gap and is highly efficient in absorption light; (ii) the band edge has to be more negative than standard hydrogen reduction potential; (iii) it should have good enough charge separation and migration to provide enough active sites.

There is a wide variety of photocatalysts but CdS is posited as a promising candidate for the generation of hydrogen. It has an optimal band gap of 2.4 eV that allows it to have photocatalytic activity under wavelengths of up to 500–600 nm in addition to a conduction band edge suitable for the generation of H_2 [134]. On the other hand, this semiconductor suffers from a rapid recombination of charges that make it inefficient. As an alternative to improve the migration and separation of charges, the generation of metal-semiconductor hybrids appears again as an alternative recently explored in the field [61,133–137]. In this particular hybrid, the rich existing chemistry of quantum dots has boosted the search for a wide variety of architectures searching the optimal coupling of energy levels necessary to improve the photocatalytic outcome towards H_2 production [34,136]. Developing anisotropic structures such as dumbbell type bimetallic particles [135,136], Janus-type particles [62], or yolk-shell structures [34,138] have reported considerable increases in the activity of the systems in comparison with more conventional core-shell type structures. No considerable activity enhancements have been shown in this latter configuration in comparison with the bare CdS semiconductor.

Zhao et al. [137], developed core-shell type hybrid structures and verified their efficiency in photocatalytic reactions for the production of hydrogen. They were able to synthesize concentric nanostructures where the plasmonic Au-core was completely covered by the semiconductor. Alternatively, they were also able to generate eccentric structures where the core of Au was shifted, having part of internal Au core exposed (see Figure 9a). For the synthesis of these nanostructures, a two-step multistage was necessary in which, first, silver-coated Au structures were generated in a core-shell configuration. Subsequently, the silver shell was sulfurized and a mixture of these nanoparticles were allowed to evolve with the cadmium precursor and certain conditions to enhance Au-CdS Janus-like/core shell-like nanoparticles. Zhao et al. reported [62] the need of eccentric nanoparticles of silver sulfide with partial crystallinity for further evolve into the generation of Janus-type structures. The catalytic studies showed that anisotropic materials with partially exposed Au core possessed much higher photocatalytic activity than completely coated core-shell configurations. A remarkable 730-fold enhancement in the H_2 evolution rates was detected for the anisotropic dimer (7.3 mmol H_2 $g^{-1}h^{-1}$ vs. 0.9 µmol H_2 $g^{-1}h^{-1}$, respectively). The authors claimed that increasing the temperature was necessary to ensure the proper presence of a well-defined crystalline phase. This phase was the niche that enabled a rational control of the anisotropy degree (see Figure 9a). In addition the partial exposure of the plasmonic Au was also determinant to maximize the SPR effects [62].

The photocatalytic activity of these hybrid structures was proved to have a highly symmetry dependence [137]. Janus-type particles were the most active, followed by analogous semiconductor particles and ending with core-shell Au-CdS nanoparticles. Furthermore, the Janus hybrids with the more pronounced anisotropy also exhibited the highest photoactivity using 300 W Xe lamp with application of a 400 nm cut-off long pass filter for water splitting reaction. This was attributed to the supporting role of Au acting as electron sink that could delay e/h$^+$ recombination rates and perform water reduction reaction. Upon excitation at wavelengths higher than 400 nm, a rapid electron transfer occurs between the conduction layers of the semiconductor (CdS) towards the Au surface thereby enabling a more efficient water reduction to generate hydrogen. However, the nanostructured bare

CdS or core-shell Au-CdS configurations prevent the separation of charges in such an efficient way and in the case of core-shell particles, the electrons remain trapped in the core of the particle and precluding the photocatalyst from an efficient hydrogen evolution reaction [137].

In an attempt to expand the photocatalytic response towards the less energetic NIR range and enhance core anisotropy, Li et al. [139] explored the possibilities of plasmonic AuNRs as inner plasmonic core. In this case, AuNRs were completely covered with silver to generate silver sulfide in a second step that would end up giving rise to CdS under certain conditions. Wang et al. [140] also developed more complex core shell-like nanostructures generating multilayered chalcogenide shells containing different combinations of Bi_2S_3 and CdS (Figure 9b,c). These samples were tested for dye photodegradation under visible region being the ternary combinations the most active in comparison with the binary counterparts (Au-Bi_2S_3 or Au-CdS). The enhancement observed in the ternary component was attributed to the transfer of electrons from the exposed semiconductor (CdS) with a 2.4 eV band gap passing through the Bi_2S_3 semiconductor intermediate layer with a 1.32 eV band gap until reaching the internal Au core. However, holes as positive charges, take longer to move from the valence layer from one semiconductor to another, enabling charges separation and, therefore, increasing photocatalytic activity. Moreover, hot electron injection from visible activated dumbbell-like AuNR to semiconductor can considerably increase the photoactivity compared with bare semiconductor or semiconductor coated conventional GNRs (see Figure 9b,c).

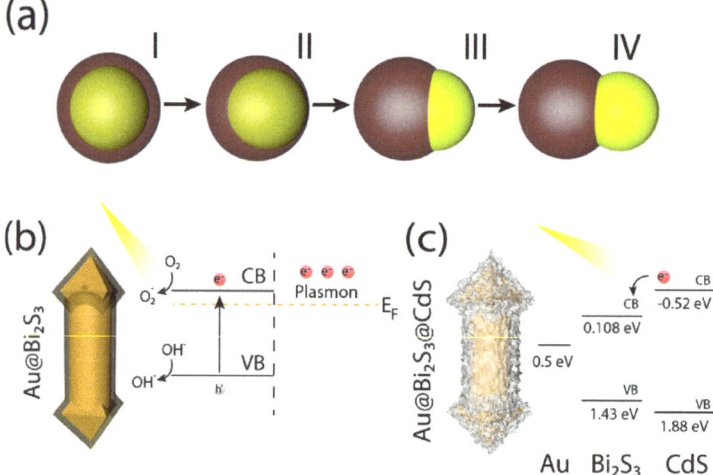

Figure 9. (**a**) Concentric core-shell (I), eccentric core-shell (II), and Janus (III, IV) hybrid structures (adapted from [137]); (**b**) AuNR@Bi_2S_3 hybrid structure and its excitation mechanism under visible light irradiation; (**c**) AuNR@Bi_2S_3@CdS core-(double) shell hybrid structure. Excitation mechanism for a more efficient carrier migration and its considerable improvement when both semiconductors were coupled in a double shell structure (adapted from [140]).

Xu et al. [136] developed different metal-CdS hybrid bimetallic structures and evaluated their photocatalytic activity to form benzaldehyde and H_2. They synthesized three types of plasmonic heteronanostructures: CdS coated AuNRs, CdS coated-Pt coated AuNRs, and CdS coated-Pt tipped AuNRs and compared their photocatalytic activity. It was observed an increasing order of activity: Au@CdS < Au@Pt@CdS < Au-Pt@CdS (50 vs. 100 vs. 150 µmol H_2 g^{-1} h^{-1}, respectively). They also carried out controls with the CdS semiconductor and metallic nanoparticles, but the activity decreased considerably in both cases. In this study, the use of AuNRs improved the harvesting of visible-NIR light and induces a local electric field (Figure 2) that is capable of promoting the generation and separation of charges from the CdS semiconductor (Figure 10). The anisotropically deposited platinum at the edge of the nanorod acted as an electron sinker or electron reservoir enabling the electrons to be directed to the places where the reduction reaction of the protons takes place to obtain H_2. In addition, the asymmetric separation of both metals served as a highway for an efficient transport of electrons that partially prevent electron-hole recombination (see Figure 10). This represents another good example of how smart designs and proper understanding of the carrier transport mechanisms facilitates an improved photo-response outperforming the simpler configurations.

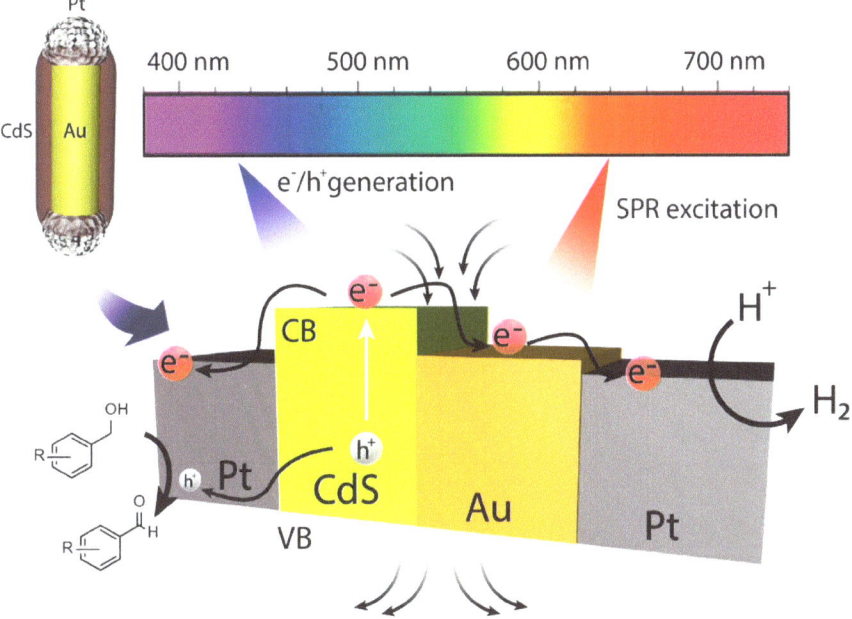

Figure 10. Schematic representation of the energy band multichannel in bimetallic@CdS heterostructures (adapted from [136]); semiconductor charge separation under visible light irradiation and its further electron migration through AuNRs to enhance tipped Pt reservoir; Induced both the resonance energy transfer (RET) and hot electron transfer processes and synergistically boosted the photocatalytic reaction (adapted from [136]).

Wu et al. [135] explored an analogous combination of Au-Pt-CdS components structured in core-shell (Au@CdS) or AuPt-CdS dumbbell like configurations with different coverage degree (see Figure 11a). These hybrids were systematically tested in the photocatalytic hydrogen generation reaction under visible or NIR light irradiation [135]. The photocatalytic activities were compared when systems were faced with wavelengths belonging to the visible or the near infrared light and it was seen that the order in the activity depended not only on the type of structure used, but also on the wavelength used because of the mechanisms enhancement. When the catalysts were excited with wavelengths within the visible range, the order in the catalytic activity was as follows: Au-Pt-CdS >> Au@CdS > CdS >> Au-Pt. In contrast, when NIR illumination was used, the activity dropped significantly, and the order of the catalytic activity was: Au-Pt > Au-Pt@CdS > Au > Au@CdS. Still, the H_2 evolution rates (in the micromolar range) were less remarkable than those reported for the heterodimers (vide supra).

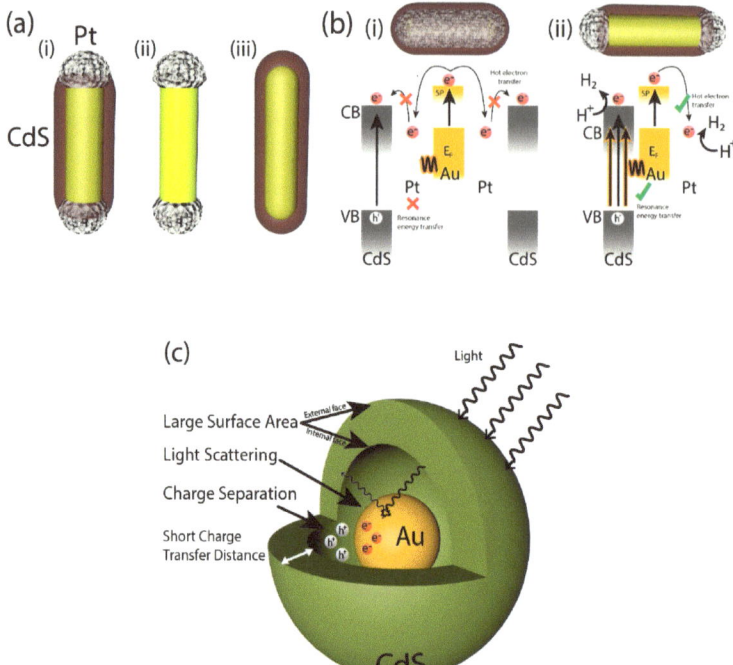

Figure 11. (**a**) Representation of Au-Pt@CdS (i), Au-Pt (ii), Au@CdS (iii) anisotropic hybrid structures; (**b**) schematic representation of energy band multichannel of Au@CdS and Au-Pt@CdS (adapted from [135]) and comparative excitation pathways: Au-assisted CdS charge separation and Au-Pt hot electron transfer is enhanced in the anisotropic structure. Hot electron transfer or RET is not easily achieved when Pt shell is deposited over AuNRs (adapted from [135]); (**c**) schematic representation of photoactivation mechanisms in yolk-shell Au-CdS nanostructures (adapted from [138]).

Under sunlight illumination, CdS semiconductor deposited on the sides of Pt-Au bimetallic nanoparticle induced both the RET and hot electron transfer processes and synergistically boosted the photocatalytic H_2 generation. However, when catalysts were activated by infrared light, Au-Pt bimetallic structure became the most active architecture indicating that excited hot electrons can further transfer to tipped Pt, but it is more complicated for electrons to enhance further migration from AuNR to CdS semiconductor (see mechanisms in Figure 11b). It was also observed how the catalytic activity of CdS coated metal tipped AuNRs (Au-Pt-CdS) was more active than its concentric counterpart Au@Pt@CdS when nanocatalyst were activated with visible light. The transversal LSPR induced electric field cannot reach the outside CdS semiconductor due to the presence of Pt NPs between the inner Au core and the external CdS layer (see Figure 11b). Pt nanoparticles around AuNR prevented the charge transfer between CdS and Au internal core. It has been observed how strategically modulating the metallic core of the structure by incorporating anisotropy or integrating several metals in a controlled way, the activity of the system was highly increased.

Likewise, yolk-shell structures have been recently reported as promising alternative hybrids with enhanced photocatalytic outcome provided by the selective control of the shell properties [138,141]. Han et al. [138] developed yolk-shell structures containing a plasmonic AuNR. The photocatalytic activity towards the generation of H_2 was compared to analogous core-shell type architectures and hollow CdS nanoparticles [138]. Using hole scavengers such as Na_2S or Na_2SO_3 to avoid photocorrosion of the CdS and exciting upon visible wavelengths, the yolk-shell structure generated a hydrogen conversion over 27 (1.7 mmol H_2 g^{-1}h^{-1}) or 12 (0.5 mmol H_2 g^{-1}h^{-1}) times greater than the core-shell type and hollow semiconductor structures (0.02 mmol H_2 $g^{-1}h^{-1}$). The authors also compared the catalytic activity of the yolk-shell Au-CdS hybrid system with an analogous system containing an insulating intermediate layer of silicon oxide Au@SiO_2-CdS [138]. No significant differences were observed when compared both photocatalysts under the same experimental conditions. For further explanation, several mechanisms were proposed.

The thermal activity of the rod was tentatively considered as one of the possible causes of the increase in activity, but the thermal evolution of the system for all photocatalysts was very similar. Since higher activity was achieved by using Au-coated plasmonic structures covered with SiO_2, it was concluded that plasmon energy transfer or plasmon-induced RET (see Figure 10) was not the mechanism by which these systems increased their activity. The researchers determined that the main cause for this process was radiative relaxation of the SPR of the AuNRs, such as near-field enhancement and resonant photon scattering (see Figures 2 and 11c). More recent studies have concurred in finding higher activity for the yolk-shell type [34]. Yolk-shell type structures have the ability to generate a very large surface-to-volume ratio, increasing the amount of surface exposed for photocatalysis and decreasing the amount of bulk material to reduce charge recombination probabilities. In addition, plasmonic Au is capable of generating better charge separation as well as potential light scattering of the light inside the nanoparticle (see Figure 11c).

We have seen several examples of hybrid architectures, where not only the type of metal incorporated (Au, Pt) was important, but its arrangement with respect to the CdS semiconductor played a crucial role in enhancing the catalytic activity. Yolk-shell type structures provided a better optical response thanks to the light scattering generated in the internal part of this void structure. Dumbbell-like Au-Pt type bimetallic structures allow better migration and separation of charges, solving one of the most important problems of this type of semiconductor. It is important to highlight the development of CdS-based complex hybrid structures where a good design and configuration of the constructed architecture allows us to considerably improve the activity of these systems when it is compared with other less complex hybrid structures, such as bare CdS or core-shell architectures.

6. Conclusions

TiO_2 and ZnO assembled with plasmonic metals are still the most explored and developed heteronanostructures obtained using semiconductors. The search for novel alternatives that can expand and take advantage of the full-solar irradiation spectra has stimulated the search for novel alternatives that can provide additional features in terms of more efficient harvesting properties, more effective carrier transport and enhanced selectivities towards specific end-products. In this regard, the need to find newer and cleaner routes for energy production, the need to improve the efficiency of decontamination of harmful pollutants and the willingness to promote the revalorization of wastes or unwanted byproducts has boosted the research in the area of photocatalysis. Nowadays, the innovation in the design of smart combination of metal-semiconductors continue to be a very promising alternative. In this review, we presented some of the most promising and versatile alternative gold-semiconductor heterostructures. The main results highlighted throughout this revision of the most recent literature suggests an important morphology dependence in the final outcome and how less-conventional designs beyond core-shell configurations may pave the way for important breakthroughs in forthcoming years. Furthermore, the selection of anisotropic metals also represents an appealing strategy to maximize the virtues of plasmon excitation. Finally, we would like to make emphasis on the need to establish exquisite control and definition of these hetero-nanoarchitectures at the nanoscale level in order to establish a clear progress in biomedical applications, where size control and reproducibility are key for clinical translation.

Author Contributions: Conceptualization, J.L.H.; writing—original draft preparation, J.I.G.-P., J.B.-A., C.J.B.-A., J.L.H.; writing—review and editing, J.L.H. All authors have read and agreed to the published version of the manuscript.

Funding: This research was funded by the European Research Council (ERC) through an Advanced Research Grant (CADENCE, grant number 742684). The APC was waived by the journal.

Acknowledgments: The authors thank the Platform of Production of Biomaterials and Nanoparticles of the NANBIOSIS-ICTS of the CIBER in BioEngineering, Biomaterials & Nanomedicine (CIBER-BBN). J.B.-A. acknowledges the Spanish Government for a PhD predoctoral grant (FPU18/04618). J.I.G.-P. thanks the Regional Government of Aragon (DGA) for granting a PhD predoctoral contract. The Regional Government of Aragon is also acknowledged.

Conflicts of Interest: The authors declare no conflict of interest.

References

1. Eibner, A.J.C.-Z. Action of light on pigments I. *Chem-Ztg* **1911**, *35*, 753–755.
2. Coronado, J.M. A historical introduction to photocatalysis. In *Design of Advanced Photocatalytic Materials for Energy and Environmental Applications*; Coronado, J.M., Fresno, F., Hernández-Alonso, M.D., Portela, R., Eds.; Springer: London, UK, 2013; pp. 1–4. [CrossRef]
3. Reyes-Coronado, D.; Rodríguez-Gattorno, G.; Espinosa-Pesqueira, M.E.; Cab, C.; de Coss, R.; Oskam, G. Phase-pure TiO2 nanoparticles: Anatase, brookite and rutile. *Nanotechnology* **2008**, *19*, 145605. [CrossRef] [PubMed]
4. Bakin, A.; El-Shaer, A.; Mofor, A.C.; Al-Suleiman, M.; Schlenker, E.; Waag, A. ZnMgO-ZnO quantum wells embedded in ZnO nanopillars: Towards realisation of nano-LEDs. *Phys. Status Solidi C* **2007**, *4*, 158–161. [CrossRef]
5. Bueno-Alejo, C.J.; Graus, J.; Arenal, R.; Lafuente, M.; Bottega-Pergher, B.; Hueso, J.L. Anisotropic Au-ZnO photocatalyst for the visible-light expanded oxidation of n-hexane. *Catal. Today* **2020**. [CrossRef]
6. Ortega-Liebana, M.C.; Hueso, J.L.; Ferdousi, S.; Arenal, R.; Irusta, S.; Yeung, K.L.; Santamaria, J. Extraordinary sensitizing effect of co-doped carbon nanodots derived from mate herb: Application to enhanced photocatalytic degradation of chlorinated wastewater compounds under visible light. *Appl. Catal. B Environ.* **2017**, *218*, 68–79. [CrossRef]
7. Suarez, H.; Ramirez, A.; Bueno-Alejo, C.J.; Hueso, J.L. Silver-copper oxide heteronanostructures for the plasmonic-enhanced photocatalytic oxidation of N-hexane in the visible-NIR range. *Materials* **2019**, *12*, 3858. [CrossRef]
8. Graus, J.; Bueno-Alejo, C.J.; Hueso, J.L. In-situ deposition of plasmonic gold nanotriangles and nanoprisms onto layered hydroxides for full-range photocatalytic response towards the selective reduction of p-nitrophenol. *Catalysts* **2018**, *8*, 354. [CrossRef]
9. Mas, N.; Hueso, J.L.; Martinez, G.; Madrid, A.; Mallada, R.; Ortega-Liebana, M.C.; Bueno-Alejo, C.; Santamaria, J. Laser-driven direct synthesis of carbon nanodots and application as sensitizers for visible-light photocatalysis. *Carbon* **2020**, *156*, 453–462. [CrossRef]
10. Kawamura, G.; Matsuda, A. Synthesis of plasmonic photocatalysts for water splitting. *Catalysts* **2019**, *9*, 982. [CrossRef]
11. Wang, J.; Wang, X.; Mu, X. Plasmonic photocatalysts monitored by tip-enhanced raman spectroscopy. *Catalysts* **2019**, *9*, 109. [CrossRef]
12. Gong, Z.; Ji, J.; Wang, J. Photocatalytic reversible reactions driven by localized surface plasmon resonance. *Catalysts* **2019**, *9*, 193. [CrossRef]
13. Wei, Z.; Janczarek, M.; Wang, K.; Zheng, S.; Kowalska, E. Morphology-governed performance of plasmonic photocatalysts. *Catalysts* **2020**, *10*, 1070. [CrossRef]
14. Fang, J.; Cao, S.-W.; Wang, Z.; Shahjamali, M.M.; Loo, S.C.J.; Barber, J.; Xue, C. Mesoporous plasmonic Au–TiO2 nanocomposites for efficient visible-light-driven photocatalytic water reduction. *Int. J. Hydrogen Energy* **2012**, *37*, 17853–17861. [CrossRef]
15. Fragua, D.M.; Abargues, R.; Rodriguez-Canto, P.J.; Sanchez-Royo, J.F.; Agouram, S.; Martinez-Pastor, J.P. Au–ZnO nanocomposite films for plasmonic photocatalysis. *Adv. Mater. Interfaces* **2015**, *2*, 1500156. [CrossRef]

16. Kim, M.; Lin, M.; Son, J.; Xu, H.; Nam, J.-M. Hot-electron-mediated photochemical reactions: Principles, recent advances, and challenges. *Adv. Opt. Mater.* **2017**, *5*, 1700004. [CrossRef]
17. Liu, J.; Ma, N.; Wu, W.; He, Q. Recent progress on photocatalytic heterostructures with full solar spectral responses. *Chem. Eng. J.* **2020**, *393*, 124719. [CrossRef]
18. Volokh, M.; Mokari, T. Metal/semiconductor interfaces in nanoscale objects: Synthesis, emerging properties and applications of hybrid nanostructures. *Nanoscale Adv.* **2020**, *2*, 930–961. [CrossRef]
19. Ghosh Chaudhuri, R.; Paria, S. Core/Shell nanoparticles: Classes, properties, synthesis mechanisms, characterization, and applications. *Chem. Rev.* **2012**, *112*, 2373–2433. [CrossRef] [PubMed]
20. Dutta, S.K.; Mehetor, S.K.; Pradhan, N. Metal semiconductor heterostructures for photocatalytic conversion of light energy. *J. Phys. Chem. Lett.* **2015**, *6*, 936–944. [CrossRef]
21. Li, B.; Gu, T.; Ming, T.; Wang, J.; Wang, P.; Wang, J.; Yu, J.C. (Gold Core)@(Ceria Shell) nanostructures for plasmon-enhanced catalytic reactions under visible light. *ACS Nano* **2014**, *8*, 8152–8162. [CrossRef]
22. Jia, H.; Du, A.; Zhang, H.; Yang, J.; Jiang, R.; Wang, J.; Zhang, C.-Y. Site-selective growth of crystalline ceria with oxygen vacancies on gold nanocrystals for near-infrared nitrogen photofixation. *J. Am. Chem. Soc.* **2019**, *141*, 5083–5086. [CrossRef] [PubMed]
23. Chen, T.-M.; Xu, G.-Y.; Ren, H.; Zhang, H.; Tian, Z.-Q.; Li, J.-F. Synthesis of Au@TiO2 core–shell nanoparticles with tunable structures for plasmon-enhanced photocatalysis. *Nanoscale Adv.* **2019**, *1*, 4522–4528. [CrossRef]
24. Liu, T.-M.; Conde, J.; Lipiński, T.; Bednarkiewicz, A.; Huang, C.-C. Revisiting the classification of NIR-absorbing/emitting nanomaterials for in vivo bioapplications. *NPG Asia Mater.* **2016**, *8*, e295. [CrossRef]
25. Ola, O.; Maroto-Valer, M.M. Review of material design and reactor engineering on TiO$_2$ photocatalysis for CO$_2$ reduction. *J. Photochem. Photobiol. C Photochem. Rev.* **2015**, *24*, 16–42. [CrossRef]
26. Habisreutinger, S.N.; Schmidt-Mende, L.; Stolarczyk, J.K. Photocatalytic reduction of CO$_2$ on TiO$_2$ and other semiconductors. *Angew. Chem. Int. Ed.* **2013**, *52*, 7372–7408. [CrossRef] [PubMed]
27. Wu, B.; Liu, D.; Mubeen, S.; Chuong, T.T.; Moskovits, M.; Stucky, G.D. Anisotropic growth of TiO$_2$ onto gold nanorods for plasmon-enhanced hydrogen production from water reduction. *J. Am. Chem. Soc.* **2016**, *138*, 1114–1117. [CrossRef] [PubMed]
28. Seh, Z.W.; Liu, S.; Low, M.; Zhang, S.Y.; Liu, Z.; Mlayah, A.; Han, M.Y. Janus Au-TiO$_2$ photocatalysts with strong localization of plasmonic near-fields for efficient visible-light hydrogen generation. *Adv. Mater. (Deerfield Beach Fla.)* **2012**, *24*, 2310–2314. [CrossRef]
29. Sun, H.; He, Q.; Zeng, S.; She, P.; Zhang, X.; Li, J.; Liu, Z. Controllable growth of Au@TiO2 yolk–shell nanoparticles and their geometry parameter effects on photocatalytic activity. *New J. Chem.* **2017**, *41*, 7244–7252. [CrossRef]
30. Li, J.-M.; Tsao, C.-W.; Fang, M.-J.; Chen, C.-C.; Liu, C.-W.; Hsu, Y.-J. TiO2-Au-Cu2O photocathodes: Au-mediated z-scheme charge transfer for efficient solar-driven photoelectrochemical reduction. *ACS Appl. Nano Mater.* **2018**, *1*, 6843–6853. [CrossRef]
31. Han, C.; Qi, M.-Y.; Tang, Z.-R.; Gong, J.; Xu, Y.-J. Gold nanorods-based hybrids with tailored structures for photoredox catalysis: Fundamental science, materials design and applications. *Nano Today* **2019**, *27*, 48–72. [CrossRef]
32. Yu, X.; Liu, F.; Bi, J.; Wang, B.; Yang, S. Improving the plasmonic efficiency of the Au nanorod-semiconductor photocatalysis toward water reduction by constructing a unique hot-dog nanostructure. *Nano Energy* **2017**, *33*, 469–475. [CrossRef]
33. Zhu, M.; Wang, Y.; Deng, Y.-H.; Peng, X.; Wang, X.; Yuan, H.; Yang, Z.-J.; Wang, Y.; Wang, H. Strategic modulation of energy transfer in Au-TiO$_2$-Pt nanodumbbells: Plasmon-enhanced hydrogen evolution reaction. *Nanoscale* **2020**, *12*, 7035–7044. [CrossRef] [PubMed]
34. Wang, L.; Chong, J.; Fu, Y.; Li, R.; Liu, J.; Huang, M. A novel strategy for the design of Au@CdS yolk-shell nanostructures and their photocatalytic properties. *J. Alloys Compd.* **2020**, *834*, 155051. [CrossRef]
35. Atta, S.; Pennington, A.M.; Celik, F.E.; Fabris, L. TiO2 on Gold Nanostars Enhances Photocatalytic Water Reduction in the Near-Infrared Regime. *Chem* **2018**, *4*, 2140–2153. [CrossRef]
36. Sun, Y.; Sun, Y.; Zhang, T.; Chen, G.; Zhang, F.; Liu, D.; Cai, W.; Li, Y.; Yang, X.; Li, C. Complete Au@ZnO core–shell nanoparticles with enhanced plasmonic absorption enabling significantly improved photocatalysis. *Nanoscale* **2016**, *8*, 10774–10782. [CrossRef] [PubMed]

37. Shao, X.; Li, B.; Zhang, B.; Shao, L.; Wu, Y. Au@ZnO core–shell nanostructures with plasmon-induced visible-light photocatalytic and photoelectrochemical properties. *Inorg. Chem. Front.* **2016**, *3*, 934–943. [CrossRef]
38. Jiang, R.; Li, B.; Fang, C.; Wang, J. Metal/Semiconductor hybrid nanostructures for plasmon-enhanced applications. *Adv. Mater.* **2014**, *26*, 5274–5309. [CrossRef]
39. Zhou, N.; López-Puente, V.; Wang, Q.; Polavarapu, L.; Pastoriza-Santos, I.; Xu, Q.-H. Plasmon-enhanced light harvesting: Applications in enhanced photocatalysis, photodynamic therapy and photovoltaics. *RSC Adv.* **2015**, *5*, 29076–29097. [CrossRef]
40. Clavero, C. Plasmon-induced hot-electron generation at nanoparticle/metal-oxide interfaces for photovoltaic and photocatalytic devices. *Nat. Photonics* **2014**, *8*, 95–103. [CrossRef]
41. Semenov, A.D.; Gol tsman, G.N.; Sobolewski, R. Hot-electron effect in superconductors and its applications for radiation sensors. *Supercond. Sci. Technol.* **2002**, *15*, R1–R16. [CrossRef]
42. Knight, M.W.; Wang, Y.; Urban, A.S.; Sobhani, A.; Zheng, B.Y.; Nordlander, P.; Halas, N.J. Embedding plasmonic nanostructure diodes enhances hot electron emission. *Nano Lett.* **2013**, *13*, 1687–1692. [CrossRef] [PubMed]
43. White, T.P.; Catchpole, K.R. Plasmon-enhanced internal photoemission for photovoltaics: Theoretical efficiency limits. *Appl. Phys. Lett.* **2012**, *101*, 073905. [CrossRef]
44. Tung, R.T. The physics and chemistry of the Schottky barrier height. *Appl. Phys. Rev.* **2014**, *1*, 011304. [CrossRef]
45. Knight, M.W.; Sobhani, H.; Nordlander, P.; Halas, N.J. Photodetection with active optical antennas. *Science* **2011**, *332*, 702. [CrossRef] [PubMed]
46. Zhang, N.; Han, C.; Fu, X.; Xu, Y.-J. Function-oriented engineering of metal-based nanohybrids for photoredox catalysis: Exerting plasmonic effect and beyond. *Chem* **2018**, *4*, 1832–1861. [CrossRef]
47. Kochuveedu, S.T.; Jang, Y.H.; Kim, D.H. A study on the mechanism for the interaction of light with noble metal-metal oxide semiconductor nanostructures for various photophysical applications. *Chem. Soc. Rev.* **2013**, *42*, 8467–8493. [CrossRef]
48. Liu, Z.; Hou, W.; Pavaskar, P.; Aykol, M.; Cronin, S.B. Plasmon resonant enhancement of photocatalytic water splitting under visible illumination. *Nano Lett.* **2011**, *11*, 1111–1116. [CrossRef]
49. Tian, Y.; Tatsuma, T. Plasmon-induced photoelectrochemistry at metal nanoparticles supported on nanoporous TiO_2. *Chem. Commun.* **2004**, 1810–1811. [CrossRef]
50. Cushing, S.K.; Li, J.; Meng, F.; Senty, T.R.; Suri, S.; Zhi, M.; Li, M.; Bristow, A.D.; Wu, N. Photocatalytic activity enhanced by plasmonic resonant energy transfer from metal to semiconductor. *J. Am. Chem. Soc.* **2012**, *134*, 15033–15041. [CrossRef]
51. Christopher, P.; Ingram, D.B.; Linic, S. Enhancing photochemical activity of semiconductor nanoparticles with optically active Ag nanostructures: Photochemistry mediated by ag surface plasmons. *J. Phys. Chem. C* **2010**, *114*, 9173–9177. [CrossRef]
52. Kelly, K.L.; Coronado, E.; Zhao, L.L.; Schatz, G.C. The optical properties of metal nanoparticles: The influence of size, shape, and dielectric environment. *J. Phys. Chem. B* **2003**, *107*, 668–677. [CrossRef]
53. Tian, N.; Zhou, Z.-Y.; Sun, S.-G.; Ding, Y.; Wang, Z.L. Synthesis of tetrahexahedral platinum nanocrystals with high-index facets and high electro-oxidation activity. *Science* **2007**, *316*, 732. [CrossRef] [PubMed]
54. Ming, T.; Feng, W.; Tang, Q.; Wang, F.; Sun, L.; Wang, J.; Yan, C. Growth of tetrahexahedral gold nanocrystals with high-index facets. *J. Am. Chem. Soc.* **2009**, *131*, 16350–16351. [CrossRef] [PubMed]
55. Wang, F.; Li, C.; Sun, L.-D.; Wu, H.; Ming, T.; Wang, J.; Yu, J.C.; Yan, C.-H. Heteroepitaxial growth of high-index-faceted palladium nanoshells and their catalytic performance. *J. Am. Chem. Soc.* **2011**, *133*, 1106–1111. [CrossRef]
56. Lu, B.; Liu, A.; Wu, H.; Shen, Q.; Zhao, T.; Wang, J. Hollow Au–Cu_2O core–shell nanoparticles with geometry-dependent optical properties as efficient plasmonic photocatalysts under visible light. *Langmuir* **2016**, *32*, 3085–3094. [CrossRef]
57. Xia, X.; Zeng, J.; McDearmon, B.; Zheng, Y.; Li, Q.; Xia, Y. Silver nanocrystals with concave surfaces and their optical and surface-enhanced raman scattering properties. *Angew. Chem. Int. Ed.* **2011**, *50*, 12542–12546. [CrossRef]

58. Yin, P.-G.; You, T.-T.; Tan, E.-Z.; Li, J.; Lang, X.-F.; Jiang, L.; Guo, L. Characterization of tetrahexahedral gold nanocrystals: A combined study by surface-enhanced raman spectroscopy and computational simulations. *J. Phys. Chem. C* **2011**, *115*, 18061–18069. [CrossRef]
59. Rodríguez-Lorenzo, L.; Álvarez-Puebla, R.A.; Pastoriza-Santos, I.; Mazzucco, S.; Stéphan, O.; Kociak, M.; Liz-Marzán, L.M.; García de Abajo, F.J. Zeptomol detection through controlled ultrasensitive surface-enhanced raman scattering. *J. Am. Chem. Soc.* **2009**, *131*, 4616–4618. [CrossRef]
60. Hong, J.W.; Wi, D.H.; Lee, S.-U.; Han, S.W. Metal–semiconductor heteronanocrystals with desired configurations for plasmonic photocatalysis. *J. Am. Chem. Soc.* **2016**, *138*, 15766–15773. [CrossRef]
61. Simon, T.; Bouchonville, N.; Berr, M.J.; Vaneski, A.; Adrović, A.; Volbers, D.; Wyrwich, R.; Döblinger, M.; Susha, A.S.; Rogach, A.L.; et al. Redox shuttle mechanism enhances photocatalytic H2 generation on Ni-decorated CdS nanorods. *Nat. Mater.* **2014**, *13*, 1013–1018. [CrossRef]
62. Zhao, Q.; Ji, M.; Qian, H.; Dai, B.; Weng, L.; Gui, J.; Zhang, J.; Ouyang, M.; Zhu, H. Controlling structural symmetry of a hybrid nanostructure and its effect on efficient photocatalytic hydrogen evolution. *Adv. Mater.* **2014**, *26*, 1387–1392. [CrossRef] [PubMed]
63. Pan, J.; Zhang, L.; Zhang, S.; Shi, Z.; Wang, X.; Song, S.; Zhang, H. Half-encapsulated Au nanorods@CeO2 Core@Shell nanostructures for near-infrared plasmon-enhanced catalysis. *ACS Appl. Nano Mater.* **2019**, *2*, 1516–1524. [CrossRef]
64. Mubeen, S.; Lee, J.; Singh, N.; Krämer, S.; Stucky, G.D.; Moskovits, M. An autonomous photosynthetic device in which all charge carriers derive from surface plasmons. *Nat. Nanotechnol.* **2013**, *8*, 247–251. [CrossRef] [PubMed]
65. Wang, F.; Cheng, S.; Bao, Z.; Wang, J. Anisotropic overgrowth of metal heterostructures induced by a site-selective silica coating. *Angew. Chem. (Int. Ed. Engl.)* **2013**, *52*, 10344–10348. [CrossRef] [PubMed]
66. Zheng, Z.; Tachikawa, T.; Majima, T. Single-particle study of Pt-modified Au nanorods for plasmon-enhanced hydrogen generation in visible to near-infrared region. *J. Am. Chem. Soc.* **2014**, *136*, 6870–6873. [CrossRef]
67. Yang, H.; Wang, Z.-H.; Zheng, Y.-Y.; He, L.-Q.; Zhan, C.; Lu, X.; Tian, Z.-Q.; Fang, P.-P.; Tong, Y. Tunable wavelength enhanced photoelectrochemical cells from surface plasmon resonance. *J. Am. Chem. Soc.* **2016**, *138*, 16204–16207. [CrossRef]
68. Bao, Z.; Sun, Z.; Li, Z.; Tian, L.; Ngai, T.; Wang, J. Plasmonic gold–superparamagnetic hematite heterostructures. *Langmuir* **2011**, *27*, 5071–5075. [CrossRef]
69. Li, A.; Zhu, W.; Li, C.; Wang, T.; Gong, J. Rational design of yolk–shell nanostructures for photocatalysis. *Chem. Soc. Rev.* **2019**, *48*, 1874–1907. [CrossRef]
70. Fang, C.; Jia, H.; Chang, S.; Ruan, Q.; Wang, P.; Chen, T.; Wang, J. (Gold core)/(titania shell) nanostructures for plasmon-enhanced photon harvesting and generation of reactive oxygen species. *Energy Environ. Sci.* **2014**, *7*, 3431–3438. [CrossRef]
71. Ortega-Liebana, M.C.; Hueso, J.L.; Arenal, R.; Santamaria, J. Titania-coated gold nanorods with expanded photocatalytic response. Enzyme-like glucose oxidation under near-infrared illumination. *Nanoscale* **2017**, *9*, 1787–1792. [CrossRef]
72. Kou, S.F.; Ye, W.; Guo, X.; Xu, X.F.; Sun, H.Y.; Yang, J. Gold nanorods coated by oxygen-deficient TiO$_2$ as an advanced photocatalyst for hydrogen evolution. *RSC Adv.* **2016**, *6*, 39144–39149. [CrossRef]
73. Paier, J.; Penschke, C.; Sauer, J. Oxygen defects and surface chemistry of ceria: Quantum chemical studies compared to experiment. *Chem. Rev.* **2013**, *113*, 3949–3985. [CrossRef] [PubMed]
74. Chueh, W.C.; Falter, C.; Abbott, M.; Scipio, D.; Furler, P.; Haile, S.M.; Steinfeld, A. High-flux solar-driven thermochemical dissociation of CO$_2$ and H$_2$O using nonstoichiometric ceria. *Science* **2010**, *330*, 1797. [CrossRef] [PubMed]
75. Cargnello, M.; Jaén, J.J.D.; Garrido, J.C.H.; Bakhmutsky, K.; Montini, T.; Gámez, J.J.C.; Gorte, R.J.; Fornasiero, P. Exceptional activity for methane combustion over modular Pd@CeO$_2$ subunits on functionalized Al$_2$O$_3$. *Science* **2012**, *337*, 713. [CrossRef] [PubMed]
76. Zhou, H.-P.; Wu, H.-S.; Shen, J.; Yin, A.-X.; Sun, L.-D.; Yan, C.-H. Thermally stable Pt/CeO$_2$ hetero-nanocomposites with high catalytic activity. *J. Am. Chem. Soc.* **2010**, *132*, 4998–4999. [CrossRef]
77. Qi, J.; Zhao, K.; Li, G.; Gao, Y.; Zhao, H.; Yu, R.; Tang, Z. Multi-shelled CeO$_2$ hollow microspheres as superior photocatalysts for water oxidation. *Nanoscale* **2014**, *6*, 4072–4077. [CrossRef]
78. Corma, A.; Atienzar, P.; García, H.; Chane-Ching, J.Y. Hierarchically mesostructured doped CeO2 with potential for solar-cell use. *Nat. Mater.* **2004**, *3*, 394–397. [CrossRef]

79. Jin-Ha, H.; Thomas, O.M. Defect chemistry and transport properties of nanocrystalline cerium oxide. *Z. Für Phys. Chem.* **1998**, *207*, 21–38. [CrossRef]
80. Wang, J.-H.; Chen, M.; Luo, Z.-J.; Ma, L.; Zhang, Y.-F.; Chen, K.; Zhou, L.; Wang, Q.-Q. Ceria-Coated Gold Nanorods for Plasmon-Enhanced Near-Infrared Photocatalytic and Photoelectrochemical performances. *J. Phys. Chem. C* **2016**, *120*, 14805–14812. [CrossRef]
81. Mitsudome, T.; Yamamoto, M.; Maeno, Z.; Mizugaki, T.; Jitsukawa, K.; Kaneda, K. One-step synthesis of core-gold/shell-ceria nanomaterial and its catalysis for highly selective semihydrogenation of alkynes. *J. Am. Chem. Soc.* **2015**, *137*, 13452–13455. [CrossRef]
82. Zhang, L.; Pan, J.; Long, Y.; Li, J.; Li, W.; Song, S.; Shi, Z.; Zhang, H. CeO_2-encapsulated hollow Ag–Au nanocage hybrid nanostructures as high-performance catalysts for cascade reactions. *Small* **2019**, *15*, 1903182. [CrossRef] [PubMed]
83. Luo, F.; Jia, C.-J.; Song, W.; You, L.-P.; Yan, C.-H. Chelating ligand-mediated crystal growth of cerium orthovanadate. *Cryst. Growth Des.* **2005**, *5*, 137–142. [CrossRef]
84. Kominami, H.; Tanaka, A.; Hashimoto, K. Mineralization of organic acids in aqueous suspensions of gold nanoparticles supported on cerium (iv) oxide powder under visible light irradiation. *Chem. Commun.* **2010**, *46*, 1287–1289. [CrossRef] [PubMed]
85. Kominami, H.; Tanaka, A.; Hashimoto, K. Gold nanoparticles supported on cerium (IV) oxide powder for mineralization of organic acids in aqueous suspensions under irradiation of visible light of λ = 530 nm. *Appl. Catal. A Gen.* **2011**, *397*, 121–126. [CrossRef]
86. Fasciani, C.; Alejo, C.J.B.; Grenier, M.; Netto-Ferreira, J.C.; Scaiano, J.C. High-temperature organic reactions at room temperature using plasmon excitation: Decomposition of dicumyl peroxide. *Org. Lett.* **2011**, *13*, 204–207. [CrossRef]
87. Heckert, E.G.; Seal, S.; Self, W.T. Fenton-like reaction catalyzed by the rare earth inner transition metal cerium. *Environ. Sci. Technol.* **2008**, *42*, 5014–5019. [CrossRef]
88. Kohantorabi, M.; Gholami, M.R. Fabrication of novel ternary Au/CeO2@g-C3N4 nanocomposite: Kinetics and mechanism investigation of 4-nitrophenol reduction, and benzyl alcohol oxidation. *Appl. Phys. A* **2018**, *124*, 441. [CrossRef]
89. Li, Y.; Sun, Q.; Kong, M.; Shi, W.; Huang, J.; Tang, J.; Zhao, X. Coupling oxygen ion conduction to photocatalysis in mesoporous nanorod-like ceria significantly improves photocatalytic efficiency. *J. Phys. Chem. C* **2011**, *115*, 14050–14057. [CrossRef]
90. Yin, D.; Zhao, F.; Zhang, L.; Zhang, X.; Liu, Y.; Zhang, T.; Wu, C.; Chen, D.; Chen, Z. Greatly enhanced photocatalytic activity of semiconductor CeO_2 by integrating with upconversion nanocrystals and graphene. *RSC Adv.* **2016**, *6*, 103795–103802. [CrossRef]
91. Nair, V.; Muñoz-Batista, M.J.; Fernández-García, M.; Luque, R.; Colmenares, J.C. Thermo-photocatalysis: Environmental and energy applications. *ChemSusChem* **2019**, *12*, 2098–2116. [CrossRef]
92. Italiano, C.; Llorca, J.; Pino, L.; Ferraro, M.; Antonucci, V.; Vita, A. CO and CO_2 methanation over Ni catalysts supported on CeO_2, Al_2O_2 and Y_2O_2 oxides. *Appl. Catal. B Environ.* **2020**, *264*, 118494. [CrossRef]
93. Hammedi, T.; Triki, M.; Alvarez, M.G.; Llorca, J.; Ghorbel, A.; Ksibi, Z.; Medina, F. Heterogeneous fenton-like oxidation of p-hydroxybenzoic acid using Fe/CeO_2-TiO_2 catalyst. *Water Sci. Technol.* **2019**, *79*, 1276–1286. [CrossRef] [PubMed]
94. Yang, C.; Yu, X.; Heißler, S.; Nefedov, A.; Colussi, S.; Llorca, J.; Trovarelli, A.; Wang, Y.; Wöll, C. Surface faceting and reconstruction of ceria nanoparticles. *Angew. Chem. Int. Ed.* **2017**, *56*, 375–379. [CrossRef] [PubMed]
95. Soler, L.; Casanovas, A.; Urrich, A.; Angurell, I.; Llorca, J. CO oxidation and COPrOx over preformed Au nanoparticles supported over nanoshaped CeO_2. *Appl. Catal. B Environ.* **2016**, *197*, 47–55. [CrossRef]
96. Lu, B.; Quan, F.; Sun, Z.; Jia, F.; Zhang, L. Photothermal reverse-water-gas-shift over Au/CeO_2 with high yield and selectivity in CO_2 conversion. *Catal. Commun.* **2019**, *129*, 105724. [CrossRef]
97. Ghoussoub, M.; Xia, M.; Duchesne, P.N.; Segal, D.; Ozin, G. Principles of photothermal gas-phase heterogeneous CO_2 catalysis. *Energy Environ. Sci.* **2019**, *12*, 1122–1142. [CrossRef]
98. Jantarang, S.; Lovell, E.C.; Tan, T.H.; Scott, J.; Amal, R. Role of support in photothermal carbon dioxide hydrogenation catalysed by Ni/Ce$_x$Ti$_y$O$_2$. *Prog. Nat. Sci. Mater. Int.* **2018**, *28*, 168–177. [CrossRef]
99. Pan, Y.; Deng, S.; Polavarapu, L.; Gao, N.; Yuan, P.; Sow, C.H.; Xu, Q.-H. Plasmon-enhanced photocatalytic properties of Cu_2O nanowire–au nanoparticle assemblies. *Langmuir* **2012**, *28*, 12304–12310. [CrossRef]

100. Ren, S.; Wang, B.; Zhang, H.; Ding, P.; Wang, Q. Sandwiched ZnO@Au@Cu$_2$O nanorod films as efficient visible-light-driven plasmonic photocatalysts. *ACS Appl. Mater. Interfaces* **2015**, *7*, 4066–4074. [CrossRef]
101. Niu, J.; Dai, Y.; Yin, L.; Shang, J.; Crittenden, J.C. Photocatalytic reduction of triclosan on Au–Cu$_2$O nanowire arrays as plasmonic photocatalysts under visible light irradiation. *Phys. Chem. Chem. Phys.* **2015**, *17*, 17421–17428. [CrossRef]
102. Sharma, K.; Maiti, K.; Kim, N.H.; Hui, D.; Lee, J.H. Green synthesis of glucose-reduced graphene oxide supported Ag-Cu$_2$O nanocomposites for the enhanced visible-light photocatalytic activity. *Compos. Part. B Eng.* **2018**, *138*, 35–44. [CrossRef]
103. Wang, X.; Dong, H.; Hu, Z.; Qi, Z.; Li, L. Fabrication of a Cu$_2$O/Au/TiO$_2$ composite film for efficient photocatalytic hydrogen production from aqueous solution of methanol and glucose. *Mater. Sci. Eng. B* **2017**, *219*, 10–19. [CrossRef]
104. McShane, C.M.; Choi, K.-S. Photocurrent enhancement of n-type Cu$_2$O electrodes achieved by controlling dendritic branching growth. *J. Am. Chem. Soc.* **2009**, *131*, 2561–2569. [CrossRef] [PubMed]
105. Kong, L.; Chen, W.; Ma, D.; Yang, Y.; Liu, S.; Huang, S. Size control of Au@Cu$_2$O octahedra for excellent photocatalytic performance. *J. Mater. Chem.* **2012**, *22*, 719–724. [CrossRef]
106. Kuo, C.-H.; Yang, Y.-C.; Gwo, S.; Huang, M.H. Facet-dependent and Au nanocrystal-enhanced electrical and photocatalytic properties of Au–Cu$_2$O core–shell heterostructures. *J. Am. Chem. Soc.* **2011**, *133*, 1052–1057. [CrossRef]
107. Yuan, G.; Lu, M.; Fei, J.; Guo, J.; Wang, Z. Morphologically controllable synthesis of core–shell structured Au@Cu$_2$O with enhanced photocatalytic activity. *RSC Adv.* **2015**, *5*, 71559–71564. [CrossRef]
108. Chiu, Y.-H.; Lindley, S.A.; Tsao, C.-W.; Kuo, M.-Y.; Cooper, J.K.; Hsu, Y.-J.; Zhang, J.Z. Hollow Au nanosphere-Cu$_2$O core–shell nanostructures with controllable core surface morphology. *J. Phys. Chem. C* **2020**, *124*, 11333–11339. [CrossRef]
109. Kuo, C.-H.; Hua, T.-E.; Huang, M.H. Au nanocrystal-directed growth of Au–Cu$_2$O core–shell heterostructures with precise morphological control. *J. Am. Chem. Soc.* **2009**, *131*, 17871–17878. [CrossRef]
110. Zhang, L.; Blom, D.A.; Wang, H. Au–Cu$_2$O core–shell nanoparticles: A hybrid metal-semiconductor heteronanostructure with geometrically tunable optical properties. *Chem. Mater.* **2011**, *23*, 4587–4598. [CrossRef]
111. Huang, W.-C.; Lyu, L.-M.; Yang, Y.-C.; Huang, M.H. Synthesis of Cu$_2$O nanocrystals from cubic to rhombic dodecahedral structures and their comparative photocatalytic activity. *J. Am. Chem. Soc.* **2012**, *134*, 1261–1267. [CrossRef]
112. Xu, W.; Jia, J.; Wang, T.; Li, C.; He, B.; Zong, J.; Wang, Y.; Fan, H.J.; Xu, H.; Feng, Y.; et al. Continuous tuning of Au-Cu(2) O janus nanostructures for efficient charge separation. *Angew. Chem. (Int. Ed. Engl.)* **2020**. [CrossRef]
113. Yu, X.; Liu, X.; Wang, B.; Meng, Q.; Sun, S.; Tang, Y.; Zhao, K. An LSPR-based "push–pull" synergetic effect for the enhanced photocatalytic performance of a gold nanorod@cuprous oxide-gold nanoparticle ternary composite. *Nanoscale* **2020**, *12*, 1912–1920. [CrossRef]
114. Liu, C.; Dong, H.; Wu, N.; Cao, Y.; Zhang, X. Plasmonic resonance energy transfer enhanced photodynamic therapy with Au@SiO2@Cu$_2$O/perfluorohexane nanocomposites. *ACS Appl. Mater. Interfaces* **2018**, *10*, 6991–7002. [CrossRef]
115. Zheng, T.; Zhou, T.; Feng, X.; Shen, J.; Zhang, M.; Sun, Y. Enhanced plasmon-induced resonance energy transfer (PIRET)-mediated photothermal and photodynamic therapy guided by photoacoustic and magnetic resonance imaging. *ACS Appl. Mater. Interfaces* **2019**, *11*, 31615–31626. [CrossRef] [PubMed]
116. Chang, Y.; Cheng, Y.; Feng, Y.; Jian, H.; Wang, L.; Ma, X.; Li, X.; Zhang, H. Resonance energy transfer-promoted photothermal and photodynamic performance of gold–copper sulfide yolk–shell nanoparticles for chemophototherapy of cancer. *Nano Lett.* **2018**, *18*, 886–897. [CrossRef] [PubMed]
117. Tao, C.; An, L.; Lin, J.; Tian, Q.; Yang, S. Surface plasmon resonance–enhanced photoacoustic imaging and photothermal therapy of endogenous H2S-triggered Au@Cu$_2$O. *Small* **2019**, *15*, 1903473. [CrossRef] [PubMed]
118. Wang, J.; Wu, X.; Ma, W.; Xu, C. Chiral AuCuAu heterogeneous nanorods with tailored optical activity. *Adv. Funct. Mater.* **2020**, *30*, 2000670. [CrossRef]
119. Luther, J.M.; Jain, P.K.; Ewers, T.; Alivisatos, A.P. Localized surface plasmon resonances arising from free carriers in doped quantum dots. *Nat. Mater.* **2011**, *10*, 361–366. [CrossRef] [PubMed]

120. Yu, X.; Bi, J.; Yang, G.; Tao, H.; Yang, S. Synergistic effect induced high photothermal performance of Au nanorod@Cu7S4 yolk–shell nanooctahedron particles. *J. Phys. Chem. C* **2016**, *120*, 24533–24541. [CrossRef]

121. Sun, M.; Fu, X.; Chen, K.; Wang, H. Dual-plasmonic gold@copper sulfide core–shell nanoparticles: Phase-selective synthesis and multimodal photothermal and photocatalytic behaviors. *ACS Appl. Mater. Interfaces* **2020**, *12*, 46146–46161. [CrossRef]

122. Zhang, Y.; Zhu, M.; Zhang, S.; Cai, Y.; Lv, Z.; Fang, M.; Tan, X.; Wang, X. Highly efficient removal of U(VI) by the photoreduction of SnO2/CdCO3/CdS nanocomposite under visible light irradiation. *Appl. Catal. B Environ.* **2020**, *279*, 119390. [CrossRef]

123. Yao, X.; Hu, X.; Zhang, W.; Gong, X.; Wang, X.; Pillai, S.C.; Dionysiou, D.D.; Wang, D. Mie resonance in hollow nanoshells of ternary TiO2-Au-CdS and enhanced photocatalytic hydrogen evolution. *Appl. Catal. B Environ.* **2020**, *276*, 119153. [CrossRef]

124. Kumar, S.G.; Kavitha, R.; Nithya, P.M. Tailoring the CdS surface structure for photocatalytic applications. *J. Environ. Chem. Eng.* **2020**, *8*, 104313. [CrossRef]

125. Peng, Y.; Kang, S.; Hu, Z. Pt nanoparticle-decorated CdS photocalysts for CO2 reduction and H2 evolution. *ACS Appl. Nano Mater.* **2020**, *3*, 8632–8639. [CrossRef]

126. Miodyńska, M.; Mikolajczyk, A.; Bajorowicz, B.; Zwara, J.; Klimczuk, T.; Lisowski, W.; Trykowski, G.; Pinto, H.P.; Zaleska-Medynska, A. Urchin-like TiO_2 structures decorated with lanthanide-doped Bi2S3 quantum dots to boost hydrogen photogeneration performance. *Appl. Catal. B Environ.* **2020**, *272*, 118962. [CrossRef]

127. Ji, Z.; Wang, H.; She, X. A novel CdS quantum dots decorated 3D Bi2O2CO3 hierarchical nanoflower with enhanced photocatalytic performance. *Catalysts* **2020**, *9*, 1046. [CrossRef]

128. Hou, H.; Zhang, X. Rational design of 1D/2D heterostructured photocatalyst for energy and environmental applications. *Chem. Eng. J.* **2020**, *395*, 125030. [CrossRef]

129. Karthikeyan, C.; Arunachalam, P.; Ramachandran, K.; Al-Mayouf, A.M.; Karuppuchamy, S. Recent advances in semiconductor metal oxides with enhanced methods for solar photocatalytic applications. *J. Alloys Compd.* **2020**, *828*, 154281. [CrossRef]

130. Wei, Y.; Jiao, J.; Zhao, Z.; Liu, J.; Li, J.; Jiang, G.; Wang, Y.; Duan, A. Fabrication of inverse opal TiO2-supported Au@CdS core–shell nanoparticles for efficient photocatalytic CO_2 conversion. *Appl. Catal. B Environ.* **2015**, *179*, 422–432. [CrossRef]

131. Wang, S.; Wang, X. Photocatalytic CO_2 reduction by CdS promoted with a zeolitic imidazolate framework. *Appl. Catal. B Environ.* **2015**, *162*, 494–500. [CrossRef]

132. Bak, T.; Nowotny, J.; Rekas, M.; Sorrell, C.C. Photo-electrochemical hydrogen generation from water using solar energy. Materials-related aspects. *Int. J. Hydrogen Energy* **2002**, *27*, 991–1022. [CrossRef]

133. Huang, J.; Mulfort, K.L.; Du, P.; Chen, L.X. Photodriven charge separation dynamics in CdSe/ZnS Core/Shell quantum Dot/Cobaloxime hybrid for efficient hydrogen production. *J. Am. Chem. Soc.* **2012**, *134*, 16472–16475. [CrossRef] [PubMed]

134. Xie, Y.P.; Yu, Z.B.; Liu, G.; Ma, X.L.; Cheng, H.-M. CdS–mesoporous ZnS core–shell particles for efficient and stable photocatalytic hydrogen evolution under visible light. *Energy Environ. Sci.* **2014**, *7*, 1895–1901. [CrossRef]

135. Wu, J.; Zhang, Z.; Liu, B.; Fang, Y.; Wang, L.; Dong, B. UV-Vis-NIR-driven plasmonic photocatalysts with dual-resonance modes for synergistically enhancing H2 generation. *Sol. RRL* **2018**, *2*, 1800039. [CrossRef]

136. Tang, W.; Fan, W.; Zhang, W.; Yang, Z.; Li, L.; Wang, Z.; Chiang, Y.-L.; Liu, Y.; Deng, L.; He, L.; et al. Wet/Sono-chemical synthesis of enzymatic two-dimensional MnO_2 nanosheets for synergistic catalysis-enhanced phototheranostics. *Adv. Mater.* **2019**, *31*, 1900401. [CrossRef] [PubMed]

137. Chen, W.-T.; Yang, T.-T.; Hsu, Y.-J. Au-CdS Core–Shell Nanocrystals with Controllable Shell Thickness and Photoinduced Charge Separation Property. *Chem. Mat.* **2008**, *20*, 7204–7206. [CrossRef]

138. Lee, S.-U.; Jung, H.; Wi, D.H.; Hong, J.W.; Sung, J.; Choi, S.-I.; Han, S.W. Metal–semiconductor yolk–shell heteronanostructures for plasmon-enhanced photocatalytic hydrogen evolution. *J. Mater. Chem. A* **2018**, *6*, 4068–4078. [CrossRef]

139. Li, M.; Yu, X.-F.; Liang, S.; Peng, X.-N.; Yang, Z.-J.; Wang, Y.-L.; Wang, Q.-Q. Synthesis of Au–CdS core–shell hetero-nanorods with efficient exciton–plasmon interactions. *Adv. Funct. Mater.* **2011**, *21*, 1788–1794. [CrossRef]

140. Ma, L.; Liang, S.; Liu, X.-L.; Yang, D.-J.; Zhou, L.; Wang, Q.-Q. Synthesis of dumbbell-like gold–metal sulfide core–shell nanorods with largely enhanced transverse plasmon resonance in visible region and efficiently improved photocatalytic activity. *Adv. Funct. Mater.* **2015**, *25*, 898–904. [CrossRef]
141. Chiu, Y.-H.; Naghadeh, S.B.; Lindley, S.A.; Lai, T.-H.; Kuo, M.-Y.; Chang, K.-D.; Zhang, J.Z.; Hsu, Y.-J. Yolk-shell nanostructures as an emerging photocatalyst paradigm for solar hydrogen generation. *Nano Energy* **2019**, *62*, 289–298. [CrossRef]

Publisher's Note: MDPI stays neutral with regard to jurisdictional claims in published maps and institutional affiliations.

© 2020 by the authors. Licensee MDPI, Basel, Switzerland. This article is an open access article distributed under the terms and conditions of the Creative Commons Attribution (CC BY) license (http://creativecommons.org/licenses/by/4.0/).

Review

Photonic Crystals for Plasmonic Photocatalysis

Tharishinny Raja-Mogan [1,2], Bunsho Ohtani [1,2] and Ewa Kowalska [1,2,*]

1. Graduate School of Environmental Science, Hokkaido University, Sapporo 060-0810, Japan; rajamogan.t@cat.hokudai.ac.jp (T.R.-M.); ohtani@cat.hokudai.ac.jp (B.O.)
2. Institute for Catalysis (ICAT), Hokkaido University, Sapporo 001-0021, Japan
* Correspondence: kowalska@cat.hokudai.ac.jp

Received: 6 July 2020; Accepted: 21 July 2020; Published: 23 July 2020

Abstract: Noble metal (NM)-modified wide-bandgap semiconductors with activity under visible light (Vis) irradiation, due to localized surface plasmon resonance (LSPR), known as plasmonic photocatalysts, have been intensively studied over the last few years. Despite the novelty of the topic, a large number of reports have already been published, discussing the optimal properties, synthesis methods and mechanism clarification. It has been proposed that both efficient light harvesting and charge carriers' migration are detrimental for high and stable activity under Vis irradiation. Accordingly, photonic crystals (PCs) with photonic bandgap (PBG) and slow photon effects seem to be highly attractive for efficient use of incident photons. Therefore, the study on PCs-based plasmonic photocatalysts has been conducted, mainly on titania inverse opal (IO) modified with nanoparticles (NPs) of NM. Although, the research is quite new and only several reports have been published, it might be concluded that the matching between LSPR and PBG (especially at red edge) by tuning of NMNPs size and IO-void diameter, respectively, is the most crucial for the photocatalytic activity.

Keywords: light harvesting; localized surface plasmon resonance; LSPR; photonic bandgap; PBG; photonic crystal; plasmonic photocatalysis; slow photons; titania inverse opal; vis-responsive photocatalysts

1. Introduction

Solar photocatalysis has been considered as one of the possible solutions for main crises facing humanity, i.e., energy, environment and water. It is believed that solar energy in the presence of photocatalyst might efficiently: (i) be converted into electricity/fuel, (ii) decompose chemical and microbiological pollutants, and (iii) purify water [1,2]. Accordingly, a large number of studies on development of efficient and stable photocatalysts have already been performed. Among various photocatalytic materials, wide-bandgap semiconductors are still the most widely investigated despite low harvesting efficiently of solar energy. It should be pointed out that though the wide bandgap is detrimental for light harvesting (absorption edge near 400 nm), it usually results in high reactivity in both oxidation and reduction reactions under UV, especially for oxide semiconductors, since valence band (VB) and conduction band (CB) are highly positive and negative, respectively [3–5]. However, it should be also remembered that all semiconductors suffer from charge carriers' recombination either in the bulk or on the surface [6]. Therefore, wide-bandgap semiconductors have been modified with various elements/compounds, and thus obtained materials exhibit much higher quantum yields of photocatalytic reactions [7]. For example, nanoparticles (NPs) of noble metals (NMs) have been used to inhibit charge carriers' recombination, as NMs work as an electron sink (higher work function than electron affinity of oxides). Kraeutler and Bard were probably the first who proved this more than 40 years ago [8]. Since then various studies on photocatalytic activity enhancement by semiconductor modification with NM have been performed, including optimization of the properties of NM/semiconductor and the mechanism clarifications [9–12]. Additionally, some complexes and clusters of NMs have shown Vis absorption, and thus semiconductors have been modified with them to obtain Vis response, e.g., PtCl$_4$ [13] and [Pt$_3$(CO)$_6$]$_6^{2-}$ clusters [14]. More than a decade ago another property of NMNPs have been used to activate wide-bandgap semiconductors towards Vis response, i.e., localized surface plasmon resonance (LSPR) [15]. Since then, many studies have been reported, and the photocatalysts containing NM deposits and being active under Vis irradiation have been named as plasmonic photocatalysts. Although the research on plasmonic photocatalysis is quite new, there are many reports on their synthesis, properties, applications, and mechanism dispute, including review papers and book chapters [16–19], and journal special issues [20,21], as shortly presented in the next sections. However, this review is not summarizing/discussing all these reports on plasmonic photocatalysis but focuses on the novel topic using semiconductors in the form of photonic crystals (PCs) as a support for NMNPs. It has been expected that such photocatalysts should possess high photocatalytic activity due to enhanced light harvesting inside PCs. The synthesis methods, properties and some application examples of PCs-based plasmonic photocatalysts are presented further.

2. Plasmonic Photocatalysis

Although, it is unknown by whom and when the term of "plasmonic photocatalysis" has been created, it is quite convenient to distinguish the activity of NM-modified wide-bandgap semiconductors under Vis irradiation from their well-known activity under UV. The activity under Vis irradiation originates from the properties of NMs, i.e., LSPR at Vis-NIR range of solar spectrum, in contrast to the activity under UV, where semiconductor is mainly responsible for the photocatalytic performance. LSPR is the result of the confinement of a surface plasmon in an NP of the size similar or smaller than the wavelength of light used to excite the plasmon, i.e., when an NP is irradiated, the oscillating electric field causes the conduction electrons to oscillate coherently. The majority of studies on plasmonic photocatalysis have been performed on titania (titanium(IV) oxide, titanium dioxide), but also other semiconductors have been tested, e.g., CeO$_2$ [22], Fe$_2$O$_3$ [23], ZnO [24], KNbO$_3$ [25], g-C$_3$N$_4$ [26], AgCl [27], Ag$_2$MoO$_4$/AgBr [28], and ZrO$_2$@CoFe$_2$O$_4$ [29]. Although plasmonic properties of NMs were found more than a century ago, and commercially used in many fields (e.g., medicine, SERS, and optical data storage), the examination of their potential for photocatalysis under Vis irradiation is quite new. Despite their novelty, many studies have already been conducted, including the improvement of the photocatalytic activity and stability as well as the explanation of the mechanism under visible light. Although the application of LSPR in the photocatalysis started at the beginning of this century, it was used only for the characterization of gold deposits on titania, i.e., the formation, properties and the stability under UV irradiation [30]. Next, Au/TiO$_2$ was used as Vis-responsive photocatalyst for generation of photocurrent [15] and oxidative degradation of organic compounds, such as methyl *tert*-butyl ether [31] and 2-propanol [32]. It should be mentioned that the direct proof confirming that LSPR is responsible for visible-light activity has been shown by action spectrum (AS) analyses [15,32], i.e., action spectra correlate with respective absorption spectra (the highest activity at max LSPR). Unfortunately, the mechanism under visible-light irradiation has not been agreed till now, and three main possibilities have been considered, as follows: (1) charge transfer, i.e., the transfer of "hot" electrons, (2) energy transfer, and (3) plasmonic heating (thermal activation).

In the case of charge transfer (1), the mechanism is similar to the sensitization, and thus NMs are also named as plasmonic photosensitizers. The incident photons are absorbed by NMNPs via LSPR excitation, and then electrons might be transferred from NM to the CB of semiconductor. The electrons in CB reduce oxygen (adsorbed on the surface), as typical for semiconductor photocatalysis under aerobic conditions (same as under UV irradiation). Whereas, the electron-deficient NMNPs might oxidize some organic compounds to recover to the original zero-valent state. Many studies have proved the possibility of this mechanism in various experiments, e.g., (i) interband absorption of electrons transferred from Au to TiO$_2$ by femtosecond transient absorption spectroscopy [33], (ii) the negative and positive shift of electrode potential and generation of anodic and cathodic photocurrent depending on the electrode structure, i.e., ITO/TiO$_2$/Au or ITO/Au/TiO$_2$, respectively [34,35], (iii) detection of different oxidation species under irradiation with UV and Vis by EPR experiments [36,37], and (iv) the conductivity under visible-light excitation only for plasmonic photocatalysts (no signal for bare titania) in the case of Au/TiO$_2$ [38] and Ag/TiO$_2$ [39] by time-resolved microwave conductivity (TRMC) study.

The energy transfer (2) between two materials might happen when their energy levels match closely, which is not expected for Au/TiO$_2$ photocatalysts as the bandgap of TiO$_2$ (ca. 3.0–3.2 eV) is much higher than LSPR of Au NPs (ca. 2.2–2.5 eV for spherical NPs). Accordingly, TiO$_2$ has been first pre-modified (in order to enable visible-light absorption), e.g., PRET (plasmon resonance energy transfer) has been suggested for Au NPs deposited on TiO$_2$ pre-modified with nitrogen and fluorine [40]. In the case of other Au-modified photocatalysts, e.g., pre-modified titania with nitrogen [41], narrow-bandgap semiconductors (with visible-light activity), e.g., CuWO$_4$ (2.0–2.5 eV) [42], self-doped TiO$_2$, i.e., with crystal defects [43] and amorphous TiO$_2$ with some disorders causing the localized states inside bandgap [44], PRET has also been postulated as the main mechanism.

The plasmonic heating (3) was first suggested by Chen et al. in 2008, reporting that plasmonic heated Au NPs could induce the oxidation of organic compounds [45]. Although plasmonic heating has been proposed as the main mechanism in some research, most reports have rejected this mechanism, mainly due to usually negligible activity of unsupported Au NPs and Au-modified insulators. For example, plasmonic heating has been rejected in research on water splitting [40], photocurrent generation [46], and hydrogen dissociation [47]. The study on the activation energy has also excluded plasmonic heating for degradation of organic compounds [48] and generation of photocurrent [49].

The plasmonic photocatalysts with various morphology have been prepared that differ significantly in the composition, physicochemical properties (light absorption and reagents' adsorption), and, thus, in the overall activities. Accordingly, it is not surprising that different mechanisms have been proposed for different nanostructures. Moreover, it is also possible that complex mechanism including electron/energy transfer and even "dark" catalytic reactions [50], might be involved in the photocatalytic process.

It should be pointed out that the activity of plasmonic photocatalysts depends on the properties of both NM and semiconductor, as well as the interactions between them, e.g., uniform distribution of NM on the photocatalyst surface is usually recommended. However, in some cases, selective deposition of metals on some surfaces/facets results in improved activity [51]. For example, negligible visible-light activity has been observed for one of the most active titania photocatalysts, i.e., decahedral anatase particles (DAP), when gold has been deposited on {101} facets, resulting probably from the fast charge carriers' recombination (electron transfer: Au → TiO$_2$ → Au), due to an intrinsic property of DAP, i.e., electron and hole reverse migration to {101} and {001} facets, respectively [52], as shown in Figure 1a,c [53]. In contrast, octahedral anatase particles (OAP) with only one type of facets {101} exhibits the highest photocatalytic activity after modification with Au NPs, despite one of the worst photoabsorption properties (narrow LSPR), among fifteen commercial and DAP samples modified with Au NPs in the same procedure, probably due to fast migration of "hot" electrons via shallow electron traps (ETs), as shown in Figure 1b,c [54].

The photoabsorption properties of plasmonic photocatalysts depend on the kind of NM, the properties of NM deposits (size and shape), and the environment (refractive index of medium). Gold and silver have been mainly used for plasmonic photocatalysis, but also other NMs have already been applied, such as palladium [55], platinum [56], and copper [57]. For example, LSPR is observed at 520–580 nm for spherical Au NPs [58] and Cu NPs [59], and at 410–430 nm for Ag NPs [60,61]). Two absorption peaks are noticed for rod-like nanostructures with transverse and longitudinal LSPR at shorter and longer wavelengths, respectively [50]. Accordingly, it has been proposed that broader LSPR peak, e.g., due to polydispersity in NMNPs, results in the higher overall activity under visible-light irradiation due to more efficient light harvesting [32,50]. In this regard, it is expected that application of photonic crystals for plasmonic photocatalysis should result in enhanced photocatalytic performance, as discussed in the Section 4.

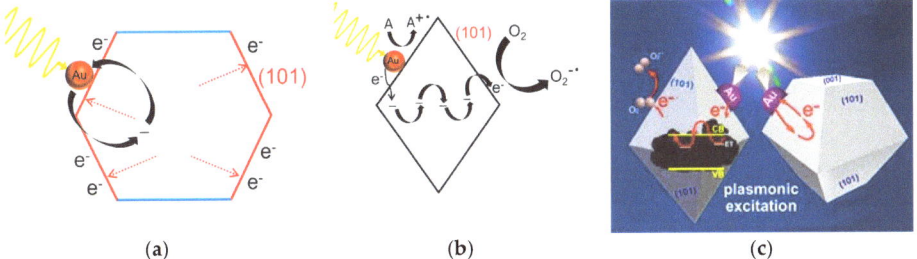

Figure 1. The mechanism of possible "hot" electron transfer on Au-modified: (**a**) decahedral anatase particles (DAP), (**b**) octahedral anatase particles (OAP), and (**c**) DAP vs. OAP; adapted with permission from [53]. Copyright 2018 Creative Commons Attribution.

3. Photonic Crystals (PCs)

Photonic crystals (PCs), periodic optical structures affecting the motion of photons, have been considered as efficient materials for light harvesting. The consistent spatial periodicity of the refractive index (n) in the well-ordered structure of PCs prevents the pathway of some wavelengths, and, thus, resulting in the formation of the stop band and the photonic bandgap (PBG) [62]. The PBG initiates the formation of "slow photons" effects at the blue and red edges of PBG. At these edges the light is scattered and reflected, resulting in the reduction of group velocity of photons [62,63]. Accordingly, it has been proposed that "slow photon" might be utilized for light harvesting of semiconductor photocatalysts.

Various nanostructures of PCs have been designed and prepared, including one-dimensional (1D), e.g., titania nanotubes-PCs (TNTs-PCs) [64,65] and three-dimensional (3D), e.g., opal PCs and inverse opal PCs (IO-PCs). In the case of 1D PCs, TNTs-PCs are prepared similarly to any other TNTs, i.e., by electrochemical methods, including one- [64] and two-step anodization [66–68]. The preparation procedure is critical for the formation and the properties of PCs [64,67]. For example, the bathochromic shift of PBG has been observed after shortening of TNTs and increasing of their diameter [65]. Moreover, the PBG position correlates directly to the distance between TNTs and their thickness.

Recently, 3D PCs of titania have been intensively examined, because of many advantages, such as large specific surface area, tunable porosity, and, thus, efficient mass transport [69–71]. IO-PCs are the exact replica structure obtained by using the opal as a template, as shown in Figure 2. The IO-PCs are usually prepared in the three-step procedure, i.e., (i) preparation of opal structure (template) by self-assembly of colloidal particles, (ii) infiltration of titania precursor (or titania) inside the pores of opal, and (iii) the removal of sacrificial template (opal). All steps are highly important for the quality (and thus optic properties) of IO-PCs to avoid the formation of any cracks and defects [72–74]. For the formation of opal structure (i), various colloidal particles have been used, such as silica (SiO_2), polystyrene (PS), and polymethyl methacrylate (PMMA). The most important for this step is the monodispersity of particles as larger variations than 5% in the size/shape inhibits the opal formation [73,75]. The self-assembly (ii) are performed by a few methods, including the vertical capillary deposition [76–78], spin coating [79], sedimentation [80], and centrifugation-assisted sedimentation [81]. Additionally, some post-thermal operations have been proposed to improve the mechanical stability and enable the good connections between the particles [82]. It has been considered that second step is the most difficult and critical, i.e., to impregnate opal with titania, obtaining the stable IO structure. Many methods have been used for this step, e.g., drop casting and capillary force [78,83], vacuum infiltration [78,83], atomic layer deposition (ALD) [84,85], spin-coating [86] and chemical vapor deposition (CVD) [87]. After infiltration, the crystallization of titania is usually carried out, causing ca. 15–30% shrinkage of the structure, and, thus, the pore size reduction [88–91]. The final step (iii) of the template removal might be performed by calcination (for polymers) [77,78,83,92] or chemical treatment [93].

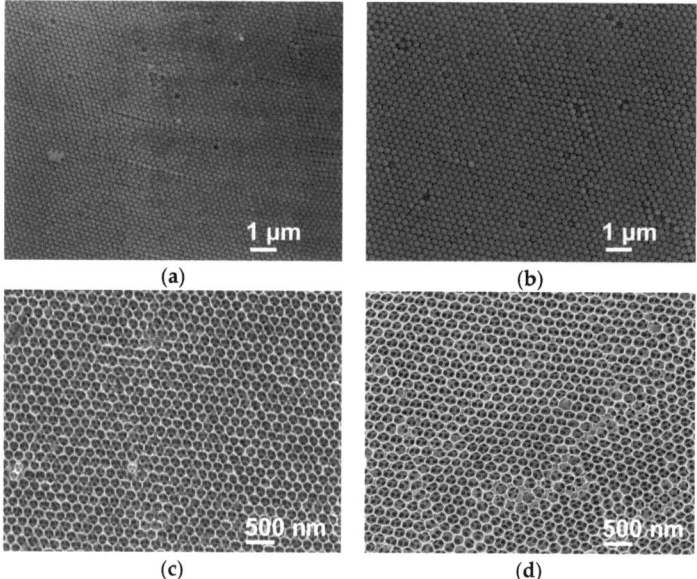

Figure 2. SEM images of opal polystyrene (PS) and inverse opal (IO) titania structures: (**a**,**b**) PS opal with: (**a**) 200-nm and (**b**) 250-nm sphere sizes, and (**c**,**d**) replicated titania IO films with: (**c**) 180-nm and (**d**) 207-nm voids; adapted with permission from [78]. Copyright 2018 Creative Commons Attribution.

The tuning of PBG might be performed by different ways, e.g., the angle of incident light and the change in the nanovoid diameter. For example, the shift of PBG towards shorter wavelengths has been observed after a decrease in nanovoid diameter caused by an increase in the calcination temperature [90]. The photocatalytic performance of titania IO-PCs have been examined under UV [66,94] and Vis [86,91,93] irradiation, as already summarized in review paper on titania-based photocatalysts [5]. It should be pointed out that probably the most important feature of IO is tunable nanovoid dimensions, which influences the scattering and reflection of light. Unfortunately, the contradictory results have been published. For example, it has been reported that reduced velocity of photons and inhibited charge carriers' recombination have been the most evident for structures with the largest voids of 610 nm [95]. In contrast, the smallest voids have been reported as the most active, giving the highest photoelectrochemical enhancements in the case of titania IO sensitized by CdS quantum dots (QDs) [84]. Therefore, the resultant photocatalytic activity might depend not only on nanovoid diameter, but also on other factors, including porosity, specific surface area and kind of co-modifications.

Besides nanovoid diameter, the slow photon effect, influenced by the angle of irradiation [83,88], impacts the photocatalytic performance. For example, hypsochromic shift of PBG has been observed after increasing the irradiation angle [83,88,90], and the highest activity has been obtained under irradiation at 40° than at different angles during degradation of stearic acid on IO with 247-nm voids [64]. Although, some reports have not shown the direct correlation between the voids' sizes and the photocatalytic activity [83], the slow photon effect, e.g., due to the overlapping between the red-edge of PBG and titania bandgap, results in the activity enhancement [83,88,96]. The comparison between the original and slightly destroyed titania IOs, e.g., by ultrasonication [94], milling [97], and grinding [89] confirms that undestroyed IO structure is necessary for high activity, e.g., 71% decrease in activity has been observed for disordered surface by milling [97].

It should be pointed out that although titania IO might exhibit PBG in Vis range of solar spectrum (depending on the void diameter), and thus is able to absorb photons with lower energy than its bandgap, this absorption should not cause the photocatalytic activity as titania is not excited (transfer of electrons from VB to CB). Accordingly, even if some reports show visible-light activity of titania IOs, these reports have been mainly performed for activity testing during dyes' degradation, and thus titania sensitization by dye must be considered as the main mechanism under visible-light irradiation, but not due to PC feature. Indeed, Curti et al. have confirmed that visible-light activity of titania IO is only observed for dyes (methylene blue; MB), whereas negligible effect is obtained for non-color molecules (acetaldehyde) [88]. Tomazatou et al. have indicated that high activity of dye-sensitized titania IO (MB/TiO_2) under visible light (in comparison with negligible activity by commercial titania P25) is probably caused by the slow-photon effect [98]. Chen et al. have proved that overlapping of PBG with bandgap of titania (TiO_2-coated SnO_2 IO) is the most recommended for activity enhancement, i.e., an increase in the activity in the following order of PBG: 2.55 < 2.98 < 3.35 eV [99]. Moreover, it has been shown that the highest efficiency for dye decomposition has been obtained when PBG overlaps with absorption band of MB (but away from its maximum to avoid the undesired light screening), confirming both dye-sensitization mechanism and slow photon effect [100].

To further increase the photocatalytic performance by enhanced light harvesting, IO structures have been modified with various compounds, including metals, e.g., gold [79,91], silver [93], and nickel [86,101], semiconductors, e.g., cadmium selenide and cadmium sulfide QDs [84,102] and zinc oxide NPs [103], and by "self-doping" (surface-disordered-engineered TiO_2 PCs [97]). The modifications with NM, resulting in mainly formation of plasmonic photocatalysts, are discussed in the next section.

4. NM-Based PCs

The heterostructural design of IOs, such as the coupling of IO-PCs with other semiconductors/materials and surface modifications with NMs, has been considered as an efficient method for enhanced performance of PCs. Although, various reasons have been proposed, including charge carriers' separation, enhanced light harvesting, synergy, higher specific surface area, it is still unclear what properties are the key-factors of high activity and which of them are the most recommended for the specific application. It is expected that for photocatalytic reactions, the most advisable modification is to use the materials with ability of photon absorption near PBG. Accordingly, the application of NM with ability of visible-light absorption seems to be highly attractive because of the possible overlapping of LSPR of NM with PBG of titania IO. Moreover, PCs-based plasmonic photocatalysts might help in the clarification on the mechanism of plasmonic photocatalysis, i.e., energy vs. electron transfer.

Recently, plasmonic PCs have been intensively investigated to combine both PBG and LSPR effects together in terms of physical appearances and electronic energies for various applications [104–109]. An interesting study has been shown for negative photoresist IO (prepared by SiO_2 opal embedding) deposited with gold, where the dual-colored micropatterns (Figure 3) open the door for possible applications in various fields, including cosmetics, optical filters and coatings [104]. Although, there are more interesting examples of plasmonic PCs for various applications, the further discussion is limited to the photocatalytic reactions.

NM-modified PCs have already been proposed for photocatalysis, including, solar cells [40], water splitting [110] and mineralization of organic pollutants [100,111]. The extended path of light in terms of duration and length from the multi-scattering phenomenon by the PCs could be well utilized by the plasmonic NPs, resulting in efficient light harvesting, and thus enhanced quantum yield of photocatalytic reactions, as discussed in the Section 4.2.

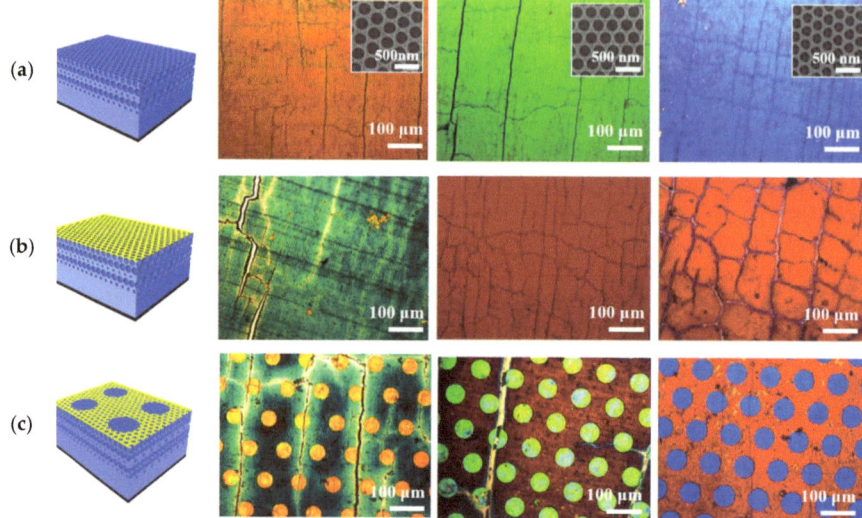

Figure 3. The models (first column) of the: (**a**) IO, (**b**) Au layer-deposited IO, and (**c**) micropatterned IO by local Au deposition; and respective optical microscopy images for different sizes of SiO_2 NPs, i.e., with diameter of 315 nm (the second column), 290 nm (the third column) and 239 nm (the fourth column), used as a template (opal); Reprinted with permission (after formatting) from [104]. Copyright (2018) WILEY.

4.1. Preparation of PCs-Based Plasmonic Photocatalysts

The fabrication of PCs-based plasmonic photocatalysts is quite challenging process to load/infiltrate plasmonic NPs on/in the PCs without destroying the Bragg's diffraction property of the PCs and to retain the effective LSPR effect of NMs. The photodeposition and thermal methods have been commonly used to deposit plasmonic NPs on IO by using the adequate precursors, such as chloroplatinic acid (H_2PtCl_6) for Pt [112], silver nitrate ($AgNO_3$) for Ag [113], and tetrachloroauric acid ($HAuCl_4$) for Au [106]. NMs have been deposited on TiO_2 IO either before [106] or after [99] its infiltration inside opal structure. Apart from that, hydrothermal method has been proposed to load Au NPs on TiO_2 PC, enhancing the stability due to the interphase sintering between TiO_2 backbone and Au NPs [91]. Erola et al. have reported a facile method for the preparation of Ag- and Au-loaded PCs via co-assembly and infiltration methods, as shown in Figure 4 [111]. Interestingly, the co-assembly method is not suitable for Ag deposition on the SiO_2-IO due to diminishing of LSPR after calcination. It has been concluded that because of less thermal stability of Ag than Au, Ag is easily oxidized during calcination. Hence, the suitable method (exemplified in Table 1) must be select carefully for a good quality of PCs-based plasmonic photocatalysts.

4.2. Photocatalytic Applications

Although, a large number of studies on plasmonic photocatalysis has been published, only few reports might be found on PCs-based plasmonic photocatalysis. NM-modified PCs have been used in the photocatalysis field, as summarized in Table 1, due to few advantages, as follows: (i) the absorption at Vis-NIR as most of the wide-bandgap semiconductors are unable to absorb light at those regions, (ii) slow photons arising from PBG/photonic effect owning to the PCs, which might enhance and strengthen the LSPR effect, and (iii) the role of NMNPs as an electron sink, which reduces the charge carriers' recombination. Although the inhibition of recombination should not be considered as plasmonic photocatalysis, various reports point this feature, which surely might enhance the overall activity under solar radiation, i.e., at UV-Vis range.

Table 1. Synthesis methods and activity tests for NM-modified PCs.

Plasmonic Based PCs	Plasmonic NPs Loading/Deposition Methods	Photocatalytic Tests	Findings	References
Pt_x/PC-TiO_2	gas bubbling-assisted membrane reduction	CO_2 reduction; $\lambda = 320$–780 nm; I	3.2 times higher activity *	[107]
TiO_2 PC/Au NPs	in situ hydrothermal reduction	degradation of 2,4-dichlorophenol; $\lambda > 420$ nm; II	PBG and SPR overlapping	[91]
AgTIO	chemical route	MB degradation; $\lambda = 254$ nm and $\lambda = 400$–760 nm; III	SPR and PBG enhancing activity	[93]
Au-TiO_2-NAA-DBRs	sputter coating	MB degradation; III	enhanced activity	[100]
TiO_2-Au-CdS	immersion and chemical bath deposition	H_2 production; $\lambda > 420$ nm; IV	enhanced H_2 generation at blue-edge PBG	[110]
i-Pt-TiO_2-o film	photodeposition	degradation of acid orange; $\lambda > 400$ nm; I	4-fold enhancement *	[112]
Au/ZnO-PCs	magnetron sputtering	photodegradation of RhB; $\lambda > 420$ nm; II	24.8-fold higher than commercial ZnO	[114]
Au-PCTNTs	magnetron sputtering	CO_2 photoreduction; $\lambda > 400$ nm; III	high selectivity of methane generation	[115]
Ag/$BiVO_4$	electrodeposition	MB degradation; $\lambda > 420$ nm; III	enhanced activity	[116]
Au/TNTs	photodeposition	IPCE (400–700 nm); photocurrent ≥ 420 nm; water splitting; II	PBG matching with LSPR—enhanced activity	[117]
Au/TiO_2 photoanode	facile ionic layer adsorption and thermal-reduction	PEC water splitting; $\lambda > 420$ nm; II	0.71% of solar energy conversion	[118]
Au/TiO_2 PC	chemical route	MO degradation; $\lambda > 420$ nm; III	7-fold increase *	[119]
rGO/Pt/3DOM TiO_2	dropwise and thermal reduction	MO degradation; $\lambda > 420$ nm; I	4-fold increase *	[120]
TiO_2-IO Au/AgNPs film	immersion	acetylene mineralization; $\lambda = 365$ nm; I	enhanced activity *	[121]
Cu/TNTs	pulsed electrochemical deposition	H_2 production $\lambda > 400$ nm; II	enhanced activity	[122]

AgTIO—TiO_2 IO films loaded with Ag NPs; i-Pt-TiO_2-o film—TiO_2 IO films loaded with Pt NPs; NAA-DBRs—functionalized nanoporous anodic alumina distributed Bragg reflectors; PEC—photoelectrochemical; PCTNTs—PCs consisting of TiO_2 nanotube arrays; rGO/Pt/3DOM TiO_2—3D-ordered macroporous TiO_2 with Pt and reduced graphene oxide; RhB—rhodamine B; * activity in the comparison with the reference sample; excitation of: I—semiconductor, II—NM, III—both (semiconductor and NM), IV—co-modifier (CdS).

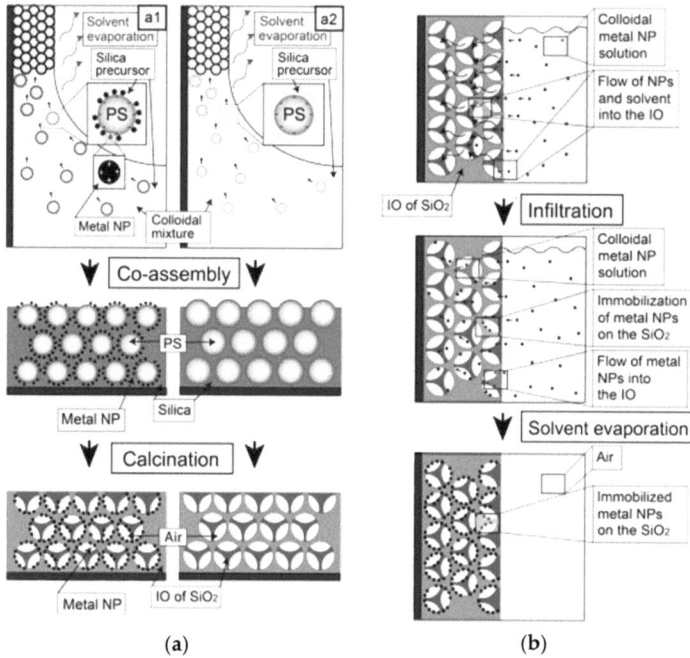

Figure 4. Fabrication of SiO_2 photonic crystal (PC) films using: (**a**) co-assembly method in which PCs were prepared with (a1) and without (a2) positively charged metal NPs and (**b**) infiltration method; Reprinted with permission (after formatting) from [111]. Copyright (2015) Elsevier.

However, the aspect mentioned at (ii) should be only expected if LSPR matches with the slow photons effect, which could be realized by the perfect tuning in the fabrication process. Indeed, even a decrease in the photocatalytic activity of TiO_2-coated nanoporous alumina IO has been observed after deposition of Au NPs or Ag NPs on its surface, where both effects do not match, i.e., LSPR at ca. 400–550 nm and PBG at ca. 800 nm [100]. In contrast, Lu et al. have succeeded the precise tuning in fabrication of Au-NP/TiO_2 PC with LSPR overlapping with the blue edge of PBG, resulting in 2.3-fold higher photocatalytic efficiency for photodegradation of 2,4-dichlorophenol in comparison to the reference Au-NP/nanocrystalline TiO_2 under visible-light irradiation [91]. Similarly, Zhang et al. [118] have shown the enhanced photoelectrochemical (PEC) water splitting by using the Au-NP-modified TiO_2 nanorod array structure (Au/TiO_2 NRPCs) by overlapping of LSPR with red edge of PBG effect (Figure 5), utilizing the slow photons (max at ca. 518 nm). Additionally, both studies [91,118] have indicated that red-edge PBG is more effective than blue-edge one. At the red-edge PBG, the light propagate with decrease group velocity, being described as a sinusoidal standing wave that has its highest amplitude in the high refractive index part of the PCs, whereas the standing wave at blue edge is localized in the low-refractive index part. Accordingly, an absorber at high dielectric part has stronger interactions with light at red-edge regions of PBG. Similarly, IO-molybdenum: bismuth vanadium (Mo:$BiVO_4$) photoelectrode with infiltrated Au NPs has shown an increase in PEC water splitting, and the enhancement has been attributed to the amplified LSPR effect when the polystyrene (PS) template of 260-nm particles results in PBG at 513 nm, synergistically matching with the LSPR of Au [123].

An important aspect that should be considered for 3D powdered PC-based plasmonic photocatalysts is the periodical/structural integrity as disordered structure could affect the enhancement of the visible-light harvesting, thus reducing the photocatalytic performance. Indeed, the crushed Au/TiO_2 biomorphic (BM)PCs have exhibited lower photodegradation efficiency (Figure 6) as compared with uncrushed ones [120]. Similarly, C. Dinh et al. have shown that the crushed Au/TiO$_2$-3D hollow nanospheres (HNSs) show about 4.3-fold lower efficiency of CO$_2$ generation from photodecomposition of isopropanol as compared with an original Au/TiO$_2$-3DHNS, indicating the probability of a decrease in photoabsorption ability by Au, due to the structure disruption [124]. It should be pointed out that although the partial crushing of IO structure results in a decrease in the activity in both studies, those activities have been much higher than that by a reference sample of commercial titania modified with Au NPs (Au/TiO$_2$-P25), highlighting the synergism between the periodic porous structure (although slightly destroyed) and plasmonic effect.

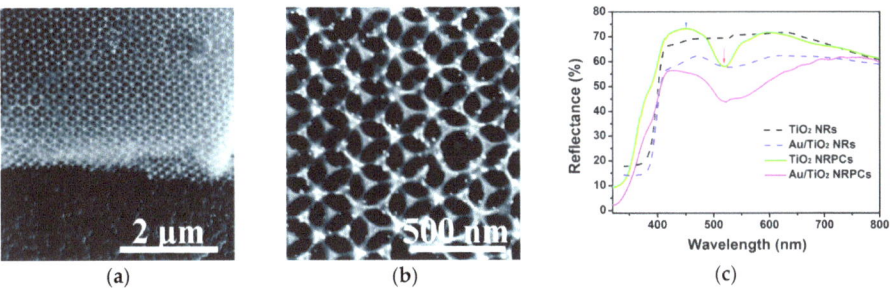

Figure 5. (a) Low-magnification and (b) high-resolution scanning electron microscopy images of Au/TiO$_2$ PCs on the fluorine-doped tin oxide (FTO) substrate, (c) Diffused reflectance spectra of TiO$_2$ nanorods (NRs), Au/TiO$_2$ NRs, TiO$_2$ NRPCs, Au/TiO$_2$ NRPCs (PBG marked with dark arrow and slow-photon region marked with red arrow); Reprinted with permission (after formatting) from [118]. Copyright (2014) RSC.

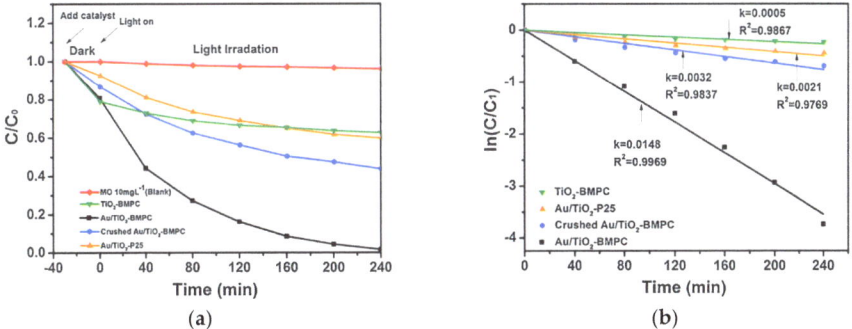

Figure 6. The time course of the photocatalytic degradation of MO under visible-light irradiation (λ > 420 nm) using bare PCs (TiO$_2$-BMPC), Au-modified PCs (Au/TiO$_2$-BMPC), crushed Au-modified PCs (Au/TiO$_2$-BMPC) and commercial titania modified with Au NPs (Au/TiO$_2$-P25): (a) a decrease in the concentration of dye during degradation (C/C$_0$), and (b) the respective logarithm plots; Reprinted with permission (after formatting) from [120]. Copyright (2016) Elsevier.

NMNPs are known for their ability to inhibit the recombination of charge carriers, by trapping electrons (as already discussed in Introduction). It should be underlined that this effect could not be described as plasmonic photocatalysis since the semiconductor is excited, but not NM. Accordingly, NMNPs, such as Pt, Au, and Ag have been infiltrated/incorporated on PCs to enhance the photocatalytic activity [93,125]. For example, Huo et al. have shown a facile preparation method of NM-modified PCs by incorporation of Pt NPs and reduced graphene oxide (r-GO) on TiO$_2$ IO macroporous structure (Figure 7a), highlighting the enhanced photodegradation of methyl orange [123]. However, it should be pointed out that this study could not be considered as plasmonic photocatalysis as TiO$_2$ instead of NM is excited, as shown in Figure 7b. In contrast, Temerov et al. have discussed the enhanced mineralization of acetylene to CO_2 on TiO_2-IO modified with Au NPs and Ag NPs (by 53% and 39%, respectively), owning to the LSPR activation (plasmonic photocatalysis), i.e., electron transfer from NM to TiO_2 with simultaneous oxidation of acetylene on the NM surface, but irradiation has been performed with UVA, and thus TiO_2 excitation must also be considered [124].

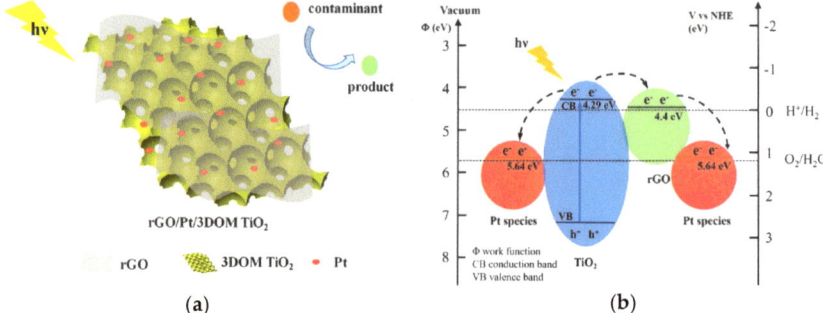

Figure 7. Schematic illustrations of: (**a**) the morphology of TiO_2 IO co-modified with Pt NPs and reduced graphene oxide (r-GO), and (**b**) the proposed charge carriers' separation and migration in rGO/Pt/TiO_2 (not plasmonic activation); Reprinted with permission from [123]. Copyright (2019) ACS.

Interestingly, plasmonic photocatalysis using both functions of NMs, i.e., plasmonic excitation and inhibition of charge carriers' recombination, has been proposed by Rahul et al. for bi-metal-modified TiO_2-IO co-doped with N/F [106]. The LSPR excitation of Au results in an electron transfer from Au to CB of TiO_2, and then to co-deposited Pt NPs, on which H_2 evolution takes place, as shown in Figure 8. It should be pointed out that co-catalyst is necessary for efficient H_2 evolution on titania, and Pt is probably the most active co-catalyst for this reaction [5,126].

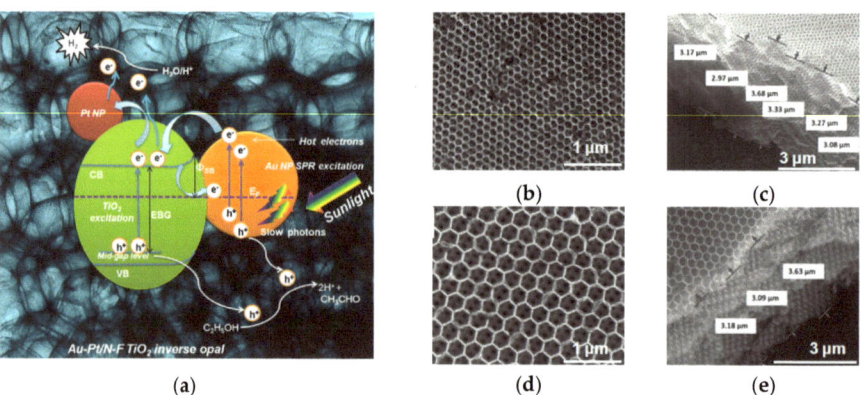

Figure 8. (**a**) Schematic illustration of the proposed mechanism of H_2 generation over Au-Pt/N-F TiO_2 IO-460 under solar light irradiation; SEM images of: (**b**,**c**) Au-Pt/ N-F TiO_2 IO-215: (**b**) top view and (**c**) cross section view, and (**d**,**e**) Au-Pt/N-F TiO_2 IO-460: (**d**) top view and (**e**) cross section view; Reprinted with permission (after formatting) from [106]. Copyright (2018) WILEY.

In the case of NM deposition on the PCs, the amount of NM is critical for the enhancement of the photocatalytic activity. For example, 2 wt.% of Pt loading on TiO_2-IO PC has been shown as the best, whereas 1.3 wt.% is not able to exhibit any amplification in the dye photodegradation [112]. Similarly, it has been shown that the lowest (2 mM) and the highest (20 mM) amount of deposited Ag (in terms of precursor concentration) on TiO_2 PC is not able to amplify the light absorption as the former is inadequate to exhibit LSPR and the latter results in Ag NPs' aggregation, diminishing the activity [93]. An interesting study has been shown for the correlation between Ag-NP deposition duration (15, 30, 45, and 60 s) by pulse current deposition on TiO$_2$-IO and the photocatalytic activity, as shown in Figure 9 [99]. Longer deposition duration has resulted in the formation of Ag aggregates (Figure 9c), which directly affects the MB photodegradation efficiency, due to the loss of LSPR effect (the aggregation) and blockage of the charge carriers' migration. Similarly, S. Zhang et al. have shown that the number of deposition cycles of Cu NPs on TiO$_2$ nanotube arrays (TNTs) affects the efficiency of photocatalytic water splitting, and 20 cycles results in 10.7-fold higher evolution of H$_2$ as compared to the reference sample (without Cu), due to the amplified plasmonic excitation [126]. Although further increase in deposition cycles (>20) has reduced the photocatalytic activity (non-clarified reason), the efficiency has been much higher than that by non-modified TNTs.

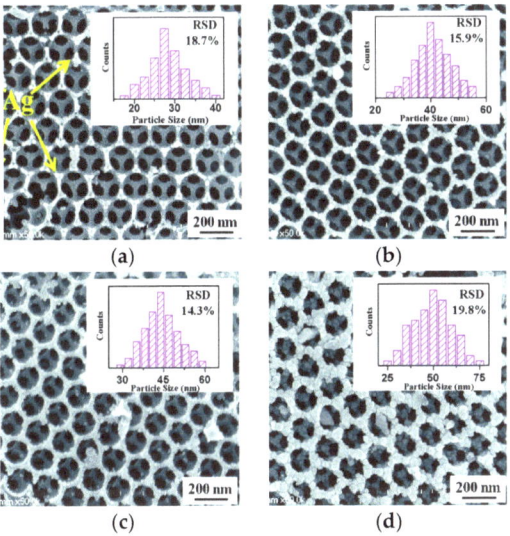

Figure 9. SEM images of PCs modified with Ag by the pulsed current deposition method during: (**a**) 15 s-Ag/TiO$_2$, (**b**) 30 s-Ag/TiO$_2$, (**c**) 45 s-Ag/TiO$_2$, and (**d**) 60 s-Ag/TiO$_2$; Reprinted with permission (after formatting) from [99]. Copyright (2014) RSC.

Moreover, it has also been postulated that the localization of NM in/on PCs might be crucial for the overall activity. For example, Lim et al. have proposed that deposition of NM layer on PCs is not recommended, as the majority of light is absorbed in the metallic coating before it reaches the bulk PCs structure, resulting in a decrease in the overall photocatalytic activity, similar to well-known "screen-effect" in other NM-TiO$_2$ structures [5,12]. Accordingly, it has been proposed that the preparation of titania IO structures with NMNPs inside each void would be the most profitable. In this regard, the size of NMNPs as well as void diameter should be tuned well to allow the matching between LSPR of NM and PBG of TiO$_2$. Our recent study on Au NPs deposited in voids of TiO$_2$ IOs has confirmed the high enhancement of activity under irradiation with wavelengths correlating with LSPR of Au NPs and PBG of TiO$_2$ IO (unpublished data [127]).

5. Conclusions

Plasmonic photocatalysis is probably one of the hottest topics in heterogeneous photocatalysis research because of possible applications under broad wavelength ranges (UV/Vis/NIR). Although, the research is quite new, there are many reports on preparation, characterization, activity and stability enhancement, mechanism clarifications, including various applications, e.g., solar energy conversion, degradation of chemical and microbiological pollutants and CO_2 conversion. However, the low quantum yields of photocatalytic reactions under visible-light irradiation are critical for the overall activity and possible applications. Accordingly, enhanced light harvesting is highly important for the activity enhancement. PCs seem to be the best materials for efficient light utilization. However, it should be pointed out that PBG and slow photon effects at Vis/NIR range of solar radiation should not be able to excite wide-bandgap semiconductors, due to insufficient energy. Although, some reports show photocatalytic activity of bare TiO_2 IO structures, those studies have been mainly performed for dyes, resulting in titania sensitization by dyes. However, PCs-based plasmonic photocatalysts due to LSPR absorption have been promising candidates for high photocatalytic performance. Accordingly, such photocatalysts have been synthesized, characterized and tested. However, the preparation of both PCs and PCs-based plasmonic photocatalysts is quite challenging, involving multi-step procedures.

It has been proposed that precise and fine tuning of photoabsorption properties, i.e., PBG matching with LSPR by nanoarchitecture of PCs void diameter and NMNPs size, respectively, is the most important aspect for the PCs-based plasmonic photocatalysts. It should be pointed out that the content of NM and structure configuration might be also a key-factor for photocatalytic activity. For example, too low content of NMs result in low LSPR, whereas too large content might cause NPs aggregation. The deposition of NMs on the surface of IO might also block the light (screening effect). Accordingly, it is proposed that the preparation of PCs with NM deposited in voids would be the most recommended for efficient light harvesting, and, thus, the overall photocatalytic effect.

Author Contributions: Conceptualization, E.K.; writing—original draft preparation, T.R.-M. and E.K.; writing—review and editing, B.O. and E.K. All authors have read and agreed to the published version of the manuscript.

Funding: This research was funded by Japanese Government (Monbukagakusho: MEXT) Scholarship. The APC was funded by E.K.

Conflicts of Interest: The authors declare no conflict of interest.

Abbreviations

1D	one-dimensional
3D	three-dimensional
AVD	atomic vapor deposition
BM	biomorphic
CB	conduction band
CVD	chemical vapor deposition
DAP	decahedral anatase particles
ETs	electron traps
FTO	fluorine-doped tin oxide
HNSs	hollow nanospheres
IO	inverse opal
IPCE	incident photon-to-electron conversion efficiency
LSPR	localized surface plasmon resonance
MB	methylene blue
NIR	near infrared
NM	noble metals
NPs	nanoparticles
NRs	nanorods
OAP	octahedral anatase particles
PCs	photonic crystals
PBG	photonic bandgap
PRET	plasmon resonance energy transfer
PMMA	polymethyl methacrylate
PS	polystyrene
QDs	quantum dots

r-GO	reduced graphene oxide
TNTs	titania nanotubes
UV	ultraviolet
VB	valence band
Vis	visible light

References

1. Hoffmann, M.R.; Martin, S.T.; Choi, W.Y.; Bahnemann, D.W. Environmental applications of semiconductor photocatalysis. *Chem. Rev.* **1995**, *95*, 69–96. [CrossRef]
2. Nakata, K.; Fujishima, A. TiO_2 photocatalysis: Design and applications. *J. Photochem. Photobiol.* **2012**, *13*, 169–189. [CrossRef]
3. Ohtani, B. Photocatalysis A to Z—What we know and what we do not know in a scientific sense. *J. Photochem. Photobiol.* **2010**, *11*, 157–178. [CrossRef]
4. Abe, R. Recent progress on photocatalytic and photoelectrochemical water splitting under visible light irradiation. *J. Photochem. Photobiol.* **2010**, *11*, 179–209. [CrossRef]
5. Wang, K.L.; Janczarek, M.; Wei, Z.S.; Raja-Mogan, T.; Endo-Kimura, M.; Khedr, T.M.; Ohtani, B.; Kowalska, E. Morphology- and crystalline composition-governed activity of titania-based photocatalysts: Overview and perspective. *Catalysts* **2019**, *9*, 1054. [CrossRef]
6. Herrmann, J.M. Heterogeneous photocatalysis: Fundamentals and applications to the removal of various types of aqueous pollutants. *Catal. Today* **1999**, *53*, 115–129. [CrossRef]
7. Zaleska, A. Doped-TiO_2: A review. *Rec. Patent. Eng.* **2008**, *2*, 157–164. [CrossRef]
8. Kraeutler, B.; Bard, A.J. Heterogeneous photocatalytic preparation of supported catalysts. Photodeposition of platinum on TiO_2 powder and other substrates. *J. Am. Chem. Soc.* **1978**, *100*, 4317–4318. [CrossRef]
9. Nishimoto, S.I.; Ohtani, B.; Kagiya, T. Photocatalytic dehydrogenation of aliphatic-alcohols by aqueous suspensions of platinized titanium-dioxide. *J. Chem. Soc. Farad. Trans.* **1985**, *81*, 2467–2474. [CrossRef]
10. Pichat, P.; Herrmann, J.M.; Disdier, J.; Courbon, H.; Mozzanega, M.N. Photocatalytic hydrogen production from aliphatic alcohols over a bifunctional platinum on titanium dioxide catalyst. *Nouv. J. Chim.* **1981**, *5*, 627–636.
11. Bahnemann, D.W.; Mönig, J.; Chapman, R. Efficient photocatalysis of the irreversible one-electron and two-electron reduction of halothane on platinized colloidal titanium dioxide in aqueous suspension. *J. Phys. Chem.* **1987**, *91*, 3782–3788. [CrossRef]
12. Wang, K.L.; Wei, Z.S.; Ohtani, B.; Kowalska, E. Interparticle electron transfer in methanol dehydrogenation on platinum-loaded titania particles prepared from P25. *Catal. Today* **2018**, *303*, 327–333. [CrossRef]
13. Macyk, W.; Burgeth, G.; Kisch, H. Photoelectrochemical properties of platinum (IV) chloride surface modified TiO_2. *Photochem. Photobiol. Sci.* **2003**, *2*, 322–328. [CrossRef]
14. Kowalska, E.; Remita, H.; Colbeau-Justin, C.; Hupka, J.; Belloni, J. Modification of titanium dioxide with platinum ions and clusters: Application in photocatalysis. *J. Phys. Chem.* **2008**, *112*, 1124–1131. [CrossRef]
15. Tian, Y.; Tatsuma, T. Mechanisms and applications of plasmon-induced charge separation at TiO_2 films loaded with gold nanoparticles. *J. Am. Chem. Soc.* **2005**, *127*, 7632–7637. [CrossRef] [PubMed]
16. Verbruggen, S.W. TiO_2 photocatalysis for the degradation of pollutants in gas phase: From morphological design to plasmonic enhancement. *J. Photochem. Photobiol.* **2015**, *24*, 64–82. [CrossRef]
17. Kowalska, E. Plasmonic photocatalysis. In *Gold Nanoparticles for Physics, Chemistry and Biology*, 2nd ed.; Louis, C., Pluchery, O., Eds.; World Scientific: Singapore, 2017; pp. 319–364.
18. Ueno, K.; Misawa, H. Surface plasmon-enhanced photochemical reactions. *J. Photochem. Photobiol.* **2013**, *15*, 31–52. [CrossRef]
19. Sarina, S.; Waclawik, E.R.; Zhu, H.Y. Photocatalysis on supported gold and silver nanoparticles under ultraviolet and visible light irradiation. *Green Chem.* **2013**, *15*, 1814–1833. [CrossRef]
20. Kowalska, E. Plasmonic Photocatalysts. Available online: https://www.mdpi.com/journal/catalysts/special_issues/plasmonic_photocatal (accessed on 2 July 2020).
21. Verbruggen, S.W. Functional Plasmonic Nanostructures. Available online: https://www.mdpi.com/journal/nanomaterials/special_issues/functional_plasmonic (accessed on 2 July 2020).

22. Kominami, H.; Tanaka, A.; Hashimoto, K. Mineralization of organic acids in aqueous suspension of gold nanoparticles supported on cerium (IV) oxide powder under visible light irradiation. *Chem. Commun.* **2010**, *46*, 1287–1289. [CrossRef]
23. Thimsen, E.; Le Formal, F.; Gratzel, M.; Warren, S.C. Influence of plasmonic Au nanoparticles on the photoactivity of Fe_2O_3 electrodes for water splitting. *Nano Lett.* **2011**, *11*, 35–43. [CrossRef]
24. Yu, H.; Ming, H.; Zhang, H.; Li, H.; Pan, K.; Liu, Y.; Wang, F.; Gong, J.; Kang, Z. Au/ZnO nanocomposite: Facile fabrication and enhanced photocatalytic activity for degradation of benzene. *Mater. Chem. Phys.* **2012**, *137*, 113–117. [CrossRef]
25. Lan, J.Y.; Zhou, X.M.; Liu, G.; Yu, J.G.; Zhang, J.C.; Zhi, L.J.; Nie, G.J. Enhancing photocatalytic activity of one-dimensional $KNbO_3$ nanowires by Au nanoparticles under ultraviolet and visible-light. *Nanoscale* **2011**, *3*, 5161–5167. [CrossRef]
26. Kashyap, T.; Biswasi, S.; Pal, A.R.; Choudhury, B. Unraveling the catalytic and plasmonic roles of g-C_3N_4 supported Ag and Au nanoparticles under selective photoexcitation. *ACS Sustain. Chem. Eng.* **2019**, *7*, 19295–19302. [CrossRef]
27. Xu, H.; Li, H.M.; Xia, J.X.; Yin, S.; Luo, Z.J.; Liu, L.; Xu, L. One-pot synthesis of visible-light-driven plasmonic photocatalyst Ag/AgCl in ionic liquid. *ACS Appl. Mater. Interfaces* **2011**, *3*, 22–29. [CrossRef] [PubMed]
28. Wang, Z.L.; Zhang, J.F.; Lv, J.L.; Dai, K.; Liang, C.H. Plasmonic Ag_2MoO_4/AgBr/Ag composite: Excellent photocatalytic performance and possible photocatalytic mechanism. *Appl. Surf. Sci.* **2017**, *396*, 791–798. [CrossRef]
29. Del Tedesco, A.; Piotto, V.; Sponchia, G.; Hossain, K.; Litti, L.; Peddis, D.; Scarso, A.; Meneghetti, M.; Benedetti, A.; Riello, P. Zirconia-based magnetoplasmonic nanocomposites: A new nanotool for magnetic-guided separations with SERS identification. *ACS Appl. Nano Mater.* **2020**, *3*, 1232–1241. [CrossRef]
30. Subramanian, V.; Wolf, E.; Kamat, P.V. Semiconductor-metal composite nanostructures. To what extent do metal nanoparticles improve the photocatalytic activity of TiO_2 films? *J. Phys. Chem.* **2001**, *105*, 11439–11446. [CrossRef]
31. Rodriguez-Gonzalez, V.; Zanella, R.; del Angel, G.; Gomez, R. MTBE visible-light photocatalytci decomposition over Au/TiO_2 and Au/TiO_2-Al_2O_3 sol-gel prepared catalysts. *J. Mol. Catal. Chem.* **2008**, *281*, 93–98. [CrossRef]
32. Kowalska, E.; Abe, R.; Ohtani, B. Visible light-induced photocatalytic reaction of gold-modified titanium(IV) oxide particles: Action spectrum analysis. *Chem. Commun.* **2009**, *2*, 241–243. [CrossRef]
33. Furube, A.; Du, L.; Hara, K.; Katoh, R.; Tachiya, M. Ulltrafast plasmon-induced electron transfer from gold nanodots into TiO_2 nanoparticles. *J. Am. Chem. Soc.* **2007**, *129*, 14852–14853. [CrossRef] [PubMed]
34. Sakai, N.; Fujiwara, Y.; Arai, M.; Yu, K.; Tatsuma, T. Electrodeposition of gold nanoparticles on ITO: Control of morphology and plasmon resonance-based absorption and scattering. *J. Electroanal. Chem.* **2009**, *628*, 7–15. [CrossRef]
35. Sakai, N.; Fujiwara, Y.; Takahashi, Y.; Tatsuma, T. Plasmon-resonance-based generation of cathodic photocurrent at electrodeposited gold nanoparticles coated with TiO_2 films. *Chem. Phys. Chem.* **2009**, *10*, 766–769. [CrossRef] [PubMed]
36. Caretti, I.; Keulemans, M.; Verbruggen, S.W.; Lenaerts, S.; Van Doorslaer, S. Light-induced processes in plasmonic gold/TiO_2 photocatalysts studied by electron paramagnetic resonance. *Top. Catal.* **2015**, *58*, 776–782. [CrossRef]
37. Priebe, J.B.; Radnik, J.; Lennox, A.J.J.; Pohl, M.M.; Karnahl, M.; Hollmann, D.; Grabow, K.; Bentrup, U.; Junge, H.; Beller, M.; et al. Solar hydrogen production by plasmonic Au-TiO_2 catalysts: Impact of synthesis protocol and TiO_2 phase on charge transfer efficiency and H_2 evolution rates. *ACS Catal.* **2015**, *5*, 2137–2148. [CrossRef]
38. Mendez-Medrano, M.G.; Kowalska, E.; Lehoux, A.; Herissan, A.; Ohtani, B.; Rau, S.; Colbeau-Justin, C.; Rodriguez-Lopez, J.L.; Remita, H. Surface modification of TiO_2 with Au nanoclusters for efficient water treatment and hydrogen generation under visible light. *J. Phys. Chem.* **2016**, *120*, 25010–25022. [CrossRef]
39. Wei, Z.; Janczarek, M.; Endo, M.; Colbeau-Justin, C.; Ohtani, B.; Kowalska, E. Silver-modified octahedral anatase particles as plasmonic photocatalyst. *Catal. Today* **2018**, *310*, 19–25. [CrossRef]
40. Liu, Z.; Hou, W.; Pavaskar, P.; Aykol, M.; Cronin, S.B. Plasmon resonance enhancement of photocatalytic water splitting under visible illumination. *Nano Lett.* **2011**, *11*, 1111–1116. [CrossRef]

41. Bouhadoun, S.; Guillard, C.; Dapozze, F.; Singh, S.; Amans, D.; Boucle, J.; Herlin-Boime, N. One step synthesis of N-doped and Au-loaded TiO$_2$ nanoparticles by laser pyrolysis: Application in photocatalysis. *Appl. Catal. Environ.* **2015**, *174*, 367–375. [CrossRef]
42. Valenti, M.; Dolat, D.; Biskos, G.; Schmidt-Ott, A.; Smith, W.A. Enhancement of the photoelectrochemical performance of CuWO$_4$ thin films for solar water splitting by plasmonic nanoparticle functionalization. *J. Phys. Chem.* **2015**, *119*, 2096–2104. [CrossRef]
43. Hou, W.; Liu, Z.; Pavaskar, P.; Hsuan Hung, W.; Cronin, S.B. Plasmonic enhancement of photocatalytic decomposition of methyl orange under visible light. *J. Catal.* **2011**, *277*, 149–153. [CrossRef]
44. Seh, Z.W.; Liu, S.W.; Low, M.; Zhang, S.-Y.; Liu, Z.; Mlayah, A.; Han, M.-Y. Janus Au-TiO$_2$ photocatalysts with strong localization of plasmonic near fields for efficient visible-light hydrogen generation. *Adv. Mater.* **2012**, *24*, 2310–2314. [CrossRef] [PubMed]
45. Chen, X.; Zhu, H.-Y.; Zhao, J.-C.; Zheng, Z.-F.; Gao, X.-P. Visible-light-driven oxidation of organic contaminants in air with gold nanoparticle catalysts on oxide supports. *Angew. Chem. Int. Ed.* **2008**, *47*, 5353–5356. [CrossRef] [PubMed]
46. Son, M.S.; Im, J.E.; Wang, K.K.; Oh, S.L.; Kim, Y.R.; Yoo, K.H. Surface plasmon enhanced photoconductance and single electron effects in mesoporous titania nanofibers loaded with gold nanoparticles. *Appl. Phys. Lett.* **2010**, *96*, 023115. [CrossRef]
47. Mukherjee, S.; Libisch, F.; Large, N.; Neumann, O.; Brown, L.V.; Cheng, J.; Lassiter, J.B.; Carter, E.A.; Nordlander, P.; Halas, N.J. Hot electrons do the impossible: Plasmon-induced dissociation of H$_2$ on Au. *Nano Lett.* **2013**, *13*, 240–247. [CrossRef] [PubMed]
48. Kominami, H.; Tanaka, A.; Hashimoto, K. Gold nanoparticles supported on cerium(IV) oxide powder for mineralization of organic acids in aqueous suspensions under irradiation of visible light of λ = 530 nm. *Appl. Catal. Gen.* **2011**, *397*, 121–126. [CrossRef]
49. Nishijima, Y.; Ueno, K.; Yokata, Y.; Murakoshi, K.; Misawa, H. Plasmon-assisted photocurrent generation from visible to near-infrared wavelength using a Au-nanorods/TiO$_2$ electrode. *J. Phys. Chem. Lett.* **2010**, *1*, 2031–2036. [CrossRef]
50. Kowalska, E.; Mahaney, O.O.P.; Abe, R.; Ohtani, B. Visible-light-induced photocatalysis through surface plasmon excitation of gold on titania surfaces. *Phys. Chem. Chem. Phys.* **2010**, *12*, 2344–2355. [CrossRef]
51. Bian, Z.F.; Tachikawa, T.; Zhang, P.; Fujitsuka, M.; Majima, T. Au/TiO$_2$ superstructure-based plasmonic photocatalysts exhibiting efficient charge separation and unprecedented activity. *J. Am. Chem. Soc.* **2014**, *136*, 458–465. [CrossRef]
52. Murakami, N.; Kurihara, Y.; Tsubota, T.; Ohno, T. Shape-controlled anatase titanium(IV) oxide particles prepared by hydrothermal treatment of peroxo titanic acid in the presence of polyvinyl alcohol. *J. Phys. Chem.* **2009**, *113*, 3062–3069. [CrossRef]
53. Wei, Z.; Janczarek, M.; Endo, M.; Wang, K.L.; Balcytis, A.; Nitta, A.; Mendez-Medrano, M.G.; Colbeau-Justin, C.; Juodkazis, S.; Ohtani, B.; et al. Noble metal-modified faceted anatase titania photocatalysts: Octahedron versus decahedron. *Appl. Catal. Environ.* **2018**, *237*, 574–587. [CrossRef]
54. Wei, Z.; Kowalska, E.; Verrett, J.; Colbeau-Justin, C.; Remita, H.; Ohtani, B. Morphology-dependent photocatalytic activity of octahedral anatase particles prepared by ultrasonication-hydrothermal reaction of titanates. *Nanoscale* **2015**, *7*, 12392–12404. [CrossRef] [PubMed]
55. Leong, K.H.; Chu, H.Y.; Ibrahim, S.; Saravanan, P. Palladium nanoparticles anchored to anatase TiO$_2$ for enhanced surface plasmon resonance-stimulated, visible-light-driven photocatalytic activity. *Beilstein J. Nanotechnol.* **2015**, *6*, 428–437. [CrossRef] [PubMed]
56. Zielinska-Jurek, A.; Wei, Z.; Wysocka, I.; Szweda, P.; Kowalska, E. The effect of nanoparticles size on photocatalytic and antimicrobial properties of Ag-Pt/TiO$_2$ photocatalysts. *Appl. Surf. Sci.* **2015**, *353*, 317–325. [CrossRef]
57. DeSario, P.A.; Pietron, J.J.; Brintlinger, T.H.; McEntee, M.; Parker, J.F.; Baturina, O.; Stroud, R.M.; Rolison, D.R. Oxidation-stable plasmonic copper nanoparticles in photocatalytic TiO$_2$ nanoarchitectures. *Nanoscale* **2017**, *9*, 11720–11729. [CrossRef]
58. Link, S.; El-Sayed, M.A. Spectral properties and relaxation dynamics of surface plasmon electronic oscillations in gold and silver nanodots and nanorods. *J. Phys. Chem. B* **1999**, *103*, 8410–8426. [CrossRef]
59. Muniz-Miranda, M.; Gellini, C.; Simonelli, A.; Tiberi, M.; Giammanco, F.; Giorgetti, E. Characterization of Copper nanoparticles obtained by laser ablation in liquids. *Appl. Phys. Mater.* **2013**, *110*, 829–833. [CrossRef]

60. Nilius, N.; Ernst, N.; Freund, H. On energy transfer processes at cluster-oxide interfaces: Silver on titania. *Chem. Phys. Lett.* **2001**, *349*, 351–357. [CrossRef]
61. Xia, Y.N.; Xiong, Y.J.; Lim, B.; Skrabalak, S.E. Shape-controlled synthesis of metal nanocrystals: Simple chemistry meets complex physics? *Angew. Chem. Int. Ed.* **2009**, *48*, 60–103. [CrossRef]
62. Yablonovitch, E. Photonic band-gap structures. *J. Opt. Soc. Am.* **1993**, *10*, 283–295. [CrossRef]
63. Lopez, C. Materials aspects of photonic crystals. *Adv. Mater.* **2003**, *15*, 1679–1704. [CrossRef]
64. Kim, W.T.; Choi, W.Y. Fabrication of TiO_2 photonic crystal by anodic oxidation and their optical sensing properties. *Sens. Actuators A Phys.* **2017**, *260*, 178–184. [CrossRef]
65. Chiarello, G.L.; Zuliani, A.; Ceresoli, D.; Martinazzo, R.; Selli, E. Exploiting the photonic crystal properties of TiO_2 nanotube arrays to enhance photocatalytic hydrogen production. *ACS Catal.* **2016**, *6*, 1345–1353. [CrossRef]
66. Zhang, Z.H.; Yang, X.L.; Hedhili, M.N.; Ahmed, E.; Shi, L.; Wang, P. Microwave-assisted self-doping of TiO_2 photonic crystals for efficient photoelectrochemical water splitting. *ACS Appl. Mater. Interfaces* **2014**, *6*, 691–696. [CrossRef] [PubMed]
67. Zhang, Z.H.; Wu, H.J. Multiple band light trapping in ultraviolet, visible and near infrared regions with TiO_2 based photonic materials. *Chem. Commun.* **2014**, *50*, 14179–14182. [CrossRef] [PubMed]
68. Li, Z.Z.; Xin, Y.M.; Wu, W.L.; Fu, B.H.; Zhang, Z.H. Phosphorus cation doping: A new strategy for boosting photoelectrochemical performance on TiO_2 nanotube photonic crystals. *ACS Appl. Mater. Interfaces* **2016**, *8*, 30972–30979. [CrossRef]
69. Likodimos, V. Photonic crystal-assisted visible light activated TiO_2 photocatalysis. *Appl. Catal. Environ.* **2018**, *230*, 269–303. [CrossRef]
70. Li, X.; Yu, J.G.; Jaroniec, M. Hierarchical photocatalysts. *Chem. Soc. Rev.* **2016**, *45*, 2603–2636. [CrossRef]
71. Chiang, C.C.; Tuyen, L.D.; Ren, C.R.; Chau, L.K.; Wu, C.Y.; Huang, P.J.; Hsu, C.C. Fabrication of titania inverse opals by multi-cycle dip-infiltration for optical sensing. *Photonics Nanostruct.* **2016**, *19*, 48–54. [CrossRef]
72. Lu, X.Y.; Zhu, Y.; Cen, T.Z.; Jiang, L. Centimeter-scale colloidal crystal belts via robust self-assembly strategy. *Langmuir* **2012**, *28*, 9341–9346. [CrossRef]
73. Jiang, P.; Bertone, J.F.; Hwang, K.S.; Colvin, V.L. Single-crystal colloidal multilayers of controlled thickness. *Chem. Mater.* **1999**, *11*, 2132–2140. [CrossRef]
74. Li, H.; Vienneau, G.; Jones, M.; Subramanian, B.; Robichaud, J.; Djaoued, Y. Crack-free 2D-inverse opal anatase TiO_2 films on rigid and flexible transparent conducting substrates: Low temperature large area fabrication and electrochromic properties. *J. Mater. Chem.* **2014**, *2*, 7804–7810. [CrossRef]
75. Mayoral, R.; Requena, J.; Moya, J.S.; Lopez, C.; Cintas, A.; Miguez, H.; Meseguer, F.; Vazquez, L.; Holgado, M.; Blanco, A. 3D long-range ordering in an SiO_2 submicrometer-sphere sintered superstructure. *Adv. Mater.* **1997**, *9*, 257–260. [CrossRef]
76. Kubrin, R.; Pasquarelli, R.M.; Waleczek, M.; Lee, H.S.; Zierold, R.; do Rosario, J.J.; Dyachenko, P.N.; Moreno, J.M.M.; Petrov, A.Y.; Janssen, R.; et al. Bottom-up fabrication of multilayer stacks of 3D photonic crystals from titanium dioxide. *ACS Appl. Mater. Interfaces* **2016**, *8*, 10466–10476. [CrossRef] [PubMed]
77. Curti, M.; Robledo, G.L.; Claro, P.C.D.; Ubogui, J.H.; Mendive, C.B. Characterization of titania inverse opals prepared by two distinct infiltration approaches. *Mater. Res. Bull.* **2018**, *101*, 12–19. [CrossRef]
78. Zhang, Y.; Li, K.; Su, F.Y.; Cai, Z.Y.; Liu, J.X.; Wu, X.W.; He, H.L.; Yin, Z.; Wang, L.H.; Wang, B.; et al. Electrically switchable photonic crystals based on liquid-crystal-infiltrated TiO_2-inverse opals. *Opt. Express* **2019**, *27*, 15391–15398. [CrossRef]
79. Kim, K.; Thiyagarajan, P.; Ahn, H.J.; Kim, S.I.; Jang, J.H. Optimization for visible light photocatalytic water splitting: Gold-coated and surface-textured TiO_2 inverse opal nano-networks. *Nanoscale* **2013**, *5*, 6254–6260. [CrossRef]
80. Zhou, Q.; Dong, P.; Liu, L.X.; Cheng, B.Y. Study on the sedimentation self-assembly of colloidal SiO_2 particles under gravitational field. *Colloids Surf.* **2005**, *253*, 169–174. [CrossRef]
81. Hua, C.X.; Xu, H.B.; Zhang, P.P.; Chen, X.Y.; Lu, Y.Y.; Gan, Y.; Zhao, J.P.; Li, Y. Process optimization and optical properties of colloidal self-assembly via refrigerated centrifugation. *Colloid. Polym. Sci.* **2017**, *295*, 1655–1662. [CrossRef]
82. Miguez, H.; Meseguer, F.; Lopez, C.; Blanco, A.; Moya, J.S.; Requena, J.; Mifsud, A.; Fornes, V. Control of the photonic crystal properties of fcc-packed submicrometer SiO_2 spheres by sintering. *Adv. Mater.* **1998**, *10*, 480–483. [CrossRef]

83. Jovic, V.; Idriss, H.; Waterhouse, G.I.N. Slow photon amplification of gas-phase ethanol photo-oxidation in titania inverse opal photonic crystals. *Chem. Phys.* **2016**, *479*, 109–121. [CrossRef]
84. Cheng, C.W.; Karuturi, S.K.; Liu, L.J.; Liu, J.P.; Li, H.X.; Su, L.T.; Tok, A.I.Y.; Fan, H.J. Quantum-dot-sensitized TiO_2 inverse opals for photoelectrochemical hydrogen generation. *Small* **2012**, *8*, 37–42. [CrossRef] [PubMed]
85. Liu, L.J.; Karuturi, S.K.; Su, L.T.; Tok, A.I.Y. TiO_2 inverse-opal electrode fabricated by atomic layer deposition for dye-sensitized solar cell applications. *Energy Environ. Sci.* **2011**, *4*, 209–215. [CrossRef]
86. Li, X.H.; Wu, Y.; Shen, Y.H.; Sun, Y.; Yang, Y.; Xie, A.J. A novel bifunctional Ni-doped TiO_2 inverse opal with enhanced SERS performance and excellent photocatalytic activity. *Appl. Surf. Sci.* **2018**, *427*, 739–744. [CrossRef]
87. Moon, J.H.; Cho, Y.S.; Yang, S.M. Room temperature chemical vapor deposition for fabrication of titania inverse opals: Fabrication, morphology analysis and optical characterization. *Korean Chem. Soc.* **2009**, *30*, 2245–2248.
88. Curti, M.; Mendive, C.B.; Grela, M.A.; Bahnemann, D.W. Stopband tuning of TiO_2 inverse opals for slow photon absorption. *Mater. Res. Bull.* **2017**, *91*, 155–165. [CrossRef]
89. Sordello, F.; Duca, C.; Maurino, V.; Minero, C. Photocatalytic metamaterials: TiO_2 inverse opals. *Chem. Commun.* **2011**, *47*, 6147–6149. [CrossRef] [PubMed]
90. Wu, M.; Liu, J.; Jin, J.; Wang, C.; Huang, S.Z.; Deng, Z.; Li, Y.; Su, B.L. Probing significant light absorption enhancement of titania inverse opal films for highly exalted photocatalytic degradation of dye pollutants. *Appl. Catal. Environ.* **2014**, *150*, 411–420. [CrossRef]
91. Lu, Y.; Yu, H.T.; Chen, S.; Quan, X.; Zhao, H.M. Integrating plasmonic nanoparticles with TiO_2 photonic crystal for enhancement of visible-light-driven photocatalysis. *Environ. Sci. Technol.* **2012**, *46*, 1724–1730. [CrossRef]
92. Curti, M.; Zvitco, G.; Grela, M.A.; Mendive, C.B. Angle dependence in slow photon photocatalysis using TiO_2 inverse opals. *Chem. Phys.* **2018**, *502*, 33–38. [CrossRef]
93. Zhao, Y.X.; Yang, B.F.; Xu, J.; Fu, Z.P.; Wu, M.; Li, F. Facile synthesis of Ag nanoparticles supported on TiO_2 inverse opal with enhanced visible-light photocatalytic activity. *Thin Solid Films* **2012**, *520*, 3515–3522. [CrossRef]
94. Srinivasan, M.; White, T. Degradation of methylene blue by three-dimensionally ordered macroporous titania. *Environ. Sci. Technol.* **2007**, *41*, 4405–4409. [CrossRef] [PubMed]
95. Wan, Y.; Wang, J.; Wang, X.; Xu, H.; Yuan, S.; Zhang, Q.; Zhang, M. Preparation of inverse opal titanium dioxide for photocatalytic performance research. *Opt. Mater.* **2019**, *96*, 109287. [CrossRef]
96. Rahul, T.K.; Sandhyarani, N. Nitrogen-fluorine co-doped titania inverse opals for enhanced solar light driven photocatalysis. *Nanoscale* **2015**, *7*, 18259–18270. [CrossRef]
97. Cai, J.M.; Wu, M.Q.; Wang, Y.T.; Zhang, H.; Meng, M.; Tian, Y.; Li, X.G.; Zhang, J.; Zheng, L.R.; Gong, J.L. Synergetic enhancement of light harvesting and charge separation over surface-disorder-engineered TiO_2 photonic crystals. *Chem* **2017**, *2*, 877–892. [CrossRef]
98. Toumazatou, A.; Arfanis, M.K.; Pantazopoulos, P.-A.; Kontos, A.G.; Falaras, P.; Stefanou, N.; Likodimos, V. Slow-photon enhancement of dye sensitized TiO_2 photocatalysis. *Mater. Lett.* **2017**, *197*, 123–126. [CrossRef]
99. Chen, Z.; Fang, L.; Dong, W.; Zheng, F.; Shen, M.; Wang, J. Inverse opal structured Ag/TiO_2 plasmonic photocatalyst prepared by pulsed current deposition and its enhanced visible light photocatalytic activity. *J. Mater. Chem.* **2014**, *2*, 824–832. [CrossRef]
100. Lim, S.Y.; Law, C.S.; Liu, L.; Markovis, M.; Abell, A.D.; Santos, A. Integrating surface plasmon resonance and slow photon effects in nanoporous anodis alumina photonic crystals for photocatalysis. *Catal. Sci. Technol.* **2019**, *9*, 3158–3176. [CrossRef]
101. Ye, J.; He, J.H.; Wang, S.; Zhou, X.J.; Zhang, Y.; Liu, G.; Yang, Y.F. Nickel-loaded black TiO_2 with inverse opal structure for photocatalytic reduction of CO_2 under visible light. *Sep. Purif. Technol.* **2019**, *220*, 8–15. [CrossRef]
102. Wang, X.Y.; Li, J.; Gao, X.N.; Shen, Y.H.; Xie, A.J. Ordered CdSe-sensitized TiO_2 inverse opal film as multifunctional surface-enhanced Raman scattering substrate. *Appl. Surf. Sci.* **2019**, *463*, 357–362. [CrossRef]
103. Zheng, X.Z.; Li, D.Z.; Li, X.F.; Chen, J.; Cao, C.S.; Fang, J.L.; Wang, J.B.; He, Y.H.; Zheng, Y. Construction of ZnO/TiO_2 photonic crystal heterostructures for enhanced photocatalytic properties. *Appl. Catal. Environ.* **2015**, *168*, 408–415. [CrossRef]

104. Lee, H.; Jeon, T.Y.; Lee, S.Y.; Lee, S.Y.; Kim, S.H. Designing multicolor micropatterns of inverse opals with photonic bandgap and surface plasmon resonance. *Adv. Funct. Mater.* **2018**, *28*, 1706664. [CrossRef]
105. Zhang, X.Y.; Zheng, Y.H.; Liu, X.; Lu, W.; Dai, J.Y.; Lei, D.Y.; MacFarlane, D.R. Hierarchical porous plasmonic metamaterials for reproducible ultrasensitive surface-enhanced Raman spectroscopy. *Adv. Mater.* **2015**, *27*, 1090–1096. [CrossRef] [PubMed]
106. Rahul, T.K.; Sandhyarani, N. Plasmonic and photonic effects on hydrogen evolution over chemically modified titania inverse opals. *Chemnanomat* **2018**, *4*, 642–648. [CrossRef]
107. Jiao, J.Q.; Wei, Y.C.; Chi, K.B.; Zhao, Z.; Duan, A.J.; Liu, J.; Jiang, G.Y.; Wang, Y.J.; Wang, X.L.; Han, C.C.; et al. Platinum nanoparticles supported on TiO_2 photonic crystals as highly active photocatalyst for the reduction of CO_2 in the presence of water. *Energy Technol.* **2017**, *5*, 877–883. [CrossRef]
108. Ding, B.Y.; Pemble, M.E.; Korovin, A.V.; Peschel, U.; Romanov, S.G. Three-dimensional photonic crystals with an active surface: Gold film terminated opals. *Phys. Rev.* **2010**, *82*, 035119. [CrossRef]
109. Paterno, G.M.; Moscardi, L.; Donini, S.; Ariodanti, D.; Kriegel, I.; Zani, M.; Parisini, E.; Scotognella, F.; Lanzani, G. Hybrid one-dimensional plasmonic-photonic crystals for optical detection of bacterial contaminants. *J. Phys. Chem. Lett.* **2019**, *10*, 4980–4986. [CrossRef]
110. Zhao, H.; Hu, Z.Y.; Liu, J.; Li, Y.; Wu, M.; Van Tendeloo, G.; Su, B.L. Blue-edge slow photons promoting visible-light hydrogen production on gradient ternary 3DOM TiO_2-Au-CdS photonic crystals. *Nano Energy* **2018**, *47*, 266–274. [CrossRef]
111. Erola, M.O.A.; Philip, A.; Ahmed, T.; Suvanto, S.; Pakkanen, T.T. Fabrication of Au- and Ag-SiO_2 inverse opals having both localized surface plasmon resonance and Bragg diffraction. *J. Solid State Chem.* **2015**, *230*, 209–217. [CrossRef]
112. Chen, J.I.L.; Loso, E.; Ebrahim, N.; Ozin, G.A. Synergy of slow photon and chemically amplified photochemistry in platinum nanocluster-loaded inverse titania opals. *J. Am. Chem. Soc.* **2008**, *130*, 5420–5421. [CrossRef]
113. Sanchez-Garcia, L.; Tserkezis, C.; Ramirez, M.O.; Molina, P.; Carvajal, J.J.; Aguilo, M.; Diaz, F.; Aizpurua, J.; Bausa, L.E. Plasmonic enhancement of second harmonic generation from nonlinear $RbTiOPO_4$ crystals by aggregates of silver nanostructures. *Opt. Express* **2016**, *24*, 8491–8500. [CrossRef]
114. Meng, S.G.; Li, D.Z.; Fu, X.L.; Fu, X.Z. Integrating photonic bandgaps with surface plasmon resonance for the enhancement of visible-light photocatalytic performance. *J. Mater. Chem.* **2015**, *3*, 23501–23511. [CrossRef]
115. Zeng, S.; Vahidzadeh, E.; VanEssen, C.G.; Kar, P.; Kisslinger, R.; Goswami, A.; Zhang, Y.; Mandi, N.; Riddell, S.; Kobryn, A.E.; et al. Optical control of selectivity of high rate CO_2 photoreduction via interbandor hot electron Z-scheme reaction pathways in Au-TiO_2 plasmonic photonic crystal photocatalyst. *Appl. Catal. Environ.* **2020**, *267*, 118644. [CrossRef]
116. Fang, L.; Nan, F.; Yang, Y.; Cao, D.W. Enhanced photoelectrochemical and photocatalytic activity in visible-light-driven Ag/$BiVO_4$ inverse opals. *Appl. Phys. Lett.* **2016**, *108*, 093902. [CrossRef]
117. Zhang, Z.; Zhang, L.; Hedhili, M.N.; Zhang, H.; Wang, P. Plasmonic gold nanocrystals coupled with photonic crystal seamlessly on TiO_2 nanotube photoelectrodes for efficient visible light photoelectrochemical water splitting. *Nano Lett.* **2013**, *13*, 14–20. [CrossRef] [PubMed]
118. Zhang, X.; Liu, Y.; Lee, S.T.; Yang, S.H.; Kang, Z.H. Coupling surface plasmon resonance of gold nanoparticles with slow-photon-effect of TiO_2 photonic crystals for synergistically enhanced photoelectrochemical water splitting. *Energy Environ. Sci.* **2014**, *7*, 1409–1419. [CrossRef]
119. Wang, Y.F.; Xiong, D.B.; Zhang, W.; Su, H.L.; Liu, Q.L.; Gu, J.J.; Zhu, S.M.; Zhang, D. Surface plasmon resonance of gold nanocrystals coupled with slow-photon-effect of biomorphic TiO_2 photonic crystals for enhanced photocatalysis under visible-light. *Catal. Today* **2016**, *274*, 15–21. [CrossRef]
120. Huo, J.W.; Yuan, C.; Wang, Y. Nanocomposites of three-dimensionally ordered porous TiO_2 decorated with Pt and reduced graphene oxide for the visible-light photocatalytic degradation of waterborne pollutants. *ACS Appl. Nano Mater.* **2019**, *2*, 2713–2724. [CrossRef]
121. Temerov, F.; Ankudze, B.; Saarinen, J.J. TiO_2 inverse opal structures with facile decoration of precious metal nanoparticles for enhanced photocatalytic activity. *Mater. Chem. Phys.* **2020**, *242*, 122471. [CrossRef]
122. Zhang, S.S.; Peng, B.Y.; Yang, S.Y.; Wang, H.G.; Yu, H.; Fang, Y.P.; Peng, F. Non-noble metal copper nanoparticles-decorated TiO_2 nanotube arrays with plasmon-enhanced photocatalytic hydrogen evolution under visible light. *Int. J. Hydrog. Energy* **2015**, *40*, 303–310. [CrossRef]
123. Zhang, L.W.; Lin, C.Y.; Valev, V.K.; Reisner, E.; Steiner, U.; Baumberg, J.J. Plasmonic enhancement in $BiVO_4$ photonic crystals for efficient water splitting. *Small* **2014**, *10*, 3970–3978. [CrossRef]

124. Dinh, C.T.; Yen, H.; Kleitz, F.; Do, T.O. Three-dimensional ordered assembly of thin-shell Au/TiO$_2$ hollow nanospheres for enhanced visible-light-driven photocatalysis. *Angew. Chem. Int. Ed.* **2014**, *53*, 6618–6623. [CrossRef] [PubMed]
125. Alessandri, I.; Ferroni, M. Exploiting optothermal conversion for nanofabrication: Site-selective generation of Au/TiO$_2$ inverse opals. *J. Mater. Chem.* **2009**, *19*, 7990–7994. [CrossRef]
126. Wei, Z.; Endo, M.; Wang, K.; Charbit, E.; Markowska-Szczupak, A.; Ohtani, B.; Kowalska, E. Noble metal-modified octahedral anatase titania particles with enhanced activity for decomposition of chemical and microbiological pollutants. *Chem. Eng. J.* **2017**, *318*, 121–134. [CrossRef] [PubMed]
127. Raja-Mogan, T.; Lehoux, A.; Takashima, M.; Kowalska, E.; Ohtani, B. A triply wavelength-tuned visible light-responsive photocatalyst: Matching of LED, PBG and LSPR. **2020**. under preparation for submission.

© 2020 by the authors. Licensee MDPI, Basel, Switzerland. This article is an open access article distributed under the terms and conditions of the Creative Commons Attribution (CC BY) license (http://creativecommons.org/licenses/by/4.0/).

Review

Review of Experimental Setups for Plasmonic Photocatalytic Reactions

Hung Ji Huang [1,*,†], Jeffrey Chi-Sheng Wu [2], Hai-Pang Chiang [3,4], Yuan-Fong Chou Chau [5], Yung-Sheng Lin [6], Yen Han Wang [2] and Po-Jui Chen [1]

1. Taiwan Instrument Research Institute, National Applied Research Laboratories, Hsinchy 30076, Taiwan; proray@narlabs.org.tw
2. Department of Chemical Engineering, National Taiwan University, Taipei 10617, Taiwan; cswu@ntu.edu.tw (J.C.-S.W.); d06524019@ntu.edu.tw (Y.H.W.)
3. Department of Optoelectronics and Materials Technology, National Taiwan Ocean University, Keelung 20224, Taiwan; hpchiang@mail.ntou.edu.tw
4. Institute of Physics, Academia Sinica, Taipei 11529, Taiwan
5. Centre for Advanced Material and Energy Sciences, Universiti Brunei Darussalam, Gadong BE1410, Negara Brunei Darussalam; chou.fong@ubd.edu.bn
6. Department of Chemical Engineering, National United University, Miaoli 36063, Taiwan; linys@nuu.edu.tw
* Correspondence: hjhuang@narlabs.org.tw or hhjhuangkimo@gmail.com
† Current address: 20, R&D Rd. VI, Hsinchu Science Park, Hsinchu 30076, Taiwan.

Received: 13 November 2019; Accepted: 26 December 2019; Published: 31 December 2019

Abstract: Plasmonic photocatalytic reactions have been substantially developed. However, the mechanism underlying the enhancement of such reactions is confusing in relevant studies. The plasmonic enhancements of photocatalytic reactions are hard to identify by processing chemically or physically. This review discusses the noteworthy experimental setups or designs for reactors that process various energy transformation paths for enhancing plasmonic photocatalytic reactions. Specially designed experimental setups can help characterize near-field optical responses in inducing plasmons and transformation of light energy. Electrochemical measurements, dark-field imaging, spectral measurements, and matched coupling of wavevectors lead to further understanding of the mechanism underlying plasmonic enhancement. The discussions herein can provide valuable ideas for advanced future studies.

Keywords: plasmonic photocatalytic reactions; photocatalytic reactors; instrumentation

1. Introduction

Applications of photocatalytic reactions have developed extensively since Fujishima and Honda presented their work in the electrochemical photolysis of water in a semiconductor electrode in 1972 [1]. The improvement of processing efficiency was started with the modification of photocatalysts. One modification approach entails loading metal nanoparticles on the surface of photocatalysts [2]. Researchers developing plasmonics perceived a possible breakthrough in plasmonic photocatalytic reactions after the work presented by Awazu et al. [3]. Several studies and reviews related to plasmonic photocatalytic reactions have been published, particularly in the past 3 years [4–21].

Plasmonic photocatalytic reactions are useful in various applications, such as antibiotics [5], photo-deposition [6], H_2 generation [22], and plasmon-induced dissociation of H_2 on Au through the generation of hot electrons [23]. The plasmonic enhancement of photocatalytic reactions is also affected by various experimental conditions. A study reported that in the photocatalytic degradation of methyl orange through a spinning disk reactor [24], the final reserved concentration ratios (C/C_0) for TiO_2 under illumination from a 4-W mercury tube lamp with and without an additional red light-emitting diode (LED; 637 nm) were 0.71 and 0.62, respectively. The red light slightly suppressed the photocatalytic reactions by interfering with the movement of generated hot charges under ultraviolet (UV) light. However, the final reserved C/C_0 observed for Au-TiO$_2$ with and without an additional red LED are 0.54 and 0.65, respectively. The Au-TiO$_2$ hybrid photocatalyst under UV light and additional 637 nm light exhibited approximately 24% greater activity than did the TiO$_2$ photocatalyst under only UV light. The red light introduced plasmons on the Au side and enhanced photocatalytic reactions. The visible light affects not only the free electrons in the metal nanoparticles, but also hot charges generated in the photocatalyst under UV light illumination. This could occasionally engender ambiguity in the identification of plasmonic effects on photocatalytic reactions.

Plasmonic effects engender not only enhanced photocatalytic reactions but also product selectivity. Cui et al. presented selectivity of benzyl alcohol oxidation reaction products, namely benzaldehyde, benzyl benzoate, and benzoic acid, in plasmonic photocatalytic reactions [25]. A 1500-nm light source yielded 99.57% selectivity toward benzaldehyde, whereas 808 and 980 nm provided 95.87% and 93.43% selectivity, respectively. Moreover, the observed selectivity toward benzyl benzoate changed substantially with light wavelengths. A selectivity of 3.90% was obtained for the 808-nm trial, 5.62% for the 980-nm trial, and only 0.28% for the 1500-nm trial.

This review presents a brief discussion of experimental designs providing plasmonic enhancements of photocatalytic reactions, particle reactors and measurements. Plasmonic enhancement must occur through the light-induced generation of plasmons, such as volume plasmons (VPs), surface plasmons (SPs), or localized surface plasmon (LSPs), to improve processing efficiency in photocatalytic reactions with various plasmonic energy transformation paths [19]. In addition, this review introduces various reactors and identification methods to provide a brief overview of models of plasmonic photocatalytic reactions. In plasmonics, the generation of plasmons, especially SPs and LSPs, must typically match specific coupling conditions to achieve high light-to-plasmon conversion efficiency. Stronger or higher concentration of plasmons, increase the corresponding plasmonic enhancement.

It is confused to identify the plasmonic enhancements of photocatalytic reactions processed chemically or physically. Typically, a chemically plasmonic-enhanced photocatalytic reaction includes the migration of induced hot charges. A physically plasmonic-enhanced photocatalytic reaction goes through "electromagnetic field-enhancement" in the process. This review also briefly identifies how various energy transformation paths enhanced the photocatalytic reaction in chemically plasmonic or physically plasmonic enhancement pathways.

2. Energy Transformation Path of Plasmonic Photocatalytic Reactions

Plasmons are collective oscillations of free electrons at a metal-dielectric interface and are induced by external electromagnetic waves [26,27]. In general, plasmons can be roughly categorized into three types; i.e., VPs, SPs, and LSPs. According to the Drude model, VPs are induced when free electrons in bulk metal resist the penetration of external electromagnetic waves within an extremely small depth. The plasma or natural frequency, $\omega_{pe} = \sqrt{n_e e^2 / m^* \varepsilon_0}$ describes the oscillation of electron density that allows wide variations in light transparency. SPs [28,29] are transformed modes of light wave propagating at the interface between negative (metal) and positive (dielectric or semiconductor) permittivity materials. A SP typically has a higher wavevector than that of light propagating in vacuum and in air. Therefore, the SPs generation require a special coupling method, such as the Otto configuration, Kretschmann configuration, grating or rough surface coupling [28]. LSPs are collective oscillation of free electrons generated and localized in a small region under external illumination. Sharp surface structures, such as nanoparticles or edges, gaps, nanometer-sized bumps, and valleys on a metal surface and silver colloid fractal clusters [30], provide a wide range of spatial frequency that can generate plasmon hot spots.

Regarding plasmonic photocatalytic reactions, this review restricts its scope to reactions that involve light harvest through the generation of various plasmons (Figure 1). In such plasmonic photocatalytic reactions, plasmons can modulate photocatalysis through the following effects: (i) strong light absorption, (ii) intensive far-field light-scattering, (iii) near-field electromagnetic field strengthening, (iii) abundant hot carrier generation and (iv) plasmonic heating effects [19,31].

Light-induced plasmons can elastically or inelastically transform to light again (Figure 1), a phenomenon that is normally viewed as a scattering process. The induced plasmon energy on a metal nanoparticle can be delivered elastically as transferring light energy to nearby bulk materials or dielectric, semiconductor, or metal nanoparticles. Light energy scattering and trapping from metal nanoparticles can increase the photon flux flow through photocatalytic nanoparticles [19,32,33] thus plasmonically enhancing the photocatalytic reaction. The plasmon energy may trigger a photocatalytic reaction as it propagates through a photocatalytic nanoparticle [3]. In this type of plasmonic photocatalytic reaction, the light that can originally excite the reaction on the photocatalyst is focused through the metal nanoparticle and has a relatively higher power intensity. The distance between the metal nanoparticles and the nano photocatalyst is approximately 100 nm or shorter, but direct contact is not necessary. Most studies on plasmonic photocatalytic reactions have used lamps also delivering UV light (e.g., mercury lamp, xenon lamp) for direct energy transformation process.

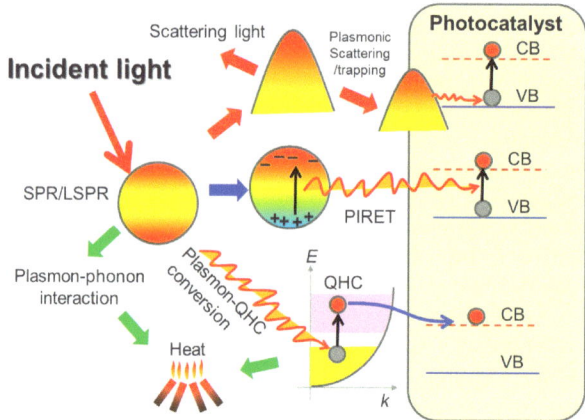

Figure 1. Various types of energy transformation in plasmonic photocatalytic reactions. Figure redrawn from reference [19].

Plasmonic energy in a metal may nonradiatively transfer to a photocatalyst in the near-field zone through a dipole-dipole interaction, and electron-hole pairs are thus generated in the photocatalyst (Figure 1) [19,34,35]. Plasmon-induced energy transformation (PIRET) [19] can occur even in the presence of an insulating space layer with a thickness of up to 25 nm between the plasmonic metal and the photocatalyst.

The plasmon energy can also nonradiatively decay to generate high-energy quantum hot charge carriers through Landau damping with intraband and interband electron transitions [19,36–38]. The hot electrons have a relatively high energy level and can be directly injected into the conduction band of the contacted photocatalyst (Figure 1) [19,39,40]. The electrons injected into the photocatalyst can still be regarded as hot carriers with an energy level higher than the conduction band level. In metal-photocatalyst heterostructures, hot electrons generated and transferred from the metal part exhibit longer excited-state lifetimes than do electrons photogenerated directly within photocatalysts through UV excitation [41]. Therefore, the metal and photocatalyst require only intimate contact for maximized electron injection. Govorov's group [31,42,43] and Kumarasinghe's group [44,45] have developed a single-electron model, involving the assumption of a noninteracting electron gas confined in a uniform background potential in metals. Their theories demonstrate that hot-electron generation and injection are highly sensitive to nanoparticle size and shape as well as "hot spot" presence.

A generated plasmon can inelastically decay to heat through a plasmon-phonon interaction. The local temperature can be substantially increased. The induced quantum hot carriers (QHCs) can also decay to convert energy to heat. The Arrhenius equation describes the positive exponential temperature-dependent processing efficiency of chemical reactions. Therefore, plasmonic heating also plays a crucial role in plasmonic photocatalytic reactions.

3. Types of Reactors for Photocatalytic Reactions

Photocatalytic reactions can be processed in various types of photocatalytic reactors (Figure 2). Slurry- and fixed-bed reactors are the most commonly used as shown in Figure 2a,c, respectively. In a slurry-bed reactor, the prepared photocatalysts are dispersed homogeneously in the target solution. Photocatalytic processing efficiency is typically further enhanced by mechanical stirring. The excitation light (e.g., mercury lamp, halogen lamp) illuminates from outside. A waterproof lamp (Figure 2d) or light guide might be placed in the solution to improve light excitation. Some photocatalysts are lost during recycling. Consequently, performing a sequential modification test on the same photocatalyst for comparison is difficult. Some studies have suggested using magnetic material, e.g., Fe, Co or Ni, to synthesize (composite) magnetic photocatalyst. Magnetic (composite) photocatalyst can be easily recycled using a magnet after use. However, this limits the material for photocatalyst fabrication. A notable modification to the slurry-bed reactors involves the use of a twin reactor (Figure 2b) to process separate photocatalytic reactions in two compartments using a membrane, [46–49]. The hydrogen and oxygen generation reactions in photocatalytic water splitting can be processed in separate compartments, thus eliminating costs associated with the separation process [46,47,50]. Moreover, advanced CO_2 reduction in the H_2 generation unit can also be processed as O_2 generation in another unit.

A fixed-bed reactor (Figure 2c) is convenient because it does not require recycling and is suitable modification testing on the same photocatalyst for comparison. A light-transparent substrate can be used to deliver excitation light directly to the photocatalyst without absorption or scattering by the test solution. Figure 2e presents a tube reactor [51]. The inside of the light-transparent tube is coated with a photocatalyst and the tube is wrapped around the lamp to increase the surface contact area of the photocatalyst. Setups in Figure 2d,e improve light use efficiency. The slurry-bed reactor has a simple experimental setup but the tasks of collecting and recycling the photocatalyst after use are difficult. The fixed-bed reactor has disadvantages in its limited contact area. Few reactants can be degraded during long travel in solution before diffusing to the fixed photocatalyst located on the substrate. The electrochemical electrode is also a type of fixed-bed with similar disadvantages.

Several modified fixed-bed reactors (Figure 2f,g) are built to improve processing efficiency in photocatalytic reactions. In the reactors in Figure 2f,g the photocatalyst is deposited on multiple glass fibers [52–54] or inside a porous medium, respectively, to increase the surface contact area. The residence time of the target pollutants to the photocatalyst increases substantially. The excitation light is coupled outside the reactor and homogeneously delivered to the photocatalyst.

The aforementioned experimental setups are simple and easy to use. However, typical photocatalytic reactors (Figure 2a–g) may have disadvantages related to the final desorption step of photocatalytic reactions [16]. In slurry-bed reactors (Figure 2a) the photocatalyst moves with the flow carried by magnetic stirring. The suspended photocatalyst has low speed relative to the flow, and the mass transfer of the reactants and products relies only on diffusion in solution. In fixed-bed reactors (Figure 2c–g) the test solution flows over a surface deposited with photocatalysts. The flow speed in different fluidic layers changes widely according to the input flow speed and shape of the reaction chamber. The nonslip boundary condition considerably reduces the flow speed in the fluidic layer adjacent to the surface deposited with photocatalyst. The low-speed layer has a limited mass transfer of reactants and products diffused in the flow. Three alternatives, namely spinning or rotating disk reactor (SDR/RDR), Taylor vortex reactor, and micro-optofluidic chip (MOFC) reactor, have been developed for high-efficiency photocatalytic reactions. These reactors increase the flow speed around the surface of the photocatalyst-deposited layer.

In SDRs (Figure 2h and Figure 3a) [24,55–61], a reaction disk rotates to drive the fluid to move on top of it. The fluidic layer near the deposited photocatalyst moves faster than other layers above. A higher rotating speed of the reaction disk also introduces larger shear force and friction between adjacent fluidic layers and generates more microvortices. The mass transfer rate of molecules dissolved in water greatly increased. The microvortices can rapidly strip off the products of the photocatalytic reactions from the surface of the deposited photocatalyst on the rotating disk. The processing efficiency of the photocatalytic reactions is greatly enhanced. In the reference work by Huang et al. [24], 10 µM methyl orange solution degraded to approximately 15% of its original concentration in only a single run of treatment. The photocatalytic degradation process for every drop of test solution was completed within less than 0.1 S. Notably, the disk was made of polycarbonate, which can be recycled from optical storage waste, making it environmentally beneficial.

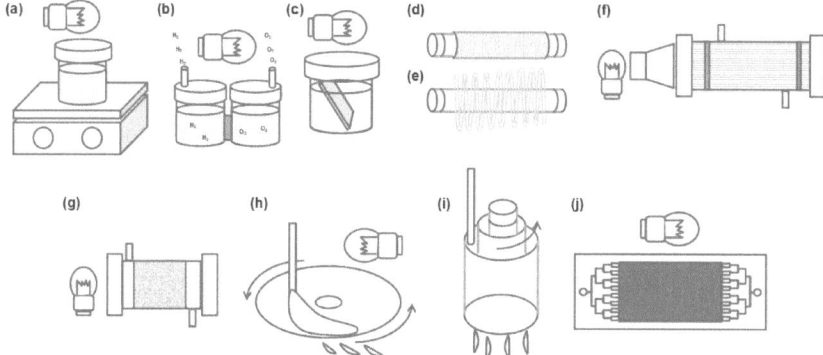

Figure 2. Various photocatalytic reactors: (**a**) slurry-bed reactor, (**b**) twin reactor, (**c**) plate fixed-bed reactor, (**d**) photocatalyst coated on the surface (or material covering the surface) of the light tube, (**e**) tube reactor, (**f**) fiber reactor, (**g**) photocatalyst deposited inside large surface area material or packed bed reactor, (**h**) spinning disk reactor, (**i**) Taylor vortex reactor, and (**j**) micro-optical fluidic chip reactor.

A Taylor vortex reactor [62–69] includes two cylinders that are arranged in symmetrically and can be oriented vertically (Figure 3a) or horizontally. Typically, the inner cylinder rotates at a high speed relative to the outer fixed cylinder. The test fluid (gas or liquid) is injected through the fixed outer cylinder and flow into the gap between the two cylinders. The rotating inner cylinder causes the test solution to also move in its rotation direction. Simultaneously, forces of centrifugal, Coriolis and gravitational are generated to move the fluid rapidly in relation to the outside cylinder. The faster the inside cylinder rotates, the more unstable the flow becomes. This thus produces an eddy current flow in the shape of rings stacked along the rotating axis. This is typically called Taylor flow. Numerous microvortices are generated from the friction and shear force in the vigorous agitated flow. The related setup is also called a Taylor vortex reactor.

In general, a Taylor vortex reactor is typical highly efficient [62–67] and the photocatalyst can be deposited on the outer surface of the rotating inner cylinder. A tube lamp can provide excitation light from the inside. The eddy current flow with numerous vortices can strip off high-affinity products generated by the photocatalytic reaction. Concurrently, reactants can be rapidly introduced, thus considerably increasing the mass transfer rate in the flow and resulting in enhanced photocatalytic reactions. Although the SDR and the Taylor vortex reactor are typically called microreactors, their processing capacity introduced by their high processing efficiency is comparable to that of a large tank reactor. These reactors characterized by a small size, low power consumption, and waste reduction properties, rendering them green technologies.

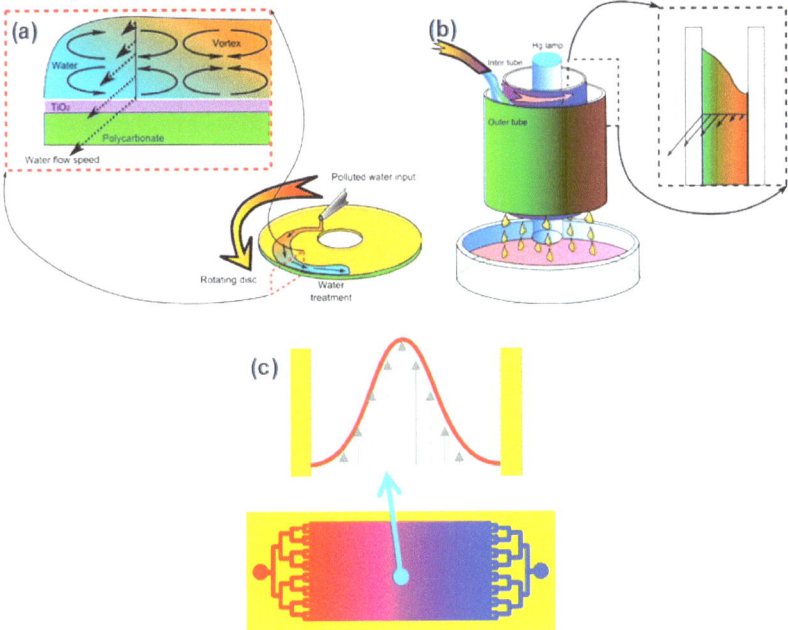

Figure 3. High-efficiency photocatalytic reactors: (**a**) spinning disk reactor, (**b**) Taylor vortex reactor, and (**c**) micro-optofluidic chip reactor.

MOFC reactors [70–85] have been used in some photocatalytic reactions. Although they have low processing volume, MOFC reactors have features that contribute to high processing efficiency in many applications: (i) Their small fluid channel size results in a high mass transfer rate, thus enabling effective transportation of reactants and products. (ii) The plasmon or light converted into heat results in increased temperature in localized area in micro fluids that further enhances mass transfer in fluidics. (iii) The external magnetic field and large variation in flow speed in the fluid channel lead to distinctive ion condensation of OH^-, thus enhancing active OH^* radical generation [85]. (iv) Their highly efficient light illumination engenders low energy loss.

4. Measurements in Plasmonic Photocatalytic Reactions

As presented in Figure 1, this review focused on photocatalytic reactions that involve light energy harvesting through generation of plasmons through several direct or indirect paths. Approaches for examining plasmonic enhancement in photocatalytic reactions can be categorized as follows: use of (i) plasmonic photocatalytic electrochemical measurements, (ii) scavengers and specially fabricated samples, (iii) optical measurements, and (iv) plasmonic light-to-heat conversion. Photocatalytic electrochemical measurements contribute the best means of examining plasmonic enhancement in photocatalytic reactions because the effect of light illumination can be easily depicted using photocurrent variations or deviations in characteristic cyclic voltammetry traces. The study of plasmonic effects is typically related to nanomaterial optical responses, and optical measurements are vital approaches for energy transformation in near-field zone. Therefore, electrochemical and optical measurements are crucial methods for characterizing plasmonic enhancements in photocatalytic reactions.

4.1. Plasmonic Photocatalytic Electrochemical Measurements

Photocatalytic technology was developed from the electrochemical photolysis of water by Fujishima and Honda in 1972 [1]. Photoelectrochemical techniques remain a powerful and direct tool for examining plasmonic enhancement in photocatalytic reactions [20]. Various types of metallic samples, such as nanoparticles [39,86,87], dendritic nanoforests [88–90], and photonic crystal [7,91], are used for examining plasmonic photocatalytic electrochemical reactions. The energy levels of reactive intermediates can be determining by monitoring overpotentials [92]. The photocurrent measurement provides real-time monitoring of the triggering enhancements [22,93]. Li et al. fabricated TiO_2 nanorods selectively planted inside an Au nanohole array [91]. The photocurrent enhancement factor was measured as a function of the wavelength and were consistent with the absorption section. Kim et al. applied a temperature-dependent photocatalytic measurement approach to acquire the activation enthalpy [87]. They observed that a reduction in activation enthalpy reduction was directly related to the photoelectrochemical potential accumulation on the Au nanoparticle under steady-state light excitation, a phenomenon analogous to electrochemical activation. Their findings are compatible with the phenomenon of additional energy compensation engendered by extra plasmonic enhancement of the photocatalytic endothermic oxidization of NH_4^+ in an SDR [94].

Plasmonic responses are typically determined using dark-field imaging (Figure 4) and optical microscopy entails monitoring scattering spectra for a plasmonic sample [38,95]. Dark-field imaging does not involve light illumination (Figure 4b) in contrast to bright-field imaging (Figure 4a). Plasmonic responses can be easily observed in dark-field imaging. Therefore, combining electrochemical measurements with dark-field microscopy [96–98], particularly hyperspectral dark-field imaging is valuable [99]. The charge carrier density can be precisely controlled by altering the applied potential and allowing for real-time optical monitoring of the affected LSP resonance responses [97,98,100]. Using hyperspectral dark-field imaging, Byers et al. [99] demonstrated that upon electrochemical tuning, a population of nanoparticles can undergo several processes ranging from nanoparticle charging to electrochemical reactions, such as chloride ion oxidation and hydrogen evolution reaction. These optical and chemical responses are altered on the base of a combination of nanoparticle or nanoparticle-substrate properties that undergo either nanoparticle charging or charge transfer in plasmon-enhanced photocatalytic reactions. Cell illumination leads to several deviations from conventional cyclic voltammetry characteristic traces, such as discrete shifts in onset potential for half reactions and an increase in the photocurrent, which provide useful information on the energy barriers for the reaction [101]. This is consistent with the quantized size dependence of light-to-heat energy transformation through a Pt thin film with a thickness of several nanometers [102]. QHC generation through the excitation of d-band electrons may be confined by selection rules.

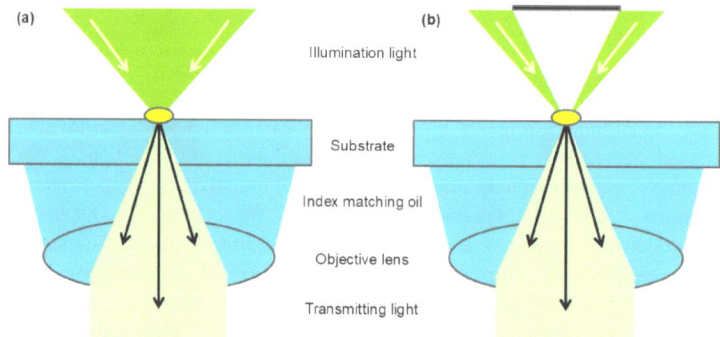

Figure 4. Optical microscopy in a (**a**) bright field and (**b**) dark field.

4.2. Optical Measurements

Optical measurements, such as transmission or absorption spectroscopy [9,22,23,25,35,37,88,103–110], Raman spectroscopy [111], and pump-probe measurements, are used for imaging optical effects in enhanced photocatalytic reactions.

The plasmonic responses of nanometallics materials are typically presented as variation in relatively broad transmission or absorption features in optical transmission spectra. The size, structure, or material change of nanometallic materials can be directly determined by feature shifts. Light-scattering spectroscopy of individual Au nanoparticles was reported to reveal the dephasing of particle plasmons in dark-field imaging [38]. A substantial reduction of the plasmon dephasing rate in nanorods as compared with small nanospheres could be attributed to a suppression of interband damping. A comparison with theory revealed that pure dephasing and interface damping negligibly contribute to the total plasmon dephasing rate.

The coating of metal nanoparticles on photocatalysts results in different optical transmission spectra that resemble those associated with typical plasmonics. Therefore, many studies have used transmission spectroscopy data to confirm the mechanism of plasmonic photocatalytic reactions. However, an action spectrum is required for further understanding how various light wavelengths are used in processing the plasmonic enhancement of photocatalytic reactions [23,89,91,112]. By using super-continuum laser, Mukherjee et al. [23] declared that "hot electrons do the impossible: plasmon-induced dissociation of H_2 on Au". Seven bandpass filters ranging from 500 to 800 nm, each with a wavelength spacing of 50 nm, were used to measure the wavelength dependence of the photocatalytic H_2 dissociation rate. The wavelength dependence of hydrogen deuteride (HD) generation was related to H_2 dissociation and was comparable to the calculated absorption cross section of Au (7 nm)-TiO_2 (30 nm) and to the diffuse reflectance spectrum. Visible light ranging from 500 to 800 nm cannot induce excitation or generation of hot charges in TiO_2 nanoparticles but can generate plasmons on Au nanoparticles. Some studies have adopted wavelength-dependent photocurrents in electrochemical measurements [88,91,93] to explain the mechanism of plasmon-enhanced photocatalytic reactions. These derived action spectra provided evidence of plasmon-induced photocatalytic reactions.

Plasmonic scattering and trapping and PIRET processes can be controlled by the spectral overlap and physical contact between a metal and photocatalyst [113]. This can help identify the PIRET and plasmonic scattering and trapping processes. For example, Awazu et al. [3] presented a plasmonic photocatalyst comprising silver nanoparticles embedded in a silica (SiO_2) spacer layer under TiO_2 layer. Localized plasmon polariton excitation on the surface of Ag nanoparticles caused a substantial increase in the near-field amplitude at well-defined wavelengths in the near-UV range under plasmonic scattering and trapping processes. The photocatalytic behavior of TiO_2 was considerably improved by the enhanced near-field amplitudes of LSP. The SiO_2 shell covered the Ag core, preventing light-induced hot electrons flowing from the directly contacted TiO_2. The optimized size of Ag NPs give rise to large LSP induced by the near-UV light can be explained by Mie scattering theory. This resulted in the seven times enhancement in photocatalytic decomposition of methylene blue. In addition, PIRET occurs in the Ag-SiO_2-TiO_2 core-shell nanoparticles because of the spectral overlap between Ag and TiO_2 [113]. Nanometals can excite charge carriers in photocatalysts through PIRET and hot-electron injection processes. In Ag-SiO_2-TiO_2 sandwich nanoparticles, the LSPs resonance band overlaps with the absorption band edge of TiO_2, enabling PIRET, and the SiO_2 barrier prevents hot-electron transfer. In Au-TiO_2, hot-electron injection occurs, but the lack of spectral overlap disables PIRET. In Ag-TiO_2, both hot-electron transfer and PIRET occur.

Photoluminescence (PL) is also used in characterizing the effect of hot charge recombination in photocatalytic reactions [114–118]. Decreased PL peaks indicate suppressed recombination of photoinduced charge carriers, which can lead to an enhanced photocatalytic reactions. Gao et al. [118] used PL spectra to measure Lorentz force–assisted charge carrier separation enhancement in TiO_2. They applied a magnetic field by placing a permanent magnet beneath the photocatalytic system. Studies on metal-decorated heterostructure photocatalyst have revealed that PL spectra of original photocatalysts had stronger signals than did those of metal-photocatalyst heterostructures [114–117]. This indicates the decreased recombination rate of photoinduced charge carriers and enhanced photocatalytic activity of metal-decorated nanoparticles. Charge carrier recombination could be effectively suppressed by introducing metal nanoparticles to the photocatalyst surface. This demonstrates a strong relationship between PL intensity and photocatalytic activity. However, additional measurements are necessary to demonstrate whether photocatalytic reaction enhancement is maintained through plasmon generation.

Raman spectroscopy is a valuable optical measurement tool for measuring the inelastic scattering of photons by matter. The exchange of energy involves the gain or loss of vibrational energy by a molecule as incident photons from a visible laser shift to lower or higher energy states. Surface-enhanced Raman scattering (SERS) [111,119–122] involves the amplification of Raman scattering signals by SPs on a metallic surface with high charge carrier density. Brooks et al. conducted an in-depth review of Raman spectroscopy applications for understanding plasmon-mediated photocatalysis [20]. As sensitive spectroscopic techniques, SERS and tip-enhanced Raman spectroscopy (TERS) offers considerably more detailed information on molecular probe structure compare with electronic spectroscopic techniques [123–135]. SERS and TERS require metal nanostructures and a sharp metal tip, respectively, to induce strong LSPs and trigger large Raman scattering signals. Therefore, Raman spectroscopy is typically used with a metal photocatalyst.

Energy transformation in plasmonic material sensing (i.e., chemical and biosensing [136–141] and SERS [119–122]) resembles that in plasmonic photocatalytic reactions. Therefore, methods of obtaining larger or enhanced signal readings in plasmonic sensing resemble to a higher enhanced processing efficiency in plasmonic photocatalytic reactions. Size dependence is typically valuable in plasmonic responses and SERS [111]. For large-scale material sensing or photocatalytic reactions, array or condensed random structures can achieve relatively high light absorption [141–143] or can even be perfect absorbers [144,145]. Furthermore, array structures help in SERS [26,27] and ultrafast real-time bioassays [140]. Metallic or heterostructure materials can facilitate the achievement of multilevel [146] and recyclable functions [147], respectively. Plasmonic incidences is strongly affected by the wavevector of incident light [95]. Therefore, prism-based light incidence systems are frequently used in plasmonic material sensing [137–139]. Photocatalyst modification is also frequently used in the study of plasmonic photocatalytic reactions and is discussed in Section 5.1.

Time-dependent property of plasmonic photocatalytic reactions is crucial for understanding plasmon progression after excitation, [19,22,35,37,38,103,148–150] as shown schematically in Figure 1. Near-field dipole-dipole interactions, light-scattering, and hot carrier responses to incident light combines as plasmonic optical response [19]. The plasmon dephasing time, as determined by radiative and nonradiative damping mechanisms, directly determines which step in the plasmon evolution is dominant in photocatalytic reaction enhancements. Metal nanoparticle reflection can be altered by the variation in plasmonic responses with particle size [37,38,103,148]. A coherent SP can be generated by introducing an incident photon source that is on-resonant with the extinction spectrum of a nanoparticle. SPs maintains its coherence for 1–10 fs and then starts to undergo energy conversion by decaying through either radiative or nonradiative pathways [151]. After the surface plasmon loses coherence, the nonradiative decay pathway produces a distribution of hot carriers that may be used in initiating photocatalytic reactions. This multistep decay process occurs primarily through Landau damping [36], where energy is transferred from a coherent plasmon to individual electron-hole pair excitations. Initially, the hot carrier distribution is nonthermal and contains charged species that are far from the Fermi level of the material [152,153]. Hot electrons and holes then rapidly undergo thermalization, reaching a Fermi-Dirac distribution that corresponds to a high effective electron temperature. This initial thermalization occurs through energy redistribution in electron-electron scattering interactions over the next several hundred femtoseconds (1–100 fs) [38,148]. During this time, the hot electrons and holes may contain energies ranging from the Fermi level to the work function. These charge carriers are sufficient in quantity and lifetime to initiate external chemical processes [152]. The hot carriers further dephase through an additional relaxation mechanism consisting of electron-phonon interactions over 1–10 ps [154,155]. During these two time intervals, the charge carriers may contain sufficient energy to transfer to a nearby chemical species to initiate a single or multistep chemical reaction. In a pump-probe measurement executed in a previous study using femtosecond diffuse reflectance spectroscopy, the transient absorption decay behavior of TiO_2 particles was determined to be independent of calcination temperature [156].

Many studies on mechanism of photocatalytic reaction enhancement have drawn conclusions after considering only a broad light absorption band of metal decorations. Such studies have not provided extra supporting experimental results. It is of concern that metal nanoparticles also responsible for accepting the light energy induced hot electrons from contacted photocatalysts (i.e., hot-electron injection model or chemical model of plasmonic photocatalytic reactions), thus reducing the recombination of the hot charge pairs.

5. Using Plasmonic Photocatalytic Reactions

Specially designed photocatalysts are typically examined using spectroscopy, selective photocatalytic reactions, pump-probe measurements, specially arranged light sources, and advanced reactors. Plasmonic light-to-heat conversion is also easily identified through temperature variation in reactions. This review considers only crucial emerging studies on plasmonic photocatalytic reactions that may provide suggestions for future studies.

5.1. Specially Fabricated Samples

Researchers have used various methods and materials to study and improve the photocatalytic processing efficiency of titanium dioxide, such as calcination temperature [156], TiO_x (x < 2) photocatalysts fabricated through flame aerosol synthesis [157], and oxygen vacancies obtained using plasma treatment [158]. Scavengers, such as EtOH [87], were used to reveal the role of hot charges in photocatalytic reactions. Bikondoa et al. [159] used a scanning tunneling microscope to directly visualize defect-mediated dissociation of water on TiO_2(110). They demonstrated that defects play a key role in oxide surface reactions. Binary photocatalytic nanoparticles may also have high photocatalytic processing efficiency. Graphene is highly conductive in two specific dimensions and is largely used like metal decoration in binary nanophotocatalysts [160–164]. For advanced methods improving photocatalytic reactions, several studies have suggested ternary nanophotocatalytst, such as Ag-TiO_2-graphene [165–167], Ag_2MoO_4-AgBr-Ag [104], Ag–rGO–Bi_2MoO_6 [105], Ag-AgCl-TiO_2 [168], Ag-AgCl-$(BiO)_2CO_3$ [106], ZnO-Ag-Ag_2WO_4 [107], Ag_2WO_4-AgBr-TiO_2 ternary nanocomposites [108], Au-$La_2Ti_2O_7$ [169], and Au-TiO_2-graphene [170]. Cu_7S_4-Pd hetero-nanostructures were also reported to exhibit near-infrared solar energy harvesting, leading to highly efficient photocatalytic reactions. Pd nanoparticles act as acceptors of light-generated holes in this system.

A noteworthy modification is the binary Pt decorated TiO_2 nanophotocatalyst, whose advantages involve the effective separation of the photogenerated electrons and holes [2,171,172]. The photogenerated electrons move from the valence band of TiO_2 to the conduction band of Pt (Figure 5a). The electrons flow through the external circuit to the Pt cathode in which water molecules are reduced to hydrogen gas; the holes remain in the TiO_2 anode in which water molecules are oxidized to oxygen. However, higher content-loaded Pt nanoparticles can absorb more incident photons that do not contribute to photocatalytic efficiency. The highest photocatalytic activity for the Pt-TiO_2 nanohybrids on MB could be achieved at 1 at % Pt loading [172]. The hot-electron injection model successfully explains the photocatalytic process of metal-decorated photocatalysts. However, in 2008, Awazu et al. [3] presented a "plasmonic photocatalyst" consisting of silver nanoparticles embedded in titanium dioxide. Their experimental results demonstrated that 30–100-nm silver nanoparticles can enhance processing efficiency under near-UV illumination even when covered by a SiO_2 layer. This signifies that metal nanoparticles can help deliver light energy by inducing a large localized electric field rather than solely mediating the accept light-generated electrons. This was a valuable breakthrough in plasmonic photocatalytic reactions. Composite nanoparticles of photocatalysts and higher plasmonic response metals of various sizes [173] and shapes (e.g., nanospheres [37,174], stars [37], and rods [174]) are recommended for advanced photocatalytic reactions. Array structures can be fabricated on a substrate to achieve high light absorption efficiency [7,91,111,175] to enhance light energy use in plasmonics. Adapting nanophotonics to photocatalysis introduces particularly slow photon effects and a strong light absorption and extinction band that enhances light harvesting [7]. Li et al. [91] fabricated TiO_2 nanorods selectively planted inside an Au nanohole array to enhance light capture and energy transfer to the plasmonic photocatalytic reaction through plasmon-induced resonant energy transfer.

In metal-photocatalyst heterojunctions, metal nanostructures may play multiple roles, including surface catalysis, surface passivation, Fermi level equilibration, and plasmonic enhancement [176]. The enhancement of photocatalytic activity or photocurrent is not solely due to the plasmonic enhancement effect of metal nanostructures. The effects of a metal nanostructure on plasmonic enhancement can be experimentally distinguished from other enhancement effects [176].

For future advanced applications in plasmonic photocatalytic reactions, a fabrication method that provides samples with stronger plasmon coupling and induction may induce a new breakthrough in plasmonic photocatalytic studies. Layered structures for creating plasmonic black absorber [93] and TiN materials for metal-like near-infrared plasmonics responses [177–182] are potential materials in relevant applications. Nanoparticles of various sizes and shapes [183,184] can achieve various plasmon modes for the target light absorption band. Dendritic nanoforests can achieve wide light absorption [185,186]. Both methods are also valuable for photochemical or photocatalytic applications.

In practical applications, it is difficult to identify a plasmonic enhancement mechanism apart trapping of light-excited hot electrons by the metal nanoparticle and longer recombination time with generated hot holes in metal-photocatalyst heterostructured nanomaterials. Therefore, it typically engenders ambiguity in determination of mechanisms underlying plasmonic enhancement in photocatalytic reactions.

5.2. Plasmonic Light-To-Heat Conversion

Light-to-heat energy conversion is a common concept, particularly under the warm sunlight at noon. When focused to a higher-order power density, this energy can even cut through steel. Light energy introduces VPs before being reflected by bulk metallic materials. However, "plasmonic" light-to-heat energy conversion is different, especially for nanomaterials [102]. Incident light excites SPs and LSPs on nanomaterials such as nanoparticles [187–191], nanorods [192], nanostars [193], nano particle arrays or networks [194,195], and nano-thin film [102] with an even higher electromagnetic field in the near-field zone. The effects of material, size, quantum, and structure result in a complex situation [102,189–191,193–197]. The size dependence of the quantum effect results in an even more complex situation when the size of the quantum well decreases to a few nanometers [102]. The harmless dosing power density for skin or tissue can be trapped or focus on a small area for detection and treatment in bioassay [140], bacterial [198], cancer cell [192,199–201], or as a nano or bulk cleavage [202]. Localized heat energy is strong and can be used to kill target cancer cells [203] while preserving the nearby cells [204].

A noteworthy application of plasmonic light-to-heat energy conversion is plasmonic polymerase chain reactions [205,206]. Enzymatic amplification of beta-globin genomic sequences, a process typically called a polymerase chain reaction, is crucial for various applications, including diagnosis, biomedical research, and criminal forensics. Most polymerase chain reaction methods rely on thermal heating and cooling cycles to induce DNA melting and enzyme-driven DNA replication. The treatment solution volume is normally small and is particularly suitable for MOFC reactor processing [207–217]. Plasmonic light-to-heat energy conversion contributed to a breakthrough ultrashort polymerase chain reaction cycle time of 30 cycles in 54 seconds [218]. This has induced more modifications with various plasmon generation methods, such as those involving the use of nanorods [219], Pt thin films [220], and Au thin films [221]. Plasmonic light-to-heat energy conversion is beneficial for rapid and precise heating in polymerase chain reaction.

Localized heating can be used in photocatalytic reactions [58,222,223]. Using a temperature-dependent photocatalytic measurement approach to acquire the activation enthalpy, Kim et al. [87] reported that reduction in activation enthalpy was directly related to the photoelectrochemical potential accumulation on the Au nanoparticle under steady-state light excitation, another phenomenon analogous to electrochemical activation. Plasmonic light-to-heat conversion is particularly crucial for endothermonic reactions such as ammonium decomposition in water. Ammonium decomposition is typically executed using wet air oxidization method at high temperatures and in a high-pressure sealed tank [224,225]. The process usually requires temperatures higher than 150 °C–200 °C to trigger efficient reactions. However, a large tank of sufficient treating volume is at risk of explosion and is characterized by high power consumption to maintain high pressure and temperature. A high-pressure environment is used to increase the reactant collision rate. Ammonium oxidization in an SDR mainly increases the mass transfer rate and collision rate of target reactants, which results in reactions under ambient conditions [58]. Providing an additional chemical-stabilized Pt thin film (8 nm) [226] could result in energy compensation in plasmonic endothermic reactions when compared with typical endothermic ammonium oxidization [223].

5.3. High-Efficiency Reactors

Slurry- and fixed-bed reactors are the widely used reactors in studies on plasmonic photocatalytic reactions. They have slightly different near-field optical responses (Figure 5a,b) that might affect plasmonic enhancement of photocatalytic reactions. Light is scattered in the elastic or inelastic decay path depending on the various matching conditions. However, typical processing setups do not consider optics for enhancing or identifying the mechanism underlying the "plasmonic enhancement" of photocatalytic reactions. Some reactors have features that enhance plasmonic responses in plasmonic photocatalytic reactions. In the study of plasmonics, the wavevector matches the plasmonic features (i.e., VPs, SPs, LSPs) and have substantial optical responses. In slurry-bed reactors, the wavevector of light is $k_{x-water} = sin\theta n_{water} \lambda$ which is the product of the water refractive index n_{water} and light wavelength λ. In this case, $sin\theta$ is typically 1 under conditions of diffuse scattering. This limits the wavevector of a specific wavelength of light to a fixed value. Light with a suitable "short" wavelength, typically UV light, for photocatalyst activation may not match the coupling condition of plasmonic features (i.e., size of the metallic decorations on a heterostructure photocatalyst). Therefore, the difference in plasmonic enhancement changed through the scattering and trapping path cannot be clearly observed. Only "long" wavelength lights, typically exhibiting visible or near-infrared absorption peaks, can couple to the decorated metallic nanoparticles and generate heat. The generated heat enhances the corresponding chemical reactions but the enhancement effect is not significant. Furthermore, the diffuse reflection of incident light by a high-density suspended photocatalyst substantially reduces the energy used in photocatalytic reactions. Diffuse light scattering can increase the chances of light absorption by photocatalysts but it may engender light decay as long-distance travel in water.

Fiber reactors [227–229] have optical advantages of delivering light with high efficiency. Glass optical fibers can deliver incident light to photocatalysts without substantial decay under diffuse scattering in water. Incident light from the glass side efficiently induces SPs and LSPs in high efficiency (Figure 5c,d). The wavevector of light propagating in an optical fiber is $k_{x-glass} = sin\theta n_{glass} \lambda$, where λ is the same as that in water; $n_{glass} \sim 1.5 > n_{water} \sim 1.3$ and only slightly increases the wavevector. However, light propagating in the optical fiber can illuminate the photocatalyst fixed on the optical fiber various angles (θ) from within the glass component. Therefore, the variation in the incident light wavevector can increase the variation in the wavevector for a specific light wavelength. The incident light can have a wavevector that matches the coupling condition of plasmon generation. This increases the probability of light inducing photocatalytic reactions on the photocatalyst or inducing enhanced plasmonic responses on the decorated metallic nanoparticles. Plasmonic scattering and trapping, PIRET, QHCs and heat transformation (Figure 1) can all have enhanced responses.

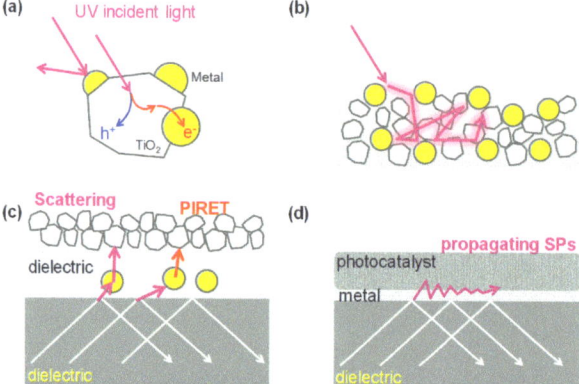

Figure 5. Near-field optical response of (**a**) binary heterogeneous photocatalyst, (**b**) binary polycrystal photocatalyst deposited on substrate, (**c**) embedded metal nanoparticles beneath the deposited polycrystal photocatalyst, and (**d**) metal thin film beneath the deposited photocatalyst layer. High-wavevector lights illuminating the substrate from below induce substantial coupling of light energy in (**c**) and (**d**).

Reactors with waveguide features or the potential to couple with light for an increased wavevector variation are particularly suitable for studying plasmonic photocatalytic reactions. Fixed-bed or MOFC reactors [71,81,85] with thin and flat glass substrates are also suitable for use as waveguides for delivering light energy. These reactors might help studies on plasmonic photocatalytic reactions. Light-transparent substrate-based SDRs are also crucial for high-efficiency plasmonic photocatalytic reaction studies. The degradation of methyl orange in water using a dual-light-source SDR (DL-SDR) and glass-embedded diffusion coupler demonstrated the mechanism underlying the enhancement of plasmonic photocatalytic reaction [24]. When visible light (637 ± 8 nm) from a spindle LED propagated in the circular polycarbonate disk waveguide of a DL-SDR, it gradually lost energy as absorbed by the Au-TiO$_2$ double-sphere nanoparticles deposited on top. This absorption enhanced the processing efficiency of the plasmonic photocatalytic reaction excited by the upper 254-nm light from a low-pressure mercury tube lamp. The additional visible light resulted in no difference in photocatalytic reactions with only TiO$_2$ nanoparticles. An experiment with multilayer reactors of aluminum disks and both visible and UV light from above reached the same conclusion [56]. Red visible light can induce plasmon on 20-nm Au nanoparticles but not in photocatalytic reactions. This is evident of plasmonic enhancement in photocatalytic reactions.

SDRs could enhance mass transfer and result in ammonium oxidization at room temperature [58] through the use of a bare-glass disk that typically executed using wet air oxidization method in environment of high temperature and high pressure [224,225]. Depositing an additional 8-nm platinum thin film using plasma-enhanced atomic layer deposition (PE-ALD) [226] has been report to further introduced plasmonic-enhanced photocatalytic ammonium oxidization in water [58,223]. The energy compensated in the plasmonic photocatalytic oxidization of ammonium is a quantized phenomenon independent of the incident light power density [94,230]. The quantized energy transformation [102] originates from the fixed quantum wavelength of the excited QHCs discussed in Section 5.2.

6. Models of Plasmonic Photocatalytic Reactions

In early studies, decorated metal nanoparticles were found to improve the photocatalytic processing efficiency of TiO$_2$. This can be explained by the lower energy level of metal conduction band than the excited band of photocatalyst that results in direct electron transfer (DET) model. However, it cannot sufficiently explain all the enhancement effects in photocatalytic reactions. The studies of optical plasmonic effect start at around 1960s and booming at 1990s provide a substantial explanation of the enhanced processing efficiency of metal-photocatalyst composited nanomaterials. Typically, a chemically plasmonic or physically plasmonic enhancement of photocatalytic reactions are related to include the migration of induced hot charges or process through "electromagnetic field-enhancement," respectively. Various energy transformation paths discussed above can enhance the photocatalytic reaction in chemically plasmonic or physically plasmonic enhancement pathways, as shown in Figure 6.

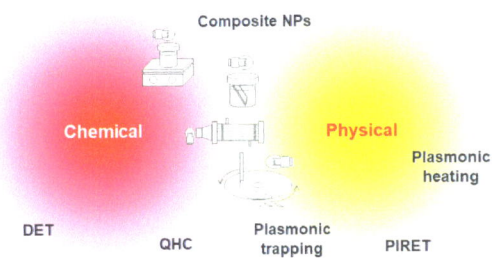

Figure 6. Chemically and physically plasmonic enhancement models of various types of energy transformation in plasmonic photocatalytic reactions.

In the metal-photocatalyst composited nanoparticles, the migrations of the ultraviolet light-induced hot-electron on the photocatalyst and plasmon-generated quantum hot electrons on the nano-metal results in reduced charge recombination rate. These are chemically plasmonic enhancements of photocatalytic reactions that need composite nanoparticles. The chemically plasmonic-enhanced photocatalytic reactions have relatively small facility-dependence and can conveniently process in various kinds of reactors.

The physically plasmonic enhancements are relatively tricky to identify clearly from the chemically plasmonic enhancements in photocatalytic reactions. Plasmonic scattering and trapping of the incident light, PIRET, and plasmonic heating were typically observed physically plasmonic enhancements in the specially arranged experiments. It usually needs some specially fabricated samples or specially arranged experimental setup to reveal the physically plasmonic enhancement in photocatalytic reactions. An insulator spacer in specially manufactured samples blocked the contact between photocatalyst and metal nanoparticles that also stop the electron migration [3]. In dual-light-source experiments [24,56], longer wavelength light has lower energy that not able to induce hot charges carriers in the photocatalysts. This can also demonstrate an extra physically plasmonic enhancement in the composited nanoparticles of no insulator spacer.

Normally, chemically plasmonic enhancements are undoubtedly processing in most of the photocatalytic reactions by metal-photocatalyst composite material. The studies in specially arranged experimental setups aiming at physically plasmonic enhancements are usually only for scientific studies. In practical applications, the plasmonic photocatalytic reactions are looking for the highest processing efficiency. It needs balanced effects from chemically plasmonic and physically plasmonic enhancements of photocatalytic reactions that also needs specially designed experimental setup. A specially designed experimental setup with high-efficiency reactor and optimized light energy delivering can thus provide further enhanced plasmonic photocatalytic reactions.

7. Conclusions

Studies on plasmonic enhancement of photocatalytic reactions are increasing in the recent years. Methods and reactors for processing the photocatalytic reactions are abundant. However, the mechanism underlying the plasmonic enhancement of photocatalytic reaction remains confusing.

Identifying a specific factor or mechanism including the greatest effects in plasmonic enhancements is difficult. This difficulty can be attributed to the complicated plasmonic optical responses as well as the complicated microchemical processes in photocatalytic reactions. The aim of this review is to highlight experimental methods used in studies in plasmonic photocatalytic reactions over the past decade and provide identifiable alternative mechanisms for theoretical modeling. To simplify plasmonic effect modeling in photocatalytic reactions, plasmonic photocatalytic reactions are defined as reactions enhanced through the light-induced generation of plasmons. The advantages of introducing plasmons in photocatalytic reactions are (i) strong light absorption, (ii) intensive far-field light-scattering, (iii) electromagnetic field strengthening in near-field zone, (iii) abundant hot carrier generation, and (iv) plasmonic heating effects. These inspired substantial research on improving the processing efficiency of photocatalytic reactions. The mechanisms of plasmonic enhancement in photocatalytic reaction systems is gradually becoming clear but remains vague in practice. This review may help to identify the plasmonic enhancement path in photocatalytic reaction from an instrumentation perspective.

Author Contributions: Conceptualization, H.J.H., J.C.-S.W., H.-P.C., Y.-F.C.C., and Y.-S.L.; reference collection, H.J.H., Y.H.W. and P.-J.C.; original draft preparation, H.J.H.; review and editing, H.J.H., J.C.-S.W., H.-P.C., Y.-F.C.C., Y.-S.L., Y.H.W. and P.-J.C. All authors have read and agreed to the published version of the manuscript.

Funding: This research was funded by the Ministry of Science and Technology of Taiwan under the project number MOST 107-2622-E-492-022-CC3, MOST 108-2112-M-492-001, MOST 108-2221-E-239-005, MOST 108-2622-E-492-023-CC3, MOST 106-2112-M-019-005-MY3, MOST 108-2221-E-002-111-MY3, 108-2119-M-002-027, University Research Grant of Universiti Brunei Darussalam (Grant No. UBD/OAVCRI/CRGWG (004) /170101), and Taiwan Academic Sinica under the project number AS-KPQ-106-DDPP.

Conflicts of Interest: The authors declare no conflict of interest. The funders had no role in the design of the study; in the collection, analyses, or interpretation of data; in the writing of the manuscript, or in the decision to publish the results.

Abbreviations

The following abbreviations are used in this manuscript:

VPs	Volume Plasmons
SPs	Surface Plasmons
LSPs	Localized Surface Plasmons
PIRET	Plasmon-Induced Energy Transformation
QHCs	Quantum Hot Charge carriers
SDR	Spinning Disk Reactor
MOFC	Micro Opto-Fluidic Chip
PL	Photo luminescence

References

1. Fujishima, A.; Honda, K. Electrochemical Photolysis of Water at a Semiconductor Electrode. *Nature* **1972**, *238*, 37–38. [CrossRef]
2. Linsebigler, A.L.; Lu, G.; Yates, Jr., J.T.; Photocatalysis on TiO_n Surfaces: Principles, Mechanisms, and Selected Results. *Chem. Rev.* **1995**, *95*, 735–758. [CrossRef]
3. Awazu, K.; Fujimaki, M.; Rockstuhl, C.; Tominaga, J.; Murakami, H.; Ohki, Y.; Yoshida, N.; Watanabe, T. A Plasmonic Photocatalyst Consisting of Silver Nanoparticles Embedded in Titanium Dioxide. *J. Am. Chem. Soc.* **2008**, *130*, 1676–1680. [CrossRef]
4. Grabowska, E. Selected perovskite oxides: Characterization, preparation and photocatalytic properties—A review. *Appl. Catal. B-Environ.* **2016**, *186*, 97–126. [CrossRef]
5. Li, D.; Shi, W. Recent developments in visible-light photocatalytic degradation of antibiotics. *Chin. J. Catal.* **2016**, *37*, 792–799. [CrossRef]
6. Wenderich, K.; Mul, G. Methods, Mechanism, and Applications of Photodeposition in Photocatalysis: A Review. *Chem. Rev.* **2016**, *116*, 14587–14619. [CrossRef]

7. Pietron, J.J.; DeSario, P.A. Review of roles for photonic crystals in solar fuels photocatalysis. *J. Photon. Energy* **2016**, *7*, 1–12. [CrossRef]
8. Singh, S.; Srivastava, V.C.; Lo, S.L. Surface Modification or Doping of WO_3 for Enhancing the Photocatalytic Degradation of Organic Pollutant Containing Wastewaters: A Review. *Mater. Sci. Forum* **2016**, *855*, 105–126. [CrossRef]
9. Lam, S.-M.; Sin, J.-C.; Mohamed, A.R. A review on photocatalytic application of g-C_3N_4/semiconductor (CNS) nanocomposites towards the erasure of dyeing waste water. *Mat. Sci. Semicon. Proc.* **2016**, *47*, 62–84. [CrossRef]
10. Devi, L.G.; Kavitha, R. A review on plasmonic metal-TiO_2 composite for generation, trapping, storing and dynamic vectorial transfer of photogenerated electrons across the Schottky junction in a photocatalytic system. *Appl. Surf. Sci.* **2016**, *360*, 601–622. [CrossRef]
11. Ni, Z.; Sun, Y.; Zhang, X.; Dong, F. Fabrication, modification and application of $(BiO)_2CO_3$-based photocatalysts: A review. *Appl. Surf. Sci.* **2016**, *365*, 314–335. [CrossRef]
12. Narang, P.; Sundararaman, R.; Atwater, H.A. Plasmonic hot carrier dynamics in solid-state and chemical systems for energy conversion. *Nanophotonics* **2016**, *5*, 96–111. [CrossRef]
13. Low, J.; Cheng, B.; Yu, J. Surface modification and enhanced photocatalytic CO_2 reduction performance of TiO_2: A review. *Appl. Surf. Sci.* **2017**, *392*, 658–686. [CrossRef]
14. Boyjoo, Y.; Sun, H.; Liu, J.; Pareek, V.K.; Wang S. A review on photocatalysis for air treatment: From catalyst development to reactor design. *Chem. Eng. J.* **2017**, *310*, 537–559. [CrossRef]
15. Reddy, P.V.L.; Kavitha, B.; Reddy, P.A.K.; Kim, K.-H. TiO_2-based photocatalytic disinfection of microbes in aqueous media: A review. *Environ. Res.* **2017**, *154*, 296–303. [CrossRef]
16. Wacławek, S.; Padil, V.V.T.; Černík, M. Major advances and challenges in heterogeneous catalysis for environmental applications: a review. *Ecolog. Chem. Eng. S* **2018**, *25*, 9–34. [CrossRef]
17. Reddy, P.V.L.; Kim, K.-H.; Kavitha, B.; Kumar, V.; Raza, N.; Kalagara, S. Photocatalytic degradation of bisphenol A in aqueous media: A review. *J. Environ. Manag.* **2018**, *213*, 189–205. [CrossRef]
18. Pirhashemi, M.; Habibi-Yangjeh, A.; Pouran, S.R. Review on the criteria anticipated for the fabrication of highly efficient ZnO-based visible-light-driven photocatalysts. *J. Ind. Eng. Chem.* **2018**, *62*, 1–25. [CrossRef]
19. Wu, N. Plasmonic metal–semiconductor photocatalysts and photoelectrochemical cells: a review. *Nanoscale* **2018**, *10*, 2679–2696. [CrossRef]
20. Brooks, J.L.; Warkentin, C.L.; Saha, D.; Keller, E.L.; Frontiera, R.R. Toward a mechanistic understanding of plasmon-mediated photocatalysis. *Nanophotonics-Berlin* **2018**, *7*, 1697–1724. [CrossRef]
21. Verbruggen, S.W. TiO_2 photocatalysis for the degradation of pollutants in gas phase: From morphological design to plasmonic enhancement. *J. Photochem. Photobiol. C* **2015**, *24*, 64–82. [CrossRef]
22. Wang, F.; Jin, Z.; Jiang, Y.; Backus, E.H.G.; Bonn, M.; Lou, S.N.; Turchinovich, D.; Amala, R. Probing the charge separation process on In_2S_3/Pt-TiO_2 nanocomposites for boosted visible-light photocatalytic hydrogen production. *Appl. Catal. B-Environ.* **2016**, *198*, 25–31. [CrossRef]
23. Mukherjee, S.; Libisch, F.; Large, N.; Neumann, O.; Brown, L.V.; Cheng, J.; Lassiter, J.B.; Carter, E.A.; Nordlander, P.; Halas, N.J. Hot electrons do the impossible: plasmon-induced dissociation of H_2 on Au. *Nano Lett.* **2013**, *13*, 240–247. [CrossRef]
24. Huang, H.J.; Huang, K.-C.; Tsai, D.P. Light absorption measurement of a plasmonic photocatalyst in the circular plane waveguide of a photocatalytic dual light source spinning disk reactor. *Opt. Rev.* **2013**, *20*, 236–240. [CrossRef]
25. Cui, J.; Li, Y.; Liu, L.; Chen, L.; Xu, J.; Ma, J.; Fang, G.; Zhu, E.; Wu, H.; Zhao, L.; Wang, L.; Huang, Y. Near-infrared plasmonic-enhanced solar energy harvest for highly efficient photocatalytic reactions. *Nano Lett.* **2015**, *15*, 6295–6301. [CrossRef]
26. Lin, W.-C.; Jen, H.-C.; Chen, C.-L.; Hwang, D.-F.; Chang, R.; Hwang, J.-S.; Chiang, H.-P. SERS study of tetrodotoxin (TTX) by using silver nanoparticle arrays. *Plasmonics* **2016**, *4*, 187–192. [CrossRef]
27. Lin, W.-C.; Huang, S.-H.; Chen, C.-L.; Chen, C.-C.; Tsai, D.P.; Chiang, H.-P. SERS study of tetrodotoxin (TTX) by using silver nanoparticle arrays. *Appl. Phys. A* **2010**, *101*, 185–189. [CrossRef]
28. Raether, H. In *Surface Plasmons on Smooth and Rough Surfaces and on Gratings*; Springer: Berlin/Heidelberg, Germany, 1988.
29. Hecht, E. Dispersion. In *Optics*, 4th ed.; Pearson Education, Inc.: San Francisco, CA, USA, 2002; pp. 67–73.

30. Tsai, D.P.; Kovacs, J.; Wang, Z.; Moskovits, M.; Shalaev, V.M.; Suh, J.S.; Botet, R. Photon scanning tunneling microscopy images of optical excitations of fractal metal colloid clusters. *Phys. Rev. Lett.* **1994**, *72*, 4149–4152. [CrossRef]
31. Govorov, A.O.; Zhang, H.; Gunko, Y.K. Theory of photoinjection of hot plasmonic carriers from metal nanostructures into semiconductors and surface molecules. *J. Phys. Chem. C* **2013**, *117*, 16616–16631. [CrossRef]
32. Schaadt, D.; Feng, B.; Yu, E. Enhanced semiconductor optical absorption via surface plasmon excitation in metal nanoparticles. *Appl. Phys. Lett.* **2005**, *86*, 063106. [CrossRef]
33. Derkacs, D.; Lim, S.; Matheu, P.; Mar, W.; Yu, E. Improved performance of amorphous silicon solar cells via scattering from surface plasmon polaritons in nearby metallic nanoparticles. *Appl. Phys. Lett.* **2006**, *89*, 093103. [CrossRef]
34. Cushing, S.K.; Li, J.; Meng, F.; Senty, T.R.; Suri, S.; Zhi, M.; Li, M.; Bristow, A.D.; Wu, N. Photocatalytic Activity Enhanced by Plasmonic Resonant Energy Transfer from Metal to Semiconductor. *J. Am. Chem. Soc.* **2012**, *134*, 15033–15041. [CrossRef]
35. Li, J.; Cushing, S.K.; Meng, F.; Senty, T.R.; Bristow, A.D.; Wu, N. Plasmon-induced resonance energy transfer for solar energy conversion. *Nat. Photonics* **2015**, *9*, 601–607. [CrossRef]
36. Li, X.; Xiao, D.; Zhang, Z. Landau damping of quantum plasmons in metal nanostructures. *New J. Phys.* **2013**, *15*, 023011. [CrossRef]
37. Sousa-Castillo, A.; Comesaña-Hermo, M.; Rodríguez-González, B.; Pérez-Lorenzo, M.; Wang, Z.; Kong, X.-T.; Govorov, A.O.; Correa-Duarte, M.A. Boosting hot electron-driven photocatalysis through anisotropic plasmonic nanoparticles with hot spots in Au-TiO$_2$ nanoarchitectures. *J. Phys. Chem. C* **2016**, *120*, 11690–11699. [CrossRef]
38. Sönnichsen, C.; Franzl, T.; Wilk, T.; Plessen, G.V.; Feldmann, J.; Wilson, O.; Mulvaney, P. Drastic reduction of plasmon damping in gold nanorods. *Phys. Rev. Lett.* **2002**, *88*, 077402. [CrossRef]
39. Tian, Y.; Tatsuma, T. Plasmon-induced photoelectrochemistry at metal nanoparticles supported on nanoporous TiO$_2$. *Chem. Commun.* **2004**, *16*, 1810–1811. [CrossRef]
40. Tian, Y.; Tatsuma, T. Mechanisms and Applications of Plasmon-Induced Charge Separation at TiO$_2$ Films Loaded with Gold Nanoparticles. *J. Am. Chem. Soc.* **2005**, *127*, 7632–7637. [CrossRef]
41. Mubeen, S1.; Lee, J.; Singh, N.; Krämer, S.; Stucky, G.D.; Moskovits, M. An autonomous photosynthetic device in which all charge carriers derive from surface plasmons. *Nat. Nanotechnol.* **2013**, *8*, 247–251. [CrossRef]
42. Zhang H.; Govorov, A.O. Optical Generation of Hot Plasmonic Carriers in Metal Nanocrystals: The Effects of Shape and Field Enhancement. *J. Phys. Chem. C* **2014**, *118*, 7606–7614. [CrossRef]
43. Govorov, A.O.; Zhang, H.; Demir H.V.; Gunko, Y.K. Photogeneration of hot plasmonic electrons with metal nanocrystals: Quantum description and potential applications. *Nano Today* **2014**, *9*, 85–101. [CrossRef]
44. Kumarasinghe, C.S.; Premaratne, M.; Bao, Q.; Agrawal, G.P. Theoretical analysis of hot electron dynamics in nanorods. *Sci. Rep.* **2015**, *5*, 12140. [CrossRef]
45. Kumarasinghe, C.S.; Premaratne, M.; Gunapala, S.D.; Agrawal, G.P. Theoretical analysis of hot electron injection from metallic nanotubes into a semiconductor interface. *Phys. Chem. Chem. Phys.* **2016**, *18*, 18227–18236. [CrossRef]
46. Lo, C.-C.; Huang, C.-W.; Liao, C.-H.; Wu, J.C.S. Novel twin reactor for separate evolution of hydrogen and oxygen in photocatalytic water splitting. *I. J. Hydrogen Energ.* **2010**, *35*, 1523–1529. [CrossRef]
47. Yu, S.-C.; Huang, C.-W.; Liao, C.-H.; Wu, J.C.S.; Chang, S.-T.; Chen, K.-H. A novel membrane reactor for separating hydrogen and oxygen in photocatalytic water splitting. *J. Membrane Sci.* **2011**, *382*, 291–299. [CrossRef]
48. Lee, W.-H.; Liao, C.-H.; Tsai, M.-F.; Huang, C.-W.; Wu, J.C.S. A novel twin reactor for CO$_2$ photoreduction to mimic artificial photosynthesis. *Appl. Catal. B-Environ.* **2013**, *132–133*, 445–451. [CrossRef]
49. Nguyen, V.-H.; Wu, J.C.S. Recent developments in the design of photoreactors for solar energy conversion from water splitting and CO$_2$ reduction. *Appl. Catal. A-Gen.* **2018**, *550*, 122–141. [CrossRef]
50. Chen, J.-J.; Wu, J.C.S.; Wu, P.C.; Tsai, D.P. Plasmonic photocatalyst for H$_2$ evolution in photocatalytic water splitting. *J. Phys. Chem. C* **2010**, *115*, 210–216. [CrossRef]
51. Ray, A.K.; Beenackers, A.A.C.M. Novel photocatalytic reactor for water purification. *AIChE J.* **1998**, *44*, 447–483. [CrossRef]

52. Marinangeli, R.E.; Ollis, D.F. Photo-assisted heterogeneous catalysis with optical fibers II. Nonisothermal single fiber and fiber bundle. *AIChE J.* **1980**, *26*, 1000–1008. [CrossRef]
53. Hofstadler, K.; Bauer, R.; Novalic, S.; Heisler, G. New reactor design for photocatalytic wastewater treatment with TiO_2 immobilized on fused-silica glass fibers: photomineralization of 4-chlorophenol. *Environ. Sci. Technol.* **1994**, *28*, 670–674. [CrossRef]
54. Peill, N.J.; Hoffmann, M.R. Development and optimization of a TiO_2-coated fiber-optic cable reactor: photocatalytic degradation of 4-chlorophenol. *Environ. Sci. Technol.* **1995**, *29*, 2974–2981. [CrossRef]
55. Yatmaz, H.C.; Wallis, C.; Howarth, C.R. The spinning disc reactor–Studies on a novel TiO_2 photocatalytic reactor. *Chemosphere* **2001**, *42*, 397–403. [CrossRef]
56. Lin, C.-N.; Chang, C.-Y.; Huang, H.J.; Tsai, D.P.; Wu, N.-L. Photocatalytic degradation of methyl orange by a multi-layer rotating disk reactor. *Environ. Sci. Pollut. R.* **2012**, *19*, 3743–3750. [CrossRef]
57. Chen, Y.L.; Kuo, L.-C.; Tseng, M.L.; Chen, H.M.; Chen, C.-K.; Huang, H.J.; Liu, R.-S.; Tsai, D.P. ZnO nanorod optical disk photocatalytic reactor for photodegradation of methyl orange. *Opt. Express* **2013**, *21*, 7240–7249. [CrossRef]
58. Huang, H.J.; Liu, B.-H.; Yeh, J.A. Ammonium oxidization at room temperature and plasmonic photocatalytic enhancement. *Catal. Comm.* **2013**, *36*, 16–19. [CrossRef]
59. Xu, X.; Lu, H.; Qian, Y.; Zhang, B.; Wang, H.; Liu, H.; Yang, Q. Gas–liquid mass transfer and bubble size distribution in a multi-cyclone separator. *AIChE J.* **2019**, *65*, 215–223. [CrossRef]
60. Marchetti, A.; Stoller, M. On the micromixing behavior of a spinning disk reactor for metallic Cu nanoparticles production. *Appl. Sci.* **2019**, *9*, 3311. [CrossRef]
61. Dell'Era, A.; Scaramuzzo, F.A.; Stoller, M.; Lupi, C.; Rossi, M.; Passeri, D.; Pasquali, M. Spinning disk reactor technique for the synthesis of nanometric sulfur TiO_2 core–shell powder for lithium batteries. *Appl. Sci.* **2019**, *9*, 1913. [CrossRef]
62. Sczechowski, J.G.; Koval, C.A.; Noble, R.D. A Taylor vortex reactor for heterogeneous photocatalysis. *Chem. Eng. Sci.* **1995**, *50*, 3163–3173. [CrossRef]
63. Sengupta, T.K.; Kabir, M.F.; Ray, A.K. A Taylor Vortex Photocatalytic Reactor for Water Purification. *Ind. Eng. Chem. Res.* **2001**, *40*, 5268–5281. [CrossRef]
64. Dutta, P.K.; Ray, A.K. Experimental investigation of Taylor vortex photocatalytic reactor for water purification. *Chem. Eng. Sci.* **2004**, *59*, 5249–5259. [CrossRef]
65. Masuda, H.; Yoshida, S.; Horie, T.; Ohmura, N.; Shimoyamada, M. Flow dynamics in Taylor–Couette flow reactor with axial distribution of temperature. *AIChE J.* **2018**, *64* 1075–1082. [CrossRef]
66. Wilkinson, N.; Dutcher, C. Axial mixing and vortex stability to in situ radial injection in Taylor–Couette laminar and turbulent flows. *J. Fluid Mech.* **2018**, *854*, 324–347. [CrossRef]
67. Jafarikojour, M.; Dabir, B.; Sohrabi, M.; Royaee, S.J. Application of a new immobilized impinging jet stream reactor for photocatalytic degradation of phenol: Reactor evaluation and kinetic modelling. *J. Photochem. Photobiol. A* **2018**, *364*, 613–624. [CrossRef]
68. Cagney, N.; Balabani, S. Taylor-Couette flow of shear-thinning fluids featured. *Phys. Fluids* **2019**, *31*, 053102. [CrossRef]
69. AlAmer, M.; Lim, A.R.; Joo, Y.L. Continuous synthesis of structurally uniform graphene oxide materials in a model Taylor–Couette flow reactor. *Ind. Eng. Chem. Res.* **2019**, *58*, 1167–1176. [CrossRef]
70. Liu, A.Q.; Huang, H.J.; Chin, L.K.; Yu, Y.F.; Li, X.C. Label-free detection with micro optical fluidic systems (MOFS): a review. *Anal. Bioanal. Chem.* **2008**, *391*, 2443–2452. [CrossRef]
71. Lei, N. Wang, X. M.; Zhang, Q.; Tai, D.P.; Tsai, Q.; Chan, H.L.W. Optofluidic planar reactors for photocatalytic water treatment using solar energy. *Biomicrofluidics* **2010**, *4*, 043004. [CrossRef]
72. Parmar, J.; Jang, S.; Soler, L.; Kim, D.P.; Sánchez, S. Nano-photocatalysts in microfluidics, energy conversion and environmental applications. *Lab Chip.* **2015**, *15*, 2352–2356. [CrossRef]
73. Meng, Z.; Zhang, X.; Qin, J. A high efficiency microfluidic-based photocatalytic microreactor using electrospun nanofibrous TiO_2 as a photocatalyst. *Nanoscale* **2013**, *5*, 4687–4690. [CrossRef]
74. Huang, X.; Wang, J.; Li, T.; Wang, J.; Xu, M.; Yu, W.; Abed, A.E.; Zhang, X. Review on optofluidic microreactors for artificial photosynthesis. *Beilstein J. Nanotechnol.* **2018**, *9*, 30–41. [CrossRef]
75. Azzouz, I.; Habba, Y.G.; Capochichi-Gnambodoe, M.; Marty, F.; Vial, J.; Leprince-Wang, Y.; Bourouina, T. Zinc oxide nano-enabled microfluidic reactor for water purification and its applicability to volatile organic compounds. *Microsyst. Nanoeng.* **2018**, *4*, 17093. [CrossRef]

76. Pradhan, S.R.; Colmenares-Quintero, R.F.; Quintero, J.C.C. Designing microflowreactors for photocatalysis using sonochemistry: a systematic review article. *Molecules* **2019**, *24*, 3315. [CrossRef]
77. Taylor, R.; Coulombe, S.; Otanicar, T.; Phelan, P.; Gunawan, A.; Lv, W.; Rosengarten, G.; Prasher, R.; Tyagi, H. Small particles, big impacts: a review of the diverse applications of nanofluids. *J. Appl. Phys.* **2013**, *113*, 011301. [CrossRef]
78. Minzioni, P.; Osellame, R.; Sada, C.; Zhao, S.; Omenetto, F.G.; Gylfason, K.B.; Haraldsson, T.; Zhang, Y.; Ozcan, A.; Wax, A.; et al. Roadmap for optofluidics. *J. Opt.* **2017**, *19*, 093003. [CrossRef]
79. Ozcelik, D.; Cai, H.; Leake, K.D.; Hawkins, A.R.; Schmidt, H. Optofluidic bioanalysis: fundamentals and applications. *Nanophotonics* **2017**, *6*, 647–661. [CrossRef]
80. Wang, N.; Tan, F.; Wan, L.; Wu, M.; Zhang, X. Microfluidic reactors for visible-light photocatalytic water purification assisted with thermolysis. *Biomicrofluidics* **2014**, *8*, 054122. [CrossRef]
81. Özbakır Y, Jonáš A, Kiraz A, Erkey C. A new type of microphotoreactor with integrated optofluidic waveguide based on solid-air nanoporous aerogels. *R. Soc. Open Sci.* **2018**, *5*, 180802. [CrossRef]
82. Lindstrom, H.; Wootton, R.; Iles, A. High surface area titania photocatalytic microfluidic reactors. *AIChE J.* **2007**, *53*, 695–702. [CrossRef]
83. Wang, N.; Tan, F.; Zhao, Y.; Tsoi, C.C.; Fan, X.; Yu, W.; Zhang, X. Optofluidic UV-Vis spectrophotometer for online monitoring of photocatalytic reactions. *Sci. Rep.* **2016**, *6*, 28928. [CrossRef]
84. Li, Y.; Lin, B.; Ge, L.; Guo, H.; Chen, X.; Lu, M. Real-time spectroscopic monitoring of photocatalytic activity promoted by graphene in a microfluidic reactor. *Sci. Rep.* **2016**, *6* 28803. [CrossRef]
85. Huang, H.J.; Wang, Y.H.; Chau, Y.-F.C.; Chiang, H.-P.; Wu, J.C.-S. Magnetic field-enhancing photocatalytic reaction in micro optofluidic chip reactor. *Nanoscale Res. Lett.* **2019**, *14*, 323. [CrossRef]
86. Lin, C.-T.; Huang, H.J.; Yang, J.-J.; Shiao, M.-H. A simple fabrication process of Pt-TiO_2 hybrid electrode for photo-assisted methanol fuel cells. *Microelectron. Eng.* **2011**, *88*, 2644–2646. [CrossRef]
87. Kim, Y.; Torres, D.D.; Jain, P.K. Activation energies of plasmonic catalysts. *Nano Lett.* **2016**, *16*, 3399–3407. [CrossRef]
88. Lin, C.-T.; Shiao, M.-H.; Chang, M.-N.; Chu, N.; Chen, Y.-W.; Peng, Y.-H.; Liao, B.-H.; Huang, H.J.; Hsiao, C.-N.; Tseng, F.-G. A facile approach to prepare silicon-based Pt-Ag tubular dendritic nano-forests (tDNFs) for solar-light-enhanced methanol oxidation reaction. *Nanoscale Res. Lett.* **2015**, *10*, 74. [CrossRef]
89. Lin, C.-T.; Chang, M.-N.; Huang, H.J.; Chen, C.-H.; Sun, R.-J.; Liao, B.-H.; Chau, Y.-F.C.; Hsiao, C.-N.; Shiao, M.-H.; Tseng, F.-G. Rapid fabrication of three-dimensional gold dendritic nanoforests for visible light-enhanced methanol oxidation. *Electrochim. Acta* **2016**, *192*, 15–21. [CrossRef]
90. Shiao, M.-H.; Lin, C.-T.; Huang, H.J.; Chen, P.-H.; Liao, B.-H.; Tseng, F.-G.; Lin, Y.-S. Novel gold dendritic nanoflowers deposited on titanium nitride for photoelectrochemical cells. *J. Solid State Electr.* **2018**, *22*, 3077–3084. [CrossRef]
91. Li, J.; Cushing, S.K.; Zheng, P.; Meng, F.; Chu, D.; Wu, N. Plasmon-induced photonic and energy-transfer enhancement of solar water splitting by a hematite nanorod array. *Nat. Commun.* **2013**, *4*, 2651. [CrossRef]
92. Lee, J.; Mubeen, S.; Ji, X.; Stucky, G.D.; Moskovits, M. Plasmonic photoanodes for solar water splitting with visible light. *Nano Lett.* **2012**, *12*, 5014–5019. [CrossRef]
93. Tan, F.; Wang, N.; Lei, D.Y.; Yu, W.; Zhang, X. Plasmonic black absorbers for enhanced photocurrent of visible-light photocatalysis. *Adv. Opt. Mater.* **2017**, *5*, 1600399. [CrossRef]
94. Huang, H.J.; Liu, B.H.; Lin, C.-T.; Su, W.-S. Plasmonic photocatalytic reactions enhanced by hot electrons in a one-dimensional quantum well. *AIP Adv.* **2015**, *5*, 117224.
95. Huang, H.J.; Yu, C.P.; Chang, H.C.; Chiu, K.P.; Liu, R.S.; Tsai, D.P. Plasmonic optical properties of a single gold nano-rod. *Opt. Express* **2007**, *15*, 7132–7139. [CrossRef]
96. Chirea, M.; Collins, S.S.; Wei, X.; Mulvaney, P. Spectroelectrochemistry of silver deposition on single gold nanocrystals. *J. Phys. Chem. Lett.* **2014**, *5*, 4331–4335. [CrossRef]
97. Hoener, B.S.; Zhang, H.; Heiderscheit, T.S.; Kirchner, S.R.; Indrasekara, A.S.D.S.; Baiyasi, R.; Cai, Y.; Nordlander, P.; Link, S.; Landes, C.F.; Chang, W.-S. Spectral response of plasmonic gold nanoparticles to capacitive charging: morphology effects. *J. Phys. Chem. Lett.* **2017**, *8*, 2681–2688. [CrossRef]
98. Novo, C.; Funston, A.M.; Gooding, A.K.; Mulvaney, P. Electrochemical charging of single gold nanorods. *J. Am. Chem. Soc.* **2009**, *131*, 14664–14666. [CrossRef]

99. Byers, C.P.; Hoener, B.S.; Chang, W.S.; Yorulmaz, M.; Link, S.; Landes, C.F. Single-particle spectroscopy reveals heterogeneity in electrochemical tuning of the localized surface plasmon. *J. Phys. Chem. B* **2014**, *118*, 14047–14055. [CrossRef]
100. Brown, A.M.; Sheldon, M.T.; Atwater, H.A. Electrochemical tuning of the dielectric function of Au nanoparticles. *ACS Photonics* **2015**, *2*, 459–464. [CrossRef]
101. Zheng, J.W.; Lu, T.H.; Cotton, T.M.; Chumanov, G. Photoinduced electrochemical reduction of nitrite at an electrochemically roughened silver surface. *J. Phys. Chem. B* **1999**, *103*, 6567–6572. [CrossRef]
102. Huang, H.J.; Liu, B.-H.; Su, J.; Chen, P.-J.; Lin, C.-T.; Chiang, H.-P.; Kao, T.S.; Chau, Y.-F.C.; Kei, C.-C.; Hwang, C.-H. Light energy transformation over a few nanometers. *J. Phys. D* **2017**, *50*, 375601. [CrossRef]
103. Link, S.; El-Sayed, A.M. Spectral properties and relaxation dynamics of surface plasmon electronic oscillations in gold and silver nanodots and nanorods. *J. Phys. Chem. B* **1999**, *103*, 8410–8426. [CrossRef]
104. Wang, Z.; Zhang, J.; Lv, J.; Dai, K.; Liang, C. Plasmonic $Ag_2MoO_4/AgBr/Ag$ composite: Excellent photocatalytic performance and possible photocatalytic mechanism. *Appl. Surf. Sci.* **2017**, *396*, 791–798. [CrossRef]
105. Meng, X.; Zhang, Z. Plasmonic ternary Ag–rGO–Bi2MoO6 composites with enhanced visible light-driven photocatalytic activity. *J. Catal.* **2016**, *344*, 616–630. [CrossRef]
106. Cui, W.; Li, X.; Gao, C.; Dong, F.; Chen, X. Ternary $Ag/AgCl-(BiO)_2CO_3$ composites as high-performance visible-light plasmonic photocatalysts. *Catal. Today* **2017**, *284*, 67–76. [CrossRef]
107. Pirhashemi, M.; Habibi-Yangjeh, A. Ultrasonic-assisted preparation of plasmonic $ZnO/Ag/Ag_2WO_4$ nanocomposites with high visible-light photocatalytic performance for degradation of organic pollutants. *J. Colloid Interf. Sci.* **2017**, *491*, 216–229. [CrossRef]
108. Feizpoor, S.; Habibi-Yangjeh, A. Integration of Ag_2WO_4 and AgBr with TiO_2 to fabricate ternary nanocomposites: Novel plasmonic photocatalysts with remarkable activity under visible light. *Mater. Res. Bull.* **2018**, *99*, 93–102. [CrossRef]
109. Liu, E.; Qi, L.; Bian, J.; Chen, Y.; Hu, X.; Fan, J.; Liu, H.; Zhu, C.; Wang, Q. A facile strategy to fabricate plasmonic Cu modified TiO_2 nano-flower films for photocatalytic reduction of CO_2 to methanol. *Mater. Res. Bull.* **2015**, *68*, 203–209. [CrossRef]
110. Santos, P.B.; Santos, J.J.; Corrêa, C.C.; Corio, P.; Andrade, G.F.S. Plasmonic photodegradation of textile dye Reactive Black 5 under visible light: a vibrational and electronic study. *J. Photochem. Photobiol. A* **2019**, *371*, 159–165. [CrossRef]
111. Lin, W.-C.; Liao, L.-S.; Chen, Y.-H.; Chang, H.-C.; Tsai, D.P.; Chiang, H.-P. Size dependence of nanoparticle-SERS enhancement from silver film over nanosphere (AgFON) substrate. *Plasmonics* **2011**, *6*, 201–206. [CrossRef]
112. Kowalska, E.; Abe, R.; Ohtani, B. Visible light-induced photocatalytic reaction of gold-modified titanium(IV) oxide particles: action spectrum analysis. *Chem. Commun.* **2009**, *2*, 241–243. [CrossRef]
113. Cushing, S.K.; Li, J.; Bright, J.; Yost, B.T.; Zheng, P.; Bristow, A.D.; Wu, N. Controlling plasmon-induced resonance energy transfer and hot electron injection processes in metal@TiO_2 Core–Shell Nanoparticles. *J. Phys. Chem. C* **2015**, *119*, 16239–16244. [CrossRef]
114. Sarma, B.; Deb, S.K.; Sarma, B.K. Photoluminescence and photocatalytic activities of Ag/ZnO metal-semiconductor heterostructure. *J. Phys. Conf. Ser.* **2016**, *765*, 012023. [CrossRef]
115. Shuang, S.; Lv, R.; Xie, Z.; Zhang, Z. Surface Plasmon Enhanced Photocatalysis of Au/Pt-decorated TiO_2 Nanopillar Arrays. *Sci. Rep.* **2016**, *6*, 26670. [CrossRef]
116. Li, B.; Wang, R.; Shao, X.; Shao, L.; Zhang, B. Synergistically enhanced photocatalysis from plasmonics and a co-catalyst in Au@ZnO–Pd ternary core–shell nanostructures. *Inorg. Chem. Front.* **2017**, *4*, 2088–2096. [CrossRef]
117. Wang, Y.; Chen, Y.; Hou, Q.; Ju, M.; Li, W. Coupling plasmonic and cocatalyst nanoparticles on N–TiO_2 for visible-light-driven catalytic organic synthesis. *Nanomaterials* **2019**, *9*, 391. [CrossRef]
118. Gao, W.; Lu, J.; Zhang, S.; Zhang, X.; Wang, Z.; Qin, W.; Wang, J.; Zhou, W.; Liu, H.; Sang, Y. Suppressing photoinduced charge recombination via the Lorentz force in a photocatalytic system. *Adv. Sci.* **2019**, *6*, 1901244. [CrossRef]
119. Nie, S.; Emory, S.R. Probing single molecules and single nanoparticles by surface-enhanced Raman scattering. *Science* **1997**, *275*, 1102–1106. [CrossRef]
120. Kneipp, K.; LemmaPatricia T.; Aroca, A.A. Single molecule detection using surface-enhanced Raman scattering (SERS). *Phys. Rev. Lett.* **1997**, *78*, 1667–1670. [CrossRef]

121. Homola, J.; Yee, S.S.; Gauglitz, G. Surface plasmon resonance sensors: review. *Sensor. Actuat. B Chem.* **1999**, *54*, 3–15. [CrossRef]
122. Wang, T.-J.; Hsu, K.-C.; Liu, Y.-C.; Lai, C.-H.; Chiang, H.-P. Nanostructured SERS substrates produced by nanosphere lithography and plastic deformation through direct peel-off on soft matter. *J. Opt.* **2016**, *18*, 055006. [CrossRef]
123. Brooks, J.L.; Frontiera, R.R. Competition between reaction and degradation pathways in plasmon-driven photochemistry. *J. Phys. Chem. C* **2016**, *120*, 20869–20876. [CrossRef]
124. Li, P.; Ma, B.; Yang, L.; Liu, J. Hybrid single nanoreactor for in situ SERS monitoring of plasmon-driven and small Au nanoparticles catalyzed reactions. *Chem. Commun.* **2015**, *51*, 11394–11397. [CrossRef]
125. Xie, W.; Walkenfort, B.; Schlucker, S. Label-free SERS monitoring of chemical reactions catalyzed by small gold nanoparticles using 3D plasmonic superstructures. *J. Am. Chem. Soc.* **2013**, *135*, 1657–1660. [CrossRef]
126. Yin, Z.; Wang, Y.; Song, C.; Zheng, L.; Ma, N.; Liu, X.; Li, S.; Lin, L.; Li, M.; Xu, Y.; et al. Hybrid Au-Ag nanostructures for enhanced plasmon-driven catalytic selective hydrogenation through visible light irradiation and surface-enhanced Raman scattering. *J. Am. Chem. Soc.* **2018**, *140*, 864–867. [CrossRef]
127. Liu, X.; Tang, L.; Niessner, R.; Ying, Y.; Haisch, C. Nitrite-triggered surface plasmon-assisted catalytic conversion of p-aminothiophenol to p,p'-dimercaptoazobenzene on gold nanoparticle: surface-enhanced Raman scattering investigation and potential for nitrite detection. *Anal. Chem.* **2015**, *87*, 499–506. [CrossRef]
128. Lee, S.J.; Kim, K. Surface-induced photoreaction of 4-nitrobenzenethiol on silver: influence of SERS-active sites. *Chem. Phys. Lett.* **2003**, *378*, 122–127. [CrossRef]
129. Kim, K.; Choi, J.Y.; Shin, K.S. Photoreduction of 4-nitrobenzenethiol on Au by hot electrons plasmonically generated from Ag nanoparticles: gap-mode surface-enhanced Raman scattering observation. *J. Phys. Chem. C* **2015**, *119*, 5187–5194. [CrossRef]
130. Dong, B.; Fang, Y.; Chen, X.; Xu, H.; Sun, M. Substrate-, wavelength-, and time-dependent plasmon-assisted surface catalysis reaction of 4-nitrobenzenethiol dimerizing to p,p'-dimercaptoazobenzene on Au, Ag, and Cu films. *Langmuir* **2011**, *27*, 10677–10682. [CrossRef]
131. Choi, H.K.; Park, W.H.; Park, C.G.; Shin, H.H.; Lee, K.S.; Kim, Z.H. Metal-catalyzed chemical reaction of single molecules directly probed by vibrational spectroscopy. *J. Am. Chem. Soc.* **2016**, *138*, 4673–4684. [CrossRef]
132. De Nijs, B.; Benz, F.; Barrow, S.J.; Sigle, D.O.; Chikkaraddy, R.; Palma, A.; Carnegie, C.; Kamp, M.; Sundararaman, R.; Narang, P.; et al. Plasmonic tunnel junctions for single-molecule redox chemistry. *Nat. Commun.* **2017**, *8*, 994. [CrossRef]
133. Sonntag, M.D.; Chulhai, D.; Seideman, T.; Jensen, L.; Van Duyne, R.P. The origin of relative intensity fluctuations in single-molecule tip-enhanced Raman spectroscopy. *J. Am. Chem. Soc.* **2013**, *135*, 17187–17192. [CrossRef]
134. Stockle, R.M.; Suh, Y.D.; Deckert, V.; Zenobi, R. Nanoscale chemical analysis by tip-enhanced Raman spectroscopy. *Chem. Phys. Lett.* **2000**, *318*, 131–136. [CrossRef]
135. Van Schrojenstein Lantman, E.M. Deckert-Gaudig, T.; Mank, A.J.G.; Deckert, V.; Weckhuysen, B.M. Catalytic processes monitored at the nanoscale with tip-enhanced Raman spectroscopy. *Nat. Nanotechnol.* **2012**, *7*, 583. [CrossRef]
136. Peng, T.-C.; Lin, W.-C.; Chen, C.-W.; Tsai, D.P.; Chiang, H.-P. Enhanced sensitivity of surface plasmon resonance phase-interrogation biosensor by using silver nanoparticles. *Plasmonics* **2011**, *6*, 29–34. [CrossRef]
137. Chiang, H.-P.; Lin, J.-L.; Chen, Z.-W. High sensitivity surface plasmon resonance sensor based on phase interrogation at optimal incident wavelengths. *Appl. Phys. Lett.* **2006**, *88*, 141105. [CrossRef]
138. Luo, W.; Wang, R.; Li, H.; Kou, J.; Zeng, X.; Huang, H.; Hu, X.; Huang, W. Simultaneous measurement of refractive index and temperature for prism-based surface plasmon resonance sensors. *Opt. Express* **2019**, *27*, 576–589. [CrossRef]
139. Chung, H.-Y.; Chen, C.-C.; Wu, P.C.; Tseng, M.L.; Lin, W.-C.; Chen, C.-W.; Chiang, H.-P. Enhanced sensitivity of surface plasmon resonance phase-interrogation biosensor by using oblique deposited silver nanorods. *Nanoscale Res. Lett.* **2014**, *9*, 476. [CrossRef]
140. Lee, J.; Cheglakov, Z.; Yi, J.; Cronin, T.M.; Gibson, K.J.; Tian, B.; Weizmann, Y. Plasmonic photothermal gold bipyramid nanoreactors for ultrafast real-time bioassays. *J. Am. Chem. Soc.* **2017**, *139*, 8054–8057. [CrossRef]

141. Kumara, N.; Chau, Y.F.C.; Huang, J.W.; Huang, H.J.; Lin, C.T.; Chiang, H.P. Plasmonic spectrum on 1D and 2D periodic arrays of rod-shape metal nanoparticle pairs with different core patterns for biosensor and solar cell applications. *J. Opt.* **2016**, *18*, 115003. [CrossRef]
142. Chau, Y.-F.C.; Wang, C.-K.; Shen, L.; Lim, C.M.; Chiang, H.-P.; Chao, C.-T.C.; Huang, H.J.; Lin, C.-T.; Kumara, N.; Voo, N.Y. Simultaneous realization of high sensing sensitivity and tunability in plasmonic nanostructures arrays. *Sci. Rep.* **2017**, *7*, 16817. [CrossRef]
143. Chau, Y.-F.C.; Chao, C.T.C.; Huang, H.J.; Wang, Y.-C.; Chiang, H.-P. Muhammad Nur Syafi'ie Md Idris, Zarifi Masri, Chee Ming Lim, Strong and tunable plasmonic field coupling and enhancement generating from the protruded metal nanorods and dielectric cores. *Results Phys.* **2019**, *15*, 102567.
144. Lu, J.Y.; Nam, S.H.; Wilke, K.; Raza, A.; Lee, Y.E.; AlGhaferi, A.; Fang, N.X.; Zhang, T. Localized surface plasmon-enhanced ultrathin film broadband nanoporous absorbers. *Adv. Opt. Mater.* **2016**, *4*, 1255–1264. [CrossRef]
145. Chau, Y.-F.C.; Chao, C.-T.C.; Lim, C.M.; Huang, H.J.; Chiang, H.-P. Depolying tunable metal-shell/dielectric core nanorod arrays as the virtually perfect absorber in the near-infrared regime. *ACS Omega* **2018**, *3*, 7508. [CrossRef]
146. Tseng, M.L.; Chang, C.M.; Cheng, B.H.; Wu, P.C.; Chung, K.S.; Hsiao, M.-K.; Huang, H.W.; Huang, D.-W.; Chiang, H.-P.; Leung, P.T.; Tsai, D.P. Multi-level surface enhanced Raman scattering using AgO$_x$ thin film. *Opt. Express* **2013**, *21*, 24460–24467. [CrossRef]
147. Deng, C.-Y.; Zhang, G.-L.; Zou, B.; Shi, H.-L.; Liang, Y.-J.; Li, Y.-C.; Fu, J.-X.; Wang, W.-Z. TiO$_2$/Ag composite nanowires for a recyclable surface enhanced raman scattering substrate. *Chin. Phys. B* **2013**, *22*, 106102. [CrossRef]
148. Link, S.; El-Sayed, M.A. Size and temperature dependence of the plasmon absorption of colloidal gold nanoparticles. *J. Phys. Chem. B* **1999**, *103*, 4212–4217. [CrossRef]
149. Cushing, S.K.; Bristow, A.D.; Wu, N. Theoretical maximum efficiency of solar energy conversion in plasmonic metal–semiconductor heterojunctions. *Phys. Chem. Chem. Phys.* **2015**, *17*, 30013–30022. [CrossRef]
150. Cushing, S.K.; Wu, N. Progress and perspectives of plasmon-enhanced solar energy conversion. *J. Phys. Chem. Lett.* **2016**, *7*, 666–675. [CrossRef]
151. Hartland, G.V. Optical studies of dynamics in noble metal nanostructures. *Chem. Rev.* **2011**, *111*, 3858–3887. [CrossRef]
152. Manjavacas, A.; Liu, J.G.; Kulkarni, V.; Nordlander, P. Plasmon-induced hot carriers in metallic nanoparticles. *ACS Nano* **2014**, *88*, 7630–7638. [CrossRef]
153. Brongersma, M.L.; Halas, N.J.; Nordlander, P. Plasmon-induced hot carrier science and technology. *Nat. Nanotechnol.* **2015**, *10*, 25–34. [CrossRef] [PubMed]
154. Ahmadi, T.S.; Logunov, S.L.; El-Sayed, M.A. Picosecond dynamics of colloidal gold nanoparticles. *J. Phys. Chem.* **1996**, *100*, 8053–8056. [CrossRef]
155. Aruda, K.O.; Tagliazucchi, M.; Sweeney, C.M.; Hannah, D.C.; Schatz, G.C.; Weiss, E.A. Identification of parameters through which surface chemistry determines the lifetimes of hot electrons in small Au nanoparticles. *Proc. Natl. Acad. Sci. USA* **2013**, *110*, 4212–4217. [CrossRef] [PubMed]
156. Salmi, M.; Tkachenko, N.; Vehmanen, V.; Lamminmäki, R.-J.; Karvinen, S.; Lemmetyinen, H. The effect of calcination on photocatalytic activity of TiO$_2$ particles: femtosecond study. *J. Photoch. Photobio. A* **2004**, *163*, 395–401. [CrossRef]
157. Dhumal, S.Y.; Daulton, T.L.; Jiang, J.; Khomami, B.; Biswas, P. Synthesis of visible light-active nanostructured TiO$_x$ ($x < 2$) photocatalysts in a flame aerosol reactor. *Appl. Catal. B-Environ.* **2009**, *86*, 145–151.
158. Nakamura, I.; Negishi, N.; Sugihara, S.; Kutsuna, S.; Takeuchi, K.; Ihara, T. Role of oxygen vacancy in the plasma-treated TiO photocatalyst with visible light activity for NO removal. *J. Mol. Catal. Chem.* **2000**, *161*, 205–212. [CrossRef]
159. Bikondoa, O.; Pang, C.L.; Ithnin, R.; Muryn, C.A.; Onishi, H.; Thornton, G. Direct visualization of defect mediated dissociation of water on TiO$_2$(110). *Nat. Mater.* **2006**, *5*, 189–192. [CrossRef]
160. Shen, J.; Yan, B.; Shi, M.; Ma, H.; Li, N.; Ye, M. One step hydrothermal synthesis of TiO$_2$-reduced graphene oxide sheets. *J. Mater. Chem.* **2011**, *21*, 3415–3421. [CrossRef]
161. Xiang, Q.; Yu, J.; Jaroniec, M. Enhanced photocatalytic H$_2$-production activity of graphene-modified titania nanosheets. *Nanoscale* **2011**, *3*, 3670–3678. [CrossRef]

162. Gao, Y.; Pu, X.; Zhang, D.; Ding, G.; Shao, X.; Ma, J. Combustion synthesis of graphene oxide-TiO_2 hybrid materials for photodegradation of methyl orange. *Carbon* **2012**, *50*, 4093–4101. [CrossRef]
163. Tan, L.-L.; Ong, W.-J.; Chai, S.-P.; Mohamed, A.R. Reduced graphene oxide-TiO_2 nanocomposite as a promising visible-light-active photocatalyst for the conversion of carbon dioxide. *Nanoscale Res. Lett.* **2013**, *8*, 465. [CrossRef] [PubMed]
164. Akhavan, O.; Ghaderi, E. Photocatalyticreduction of graphene oxide nanosheets on TiO_2 thin film for photoinactivation of bacteria in solar light irradiation. *J. Phys. Chem. C* **2009**, *113*, 20214–20220. [CrossRef]
165. Leong, K.H.; Sim, L.C.; Bahnemann, D.; Jang, M.; Ibrahim, S.; Saravanan, P. Reduced graphene oxide and Ag wrapped TiO_2 photocatalyst for enhanced visible light photocatalysis. *APL Mater.* **2015**, *3*, 104503. [CrossRef]
166. Wen, Y.; Ding, H.; Shan, Y. Preparation and visible light photocatalytic activity of Ag/TiO_2/graphene nanocomposite. *Nanoscale* **2011**, *3*, 4411–4417. [CrossRef]
167. Huang, H.J.; Zhen, S.Y.; Li, P.Y.; Tzeng, S.D.; Chiang, H.P. Confined migration of induced hot electrons in Ag/graphene/TiO_2 composite nanorods for plasmonic photocatalytic reaction. *Opt. Express* **2016**, *24*, 15603–15608. [CrossRef]
168. Yu, J.; Dai, G.; Huang, B. Fabrication and characterization of visible-light-driven plasmonic photocatalyst Ag/AgCl/TiO_2 nanotube arrays. *J. Phys. Chem. C* **2009**, *113*, 16394–16401. [CrossRef]
169. Meng, F.; Cushing, S.K.; Li, J.; Hao, S.; Wu, N. Enhancement of solar hydrogen generation by synergistic interaction of $La_2Ti_2O_7$ photocatalyst with plasmonic gold nanoparticles and reduced graphene oxide nanosheets. *ACS Catal.* **2015**, *5*, 1949–1955. [CrossRef]
170. Wang, Y.; Yu, J.; Xiao, W.; Lia, Q. Microwave-assisted hydrothermal synthesis of graphene based Au–TiO_2 photocatalysts for efficient visible-light hydrogen production. *J. Mater. Chem. A* **2014**, *2*, 3847–3855. [CrossRef]
171. Jaeger, C.D.; Bard, A.J. Spin trapping and electron spin resonance detection of radical intermediates in the photodecomposition of water at titanium dioxide particulate systems. *J. Phys. Chem.* **1979**, *83* 3146–3152. [CrossRef]
172. Wang, C.; Yin, L.; Zhang, L.; Liu, N.; Lun, N.; Qi, Y. Platinum-nanoparticle-modified TiO_2 nanowires with enhanced photocatalytic property. *ACS Appl. Mater. Interfaces* **2010**, *2*, 3373–3377. [CrossRef]
173. Morikawa, T.; Irokawa, Y.; Ohwaki, T. Enhanced photocatalytic activity of TiO_2 xNx loaded with copper ions under visible light irradiation. *Appl. Catal. A-Gen.* **2006**, *314*, 123–127. [CrossRef]
174. Reineck, P.; Brick, D.; Mulvaney, P.; Bach, U. Plasmonic hot electron solar cells: the effect of nanoparticle size on quantum efficiency. *J. Phys. Chem. Lett.* **2016**, *7*, 4137–4141. [CrossRef] [PubMed]
175. Koynov, S.; Brandt, M.S.; Stutzmann, M. Black nonreflecting silicon surfaces for solar cells. *Appl. Phys. Lett.* **2006**, *88*, 203107. [CrossRef]
176. Li, J.; Cushing, S.K.; Chu, D.; Zheng, P.; Bright, J.; Castle, C.; Manivannan A.; Wu, N. Distinguishing surface effects of gold nanoparticles from plasmonic effect on photoelectrochemical water splitting by hematite. *J. Mater. Res.* **2016**, *31*, 1608–1615. [CrossRef]
177. Naik, G.V.; Schroeder, J.L.; Ni, X.; Kildishev, A.V.; Sands, T.D.; Boltasseva, A. Titanium nitride as a plasmonic material for visible and near-infrared wavelengths. *Opt. Mater. Express* **2012**, *2*, 478–489. [CrossRef]
178. Naik, G.V.; Kim, J.; Boltasseva, A. Oxides and nitrides as alternative plasmonic materials in the optical range. *Opt. Mater. Express* **2011**, *1*, 1090–1099. [CrossRef]
179. Chen, N.C.; Lien, W.C.; Liu, C.R.; Huang, Y.L.; Lin, Y.R.; Chou, C.; Chang, S.Y.; Ho, C.W. Excitation of surface plasma wave at TiN/air interface in the Kretschmann geometry. *J. Appl. Phys.* **2011**, *109*, 043104. [CrossRef]
180. Kumar, K.K.; Raole, P.M.; Rayjada, P.A.; Chauhan, N.L.; Mukherjee, S. Study of structure development of Titanium Nitride on inclined substrates. *Surf. Coat. Tech.* **2011**, *205*, S187–S191. [CrossRef]
181. Chen, C.; Wang, Z.; Wu, K.; Chong, H.; Xu, Z.; Ye, H. ITO–TiN–ITO Sandwiches for near-infrared plasmonic materials. *ACS Appl. Mater. Interfaces* **2018**, *10*, 14886–14893. [CrossRef]
182. Zakomirnyi, V.I.; Rasskazov, I.L.; Gerasimov, V.S.; Ershov, A.E.; Polyutov, S.P.; Karpov, S.V.; Ågren, H. Titanium nitride nanoparticles as an alternative platform for plasmonic waveguides in the visible and telecommunication wavelength ranges. *Photonics Nanostruct.* **2018**, *30*, 50–56. [CrossRef]
183. Pileni M.P. Control of the size and shape of inorganic nanocrystals at various scales from nano to macrodomains *J. Phys. Chem. C* **2007**, *111*, 9019. [CrossRef]
184. Stokes, N.; McDonagh A.M. Cortie M.B. Preparation of nanoscale gold structures by nanolithography. *Gold Bull.* **2007**, *40*, 310. [CrossRef]

185. Shiao, M.H.; Lin, C.T.; Zeng, J.J.; Lin, Y.S. Novel gold dendritic nanoforests combined with titanium nitride for visible-light-enhanced chemical degradation. *Nanomaterials* **2018**, *8*, 282. [CrossRef] [PubMed]
186. Shiao, M.H.; Zeng, J.J.; Huang, H.J.; Liao, B.H.; Tang, Y.H.; Lin, Y.S. Growth of gold dendritic nanoforests on titanium nitride-coated silicon substrates. *J. Visualized Exp.* **2019**, *148*, e59603. [CrossRef] [PubMed]
187. Govorov, A.O.; Richardson H.H. Generating heat with metal nanoparticles. *Nano Today* **2007**, *2*, 30–38. [CrossRef]
188. Richardson, H.H.; Carlson, M.T.; Tandler, P.J.; Hernandez, P.; Govorov, A.O. Experimental and theoretical studies of light-to-heat conversion and collective heating effects in metal nanoparticle solutions. *Nano Lett.* **2009**, *9*, 1139–1146. [CrossRef]
189. Link, S.; El-Sayed, M.A. Shape and size dependence of radiative, non-radiative and photothermal properties of gold nanocrystals. *Int. Rev. Phys. Chem.* **2000**, *19*, 409–453. [CrossRef]
190. Zhang, W.; Li, Q.; Qiu, M. A plasmon ruler based on nanoscale photothermal effect. *Opt. Express* **2013**, *21*, 172–181. [CrossRef]
191. Jiang, K.; Smith, D.A.; Pinchuk, A. Size-dependent photothermal conversion efficiencies of plasmonically heated gold nanoparticles. *J. Phys. Chem. C* **2013**, *117*, 27073–27080. [CrossRef]
192. Mostafa, A.E.-S. Plasmonic photochemistry and photon confinement to the nanoscale. *J. Photochem. Photobiol. A* **2011**, *221*, 138–142.
193. Rodríguez-Oliveros, R.; Sánchez-Gil, J.A. Gold nanostars as thermoplasmonic nanoparticles for optical heating. *Opt. Express* **2012**, *20*, 621–626. [CrossRef] [PubMed]
194. Genov, D.A.; Zhang, S.; Zhang, X. Mimicking celestial mechanics in metamaterials. *Nat. Phys.* **2009**, *5*, 687–692. [CrossRef]
195. Sanchot, A.; Baffou, G.; Marty, R.; Arbouet, A.; Quidant, R.; Girard, C.; Dujardin, E. Plasmonic nanoparticle networks for light and heat concentration. *ACS Nano* **2012**, *6*, 3434–3440. [CrossRef] [PubMed]
196. Mehrabova M.A. The modeling of calculations of thermodynamic and electronic parameters of hot electrons in a quantum well. *Intern. J. Energy* **2010**, *4*, 63–70.
197. Baffou, G.; Kreuzer, M.P.; Kulzer, F.; Quidant R. Temperature mapping near plasmonic nanostructures using fluorescence polarization anisotropy. *Opt. Express* **2009**, *17*, 3291–3298. [CrossRef]
198. Fujishima, A.; Hashimoto, K.; Kikuchi, Y.; Sunada, K. Bactericidal and detoxification effects of TiO_2 thin film. *Photocatal. Environ. Sci. Technol.* **1998**, *32*, 726–728.
199. El-Sayed, I.H.; Huang, X.; El-Sayed, M.A. Surface plasmon resonance scattering and absorption of anti-EGFR antibody conjugated gold nanoparticles in cancer diagnostics: Applications in oral cancer. *Nano Lett.* **2005**, *5*, 829–834. [CrossRef]
200. Jain, P.K.; El-Sayed, I.H.; El-Sayed, M.A. Au nanoparticles target cancer. *Nano Today* **2007**, *2*, 18–29. [CrossRef]
201. Zhao, W.; Karp, J.M. Tumour targeting: nanoantennas heat up. *Nat. Mater.* **2009**, *8*, 453–454. [CrossRef]
202. Borgarello, E.; Kiwi, J.; Graetzel, M.; Pelizzetti, E.; Visca, M. Visible light induced water cleavage in colloidal solutions of chromium-doped titanium dioxide particles. *J. Am. Chem. Soc.* **1982**, *104*, 2996–3002. [CrossRef]
203. Wu T.-H.; Teslaa,T.; Teitell, M.A.; Chiou, P.-Y. Photothermal nanoblade for patterned cell membrane cutting. *Opt. Express* **2010**, *18*, 23153–23160. [CrossRef] [PubMed]
204. Huang, X.; El-Sayed, I.H.; Qian, W.; El-Sayed, M.A. Cancer cell imaging and photothermal therapy in the near-infrared region by using gold nanorods. *J. Am. Chem. Soc.* **2006**, *128*, 2115–2120. [CrossRef] [PubMed]
205. Saiki, R. Enzymatic amplification of beta-globin genomic sequences and restriction site analysis for diagnosis of sickle cell anemia. *Science* **1985**, *230*, 1350–1354. [CrossRef] [PubMed]
206. Saiki, R. Primer-directed enzymatic amplification of DNA with a thermostable DNA polymerase. *Science* **1988**, *239*, 487–491. [CrossRef]
207. Park, S.; Zhang, Y.; Lin, S.; Wang, T.H.; Yang, S. Advances in microfluidic PCR for point-of-care infectious disease diagnostics. *Biotechnol. Adv.* **2011**, *29*, 830–839. [CrossRef]
208. Pasquardini, L.; Potrich, C.; Quaglio, M.; Lamberti, A.; Guastella, S.; Lunelli, L.; Cocuzza, M.; Vanzetti, L.; Pirri, C.F.; Pederzolli, C. Solid phase DNA extraction on PDMS and direct amplification. *Lab Chip* **2011**, *11*, 4029–4035. [CrossRef]
209. Ahmad, F.; Hashsham, S.A. Miniaturized nucleic acid amplification systems for rapid and point-of-care diagnostics: a review. *Anal. Chim. Acta* **2012**, *733*, 1–15. [CrossRef]
210. Jiang, X.; Shao, N.; Jing, W.; Tao, S.; Liu, S.; Sui, G. Microfluidic chip integrating high throughput continuous-flow PCR and DNA hybridization for bacteria analysis. *Talanta* **2014**, *122*, 246–250. [CrossRef]

211. Xu, J.; Lv, X.; Wei, Y.; Zhang, L.; Li, R.; Deng, Y.; Xu, X. Air bubble resistant and disposable microPCR chip with a portable and programmable device for forensic test. *Sens. Actuators B* **2015**, *212*, 472–480. [CrossRef]
212. Chen, J.J.; Liao, M.H.; Li, K.T.; Shen, C.M. One-heater flow-through polymerase chain reaction device by heat pipes cooling. *Biomicrofluidics* **2015**, *9*, 1–21. [CrossRef]
213. Tachibana, H.; Saito, M.; Shibuya, S.; Tsuji, K.; Miyagawa, N.; Yamanaka, K.; Tamiya, E. On-chip quantitative detection of pathogen genes by autonomous microfluidic PCR platform. *Biosens. Bioelectron.* **2015**, *74*, 725–730. [CrossRef] [PubMed]
214. Chen, J.J.; Hsieh, I.H. Using an IR lamp to perform DNA amplifications on an oscillatory thermocycler. *Appl. Therm. Eng.* **2016**, *106*, 1–12. [CrossRef]
215. Li, T.J.; Chang, C.M.; Chang, P.Y.; Chuang, Y.C.; Huang, C.C.; Su, W.C.; Shieh, D.B. Handheld energy-efficient magneto-optical real-time quantitative PCR device for target DNA enrichment and quantification. *NPG Asia Mater.* **2016**, *8*, e277. [CrossRef]
216. Zhang, Y.; Jiang, H.R. A review on continuous-flow microfluidic PCR in droplets: Advances, challenges and future. *Anal. Chim. Acta* **2016**, *914*, 7–16. [CrossRef] [PubMed]
217. Cavanaugh, S.E.; Bathrick, A.S. Direct PCR amplification of forensic touch and other challenging DNA samples: A review. *Forensic Sci. Int. Genet.* **2018**, *32*, 40–49. [CrossRef]
218. Roche, P.J.R.; Najih, M.; Lee, S.S.; Beitel, L.K.; Carnevale, M.L.; Paliouras, M.; Kirk, A.G.; Trifiro, M.A. Real time plasmonic qPCR: How fast is ultra-fast? 30 cycles in 54 seconds. *Analyst* **2017**, *142*, 1746–1755. [CrossRef]
219. Kim, J.; Kim, H.; Park, J.H.; Jon, S. Gold nanorod-based photo-PCR system for one-step, rapid detection of bacteria. *Nanotheranostics* **2017**, *1*, 178–185. [CrossRef]
220. Jeong, S.; Lim, J.; Kim, M.Y.; Yeom, J.H.; Cho, H.; Lee, H.; Shin, Y.B.; Lee, J.H. Portable low-power thermal cycler with dual thin-film Pt heaters for a polymeric PCR chip. *Biomed. Microdev.* **2018**, *20*, 14. [CrossRef]
221. Son, J.H.; Cho, B.; Hong, S.; Lee, S.H.; Hoxha, O.; Haack, A.J.; Lee, L.P. Ultrafast photonic PCR. *Light Sci. Appl.* **2015**, *4*, e280. [CrossRef]
222. Adleman, J.R.; Boyd, D.A.; Goodwin, D.G.; Psaltis, D. Heterogenous catalysis mediated by plasmon heating. *Nano Lett.* **2009**, *9*, 4417–4423. [CrossRef]
223. Huang, H.J.; Liu, B.-H. Plasmonic energy transformation in the photocatalytic oxidation of ammonium. *Catal. Comm.* **2014**, *43*, 136–140. [CrossRef]
224. Taguchi, J.; Okuhara, T. Selective oxidative decomposition of ammonia in neutral water to nitrogen over titania-supported platinum or palladium catalyst. *Appl. Catal. A-Gen.* **2000**, *194–195*, 89–97. [CrossRef]
225. Cao, S.; Chen, G.; Hu, X.; Yue, P.L. Catalytic wet air oxidation of wastewater containing ammonia and phenol over activated carbon supported Pt catalysts. *Catal. Today* **2003**, *88*, 37–47. [CrossRef]
226. Liu, B.-H.; Huang, H.J.; Huang, S.-H.; Hsiao, C.-N. Platinum thin films with good thermal and chemical stability fabricated by inductively coupled plasma-enhanced atomic layer deposition at low temperatures. *Thin Solid Films* **2014**, *566*, 93–98. [CrossRef]
227. Wu, J.C.S.; Wu, T.H.; Chu, T.C.; Huang, H.J.; Tsai, D.P. Application of optical-fiber Photoreactor for CO_2 photocatalytic reduction. *Top. Catal.* **2008**, *47*, 131–136. [CrossRef]
228. Wu, J.C.S.; Lin, H.-M.; Lai, C.-L.; Photo reduction of CO_2 to methanol using optical-fiber photoreactor. *Appl. Catal. A Gen.* **2005**, *296*, 194–200.
229. Nguyen, T.-V.; Wu, J.C.S. Photoreduction of CO_2 to fuels under sunlight using optical-fiber reactor. *Sol. Energ. Mat. Sol. C.* **2008**, *92*, 864–872 [CrossRef]
230. Huang, H.J.; Liu, B.-H. Energy transformation of plasmonic photocatalytic oxidation on 1D quantum well of platinum thin film. *Appl. Phys. A* **2015**, *121*, 1347–1351. [CrossRef]

 © 2020 by the authors. Licensee MDPI, Basel, Switzerland. This article is an open access article distributed under the terms and conditions of the Creative Commons Attribution (CC BY) license (http://creativecommons.org/licenses/by/4.0/).

Review

Synthesis of Plasmonic Photocatalysts for Water Splitting

Go Kawamura * and Atsunori Matsuda

Department of Electrical and Electronic Information Engineering, Toyohashi University of Technology, Hibarigaoka 1-1, Tempaku, Toyohashi, Aichi 441-8580, Japan; matsuda@ee.tut.ac.jp
* Correspondence: gokawamura@ee.tut.ac.jp; Tel.: +81-532-44-6796

Received: 11 November 2019; Accepted: 22 November 2019; Published: 22 November 2019

Abstract: Production of H_2, O_2, and some useful chemicals by solar water splitting is widely expected to be one of the ultimate technologies in solving energy and environmental problems worldwide. Plasmonic enhancement of photocatalytic water splitting is attracting much attention. However, the enhancement factors reported so far are not as high as expected. Hence, further investigation of the plasmonic photocatalysts for water splitting is now needed. In this paper, recent work demonstrating plasmonic photocatalytic water splitting is reviewed. Particular emphasis is given to the fabrication process and the morphological features of the plasmonic photocatalysts.

Keywords: surface plasmon resonance; metal nanoparticle; metal nanostructure; semiconductor; solar water splitting

1. Introduction

Creating renewable energy from solar, wind, and water power has been a hot topic for a long time, because natural resources such as oil, coal, and natural gas are limited, while global energy demand is rapidly increasing. Although the growth in renewable energy generation has also been rapid, it could only provide a third of the required increase in energy generation in 2018 [1]. Solar water splitting, using a semiconductor photocatalyst, is one of the ultimate solutions to the energy problem because in principal, it only requires inexhaustible solar energy and water. Both hydrogen and oxygen generated by water splitting can be used to generate electricity using fuel cells and produce useful chemical compounds [2,3].

Studies have also localized surface plasmon resonance (LSPR) as a means to enhance the efficiency of photocatalytic performances of semiconductors [4]. Specific metal nanostructures show LSPR acts as a photosensitizer, transferring plasmonic energy to vicinal semiconductors via hot-electron transfer and resonant energy transfer [5–7]. Light scattering by LSPR can also contribute to the enhancement of photocatalysis because it elongates the optical pathways [8]. This LSPR enhancement of photocatalytic performance has huge potential, because the morphology dependence of LSPR makes it possible to absorb sunlight over the entire spectrum, though the reported enhancement factors have not been so attractive until now, as recent review articles have noted [9–11]. Therefore, in this article, the reasons for the difficulties in utilizing LSPR enhancement for photocatalysis are also discussed after reviewing the recent developments in plasmonic photocatalysts fabricated especially for solar water splitting.

2. Photocatalytic Water Splitting

Before reviewing the recent studies on plasmonic photocatalysts for water splitting, some important fundamentals are briefly discussed first. Detailed tutorial and advanced reviews on photocatalytic water splitting are available in recent articles published by Domen's group [2,12,13].

In order to split water, electrons and holes with certain reduction and oxidation potentials − < +0 and > +1.23 vs. normal hydrogen electrode (NHE), respectively, are indispensable. TiO_2, CdSe, and some other wide-bandgap semiconductors can generate such carriers when bandgap excitation occurs. However, water splitting is carried out by the carriers only if the carrier injections into the water are thermodynamically preferable. Since carrier recombination is generally the predominant process after bandgap excitation, co-catalysts are normally deposited on the surface of semiconductors to capture the carrier(s) for the enhancement of charge separation (Figure 1A).

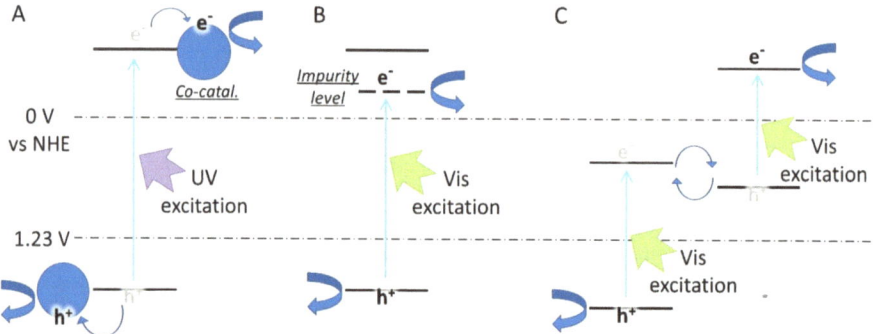

Figure 1. Schematic of energy diagrams of wide-bandgap semiconductors with (**A**) cocatalysts and (**B**) impurity energy level. (**C**) is a typical Z-scheme using two narrow bandgap semiconductors.

As carriers with the required potentials are generated by an excitation of wide-bandgap semiconductors, TiO_2 was initially [14] and the most widely investigated for water splitting [15,16]. TiO_2 absorbs only UV-rays, which occupy only ~4% of sunlight energy. Hence, the doping of metal and nonmetal species to TiO_2 has been extensively carried out to add visible light sensitivity to TiO_2 [17]. In these cases, impurity energy levels were inserted into the bandgap, so the doped TiO_2 could absorb visible light (Figure 1B). An additional advantage of doping is that some of the incorporated impurity species also work as carrier traps, which lead to carrier lifetime elongation [18,19].

Although there are many visible-light-sensitive narrow-bandgap photocatalysts with redox potential, they are unfortunately insufficient to split water. A two-step photocatalytic system, the so-called "Z-scheme", allows these narrow-bandgap photocatalysts to be used for water splitting (Figure 1C) [20,21]. In this system, two photocatalysts are connected physically, or with reversible redox mediators such as IO_3^-/I^-, I_3^-/I^-, and Fe^{3+}/Fe^{2+}. The key point in achieving highly efficient photocatalysis is to select appropriate combinations of photocatalysts and redox mediators together with interface engineering to facilitate the charge transfers. Research and development on water splitting photocatalysts including oxide and non-oxide new materials for Z-scheme systems has also been thoroughly conducted [2,12,22–24].

Among the challenges in achieving higher efficiency in solar water splitting, the use of plasmonic photocatalysts has been attracting increasing attention since plasmonic photocatalysts were first announced in 2008 [25]. In the next section, general ideas on the plasmonic enhancement of photocatalysis will be briefly introduced.

3. Plasmonic Enhancement of Photocatalysis

Plasmonic photocatalysts are generally composed of noble metal nanoparticle-deposited semiconductors. When the semiconductors are n-type, a Schottky junction is formed at the interface of the metals and semiconductors, which leads to enhanced charge separation and therefore the improved efficiency of photocatalysis (Figure 2A). In addition to the Schottky junction effect, plasmonic enhancement of photocatalysis occurs in some cases. LSPR is a collective oscillation of metal nanoparticle free electrons in resonance with incident light, and the optical absorption coefficient is large when compared to organic dyes and semiconductors [26]. The strong optical absorption leads to the generation of hot electrons and holes in metal nanoparticles, and can contribute to the enhancement of photocatalysis efficiency when the generated charges are separated via hot electron transfer from metals to semiconductors [5,6,27]. A moderate Schottky barrier energy of ~1 eV is preferable for hot electron transfer because the energies of hot electrons generally range between 1 and 4 eV (Figure 2B) [5]. The other advantageous features of hot electron transfer are that (1) wide bandgap semiconductors like TiO_2 can be endowed with visible light sensitivity if deposited metal nanoparticles exhibit LSPR under visible light illumination, and (2) both Schottky junction and hot electron transfer enhancements can be used simultaneously without any loss, even though the directions of electron transfer are opposite [28]. Aside from hot electron transfer, plasmon-induced resonance energy transfer is another major phenomenon contributing to photocatalysis efficiency enhancement. A few nonradiative and radiative energy transfer routes have been proposed, but the main and most important one for photocatalysis is energy transfer via near-field enhancement. Using this phenomenon, although the energy of metal nanoparticle LSPR must overlap with the bandgap energy of the semiconductor, and the valence band electrons in the semiconductor adjacent to the metal nanoparticles are excited by the strong electric field caused by LSPR (Figure 2C). The details of energy transfer in plasmonic photocatalysts have been widely studied and detailed review articles are available [4,5,29]. In short, the three phenomena summarized in Figure 2 should be appropriately used to achieve highly LSPR-enhanced photocatalysis including plasmonic solar water splitting.

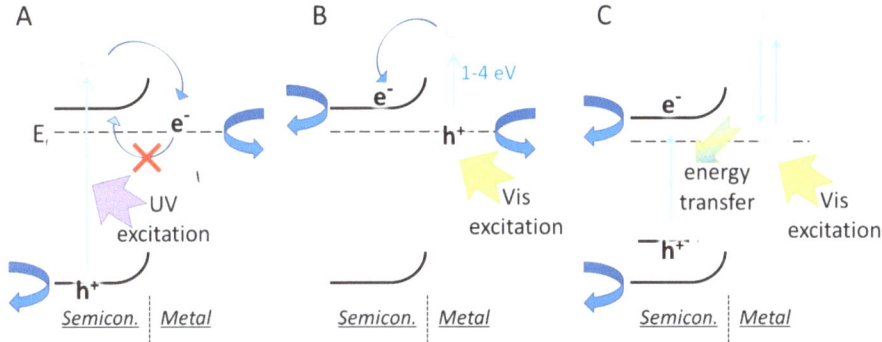

Figure 2. Schematic of energy diagrams of plasmonic photocatalyst under (**A**) UV and (**B**,**C**) Vis excitations.

4. Water Splitting by Plasmonic Photocatalysts

4.1. Titanium Dioxide

As above-mentioned, TiO_2 generates photo-induced electrons and holes with sufficient redox potential for water splitting, whereas it can absorb only UV-rays. Therefore, hot electron transfer, which endows visible light sensitivity to wide bandgap photocatalysts, has been employed to enhance the water splitting efficiency of TiO_2 by expanding the usable wavelength range as well as the Schottky junction effect, which improves reaction efficiency under UV radiation. The deposition of Au or Ag nanoparticles on TiO_2 has most often been carried out and shows enhanced photocurrents in water splitting, especially under visible light illumination [30–35]. An electric bias is applied in some cases to facilitate and enhance the water splitting reaction. As the morphology of metal/TiO_2 nanocomposites strongly affects the reaction efficiency, several researchers have reported interesting studies on the effects of morphology on the efficiency. Moskovits et al. fabricated TiO_2-deposited Au nanorod arrays using an anodic aluminum oxide template. TiO_2 was deposited on the top of the Au nanorods, and an oxygen evolution catalyst (OEC) was separately deposited on the side (Figure 3). As the LSPR wavelength of the Au nanorod can be effectively tuned by controlling the aspect ratio of the nanorod, it was expected that the solar spectrum would be mostly covered by an LSPR peak if an appropriate morphology could be fabricated [36–38]. Results showed a surprising 20-fold higher efficiency for visible light illumination compared with UV radiation because of the unique nano-architecture of the fabricated device. The authors also found that there were fast and slow components in H_2 production via hot electron transfer. This phenomenon needs to be further investigated, as it could prove important for the future design of plasmonic photocatalyst nano-architectures [39,40].

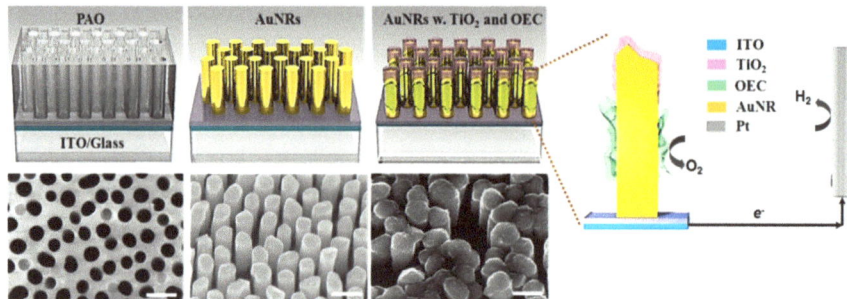

Figure 3. Schematic of the fabrication of Au nanorod arrays deposited with TiO_2. Reprinted with permission from [39], copyright (2012) American Chemical Society.

Misawa et al. prepared Au nanoparticle-deposited TiO_2 nanotube arrays by anodization of a Ti plate followed by the chemical reduction of Au precursors (Figure 4). The nanotube array structure of TiO_2 provides a large surface area for Au deposition and water splitting reaction. The quantity and dispersion state of Au nanoparticles deposited on TiO_2 nanotube arrays were carefully controlled by changing Au precursor conditions. The results found a large quantity of highly dispersed Au nanoparticles that exhibited the largest incident photon-to-current conversion efficiency. This was presumably because of the large optical absorption and effective hot electron transfer caused by the highly dispersed large number of Au nanoparticles on the TiO_2 nanotube arrays. Even though they used no sacrificial electron donors or acceptors, the solar energy was converted to chemical reaction by 0.10%, which was about 10-fold larger than that obtained in the previous study using monolith type samples composed of $SrTiO_3$ loaded with Au nanoparticles [41,42].

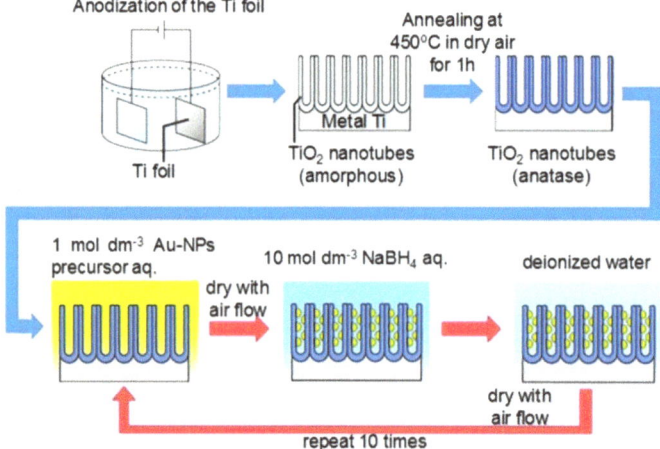

Figure 4. Schematic of the fabrication of Au nanoparticle-deposited TiO_2 nanotube arrays. Reprinted with permission from [42], copyright (2017), Royal Society of Chemistry.

Yang et al. investigated a synergistic interaction between plasmonic Au nanoparticles and oxygen vacancies in amorphous black TiO_2. TiO_2 nanoporous film was first prepared by the anodization of a Ti foil that was then reduced by $NaBH_4$ in order to introduce oxygen vacancies. Au nanoparticles were deposited by means of magnetron sputtering, followed by thermal dewetting in the N_2 atmosphere (Figure 5). The incorporated impurity energy levels in the TiO_2 facilitated hot electron transfer from the Au nanoparticles to the TiO_2. Thus, the generated plasmonic hot electron/hole pairs were efficiently separated. This interestingly resulted in a largely enhanced photoelectrochemical water splitting, compared to bare amorphous black TiO_2 and Au nanoparticle-deposited TiO_2. This kind of synergistic interaction between LSPR and impurity energy levels could lead to new strategies to design novel, highly efficient plasmonic photocatalysts [43].

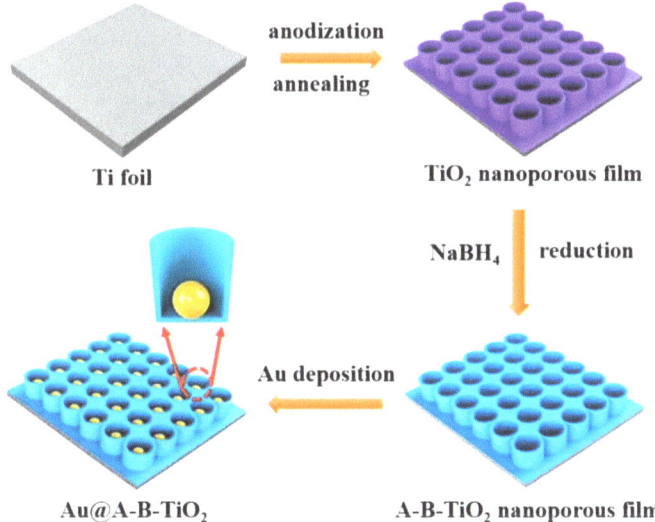

Figure 5. Schematic of the fabrication of the Au nanoparticle-deposited reduced-TiO_2 nano-porous film. Reprinted with permission from [43], copyright (2018) Royal Society of Chemistry.

TiN, ZrN, and some other transition metal nitrides have recently attracted attention because of their potential use as plasmonic materials, instead of as conventional noble metals such as Au and Ag. Naldoni et al. demonstrated that TiO_2 nanowires decorated with TiN nanoparticles showed better water splitting ability than when decorated with Au nanoparticles. The higher performance of TiN was explained mainly by the following two factors: wider absorption wavelength, and Ohmic junction formation with TiO_2. Nalgoni also proposed that a further enhancement of solar water splitting could be expected because TiN nanoparticles show plasmon resonance in the red, so that the entire solar spectrum would be covered when complemented with Au nanoparticles [44].

4.2. Iron Oxide

As stated above, TiO_2 has been widely investigated as a potential photocatalytic semiconductor because of its redox potential for water splitting. Another chemical compound, iron oxide, especially in the form hematite (α-Fe_2O_3), has been recognized as a promising photocatalyst, since it is inexpensive, non-toxic, in plentiful supply, and has an appealing band gap (~2.1 eV) for sunlight absorption. The reason why α-Fe_2O_3 has not been used practically so far is that it has several critical drawbacks such as short lifetime of charge carriers (<10 ps) and short hole diffusion length (2–4 nm) compared to deep light penetration depth (~120 nm) [45–49]. It needs to be available as a nanoparticle or thin film to enable the most generated holes to reach the surface, while a large dimension is required to effectively absorb incident light [50–52]. In order to improve the short lifetime of carriers in α-Fe$_2$O$_3$, a hole collector is often used to enhance charge separation [53]. The hole collector is called OEC when it is used for water splitting. Several kinds of OECs including Ir- and Ru-based oxides [54], and a variety of cobalt-based complexes have been investigated very recently and showed an excellent enhancement of photoelectrochemical water splitting efficiency with α-Fe$_2$O$_3$ [55–57].

Plasmonic enhancement is also soon expected to overcome these problems because of the facts stated mainly in Section 3. In short, strong optical scattering and absorption by LSPR can effectively enhance the utilization of incoming photons [8,47]. Li et al. studied the plasmonic enhancement of solar water splitting by employing incorporated α-Fe$_2$O$_3$ nanorod arrays into a gold nanohole pattern. The gold nanohole pattern was fabricated by e-beam lithography with a polystyrene nanosphere mask, followed by the growth of an α-Fe$_2$O$_3$ nanorod array using a hydrothermal method. The nanorod shape of α-Fe$_2$O$_3$ worked as an optical fiber, creating confined modes that enhanced the optical absorption of α-Fe$_2$O$_3$. At the same time, plasmonic energy transfer occurred and enabled the usage of photons with energies below the band gap. As a result, very high enhancements were observed in a wide wavelength range from 350 to 700 nm [47]. Ramadurgam et al. designed core-shell nanowire photoelectrodes composed of a Si-core, Al-mid-shell, and an α-Fe$_2$O$_3$-outer-shell (Figure 6). The theoretical comparison of the photoelectrode performances of Si–Al–Fe$_2$O$_3$, Si–Ag–Fe$_2$O$_3$, and Si–Au–Fe$_2$O$_3$ revealed that Si–Al-Fe$_2$O$_3$ had the potential to exhibit a very high efficiency rate of 14.5% solar to hydrogen, and about 93% of the theoretical maximum for bulk α-Fe$_2$O$_3$. Since Au and Ag, which are widely used as plasmon sources, are precious metals, Al has been shown to be an excellent alternative in achieving enhanced absorption of α-Fe$_2$O$_3$ and reduced the recombination of the generated charges, resulting in a high efficiency of iron-oxide-based solar water splitting [58].

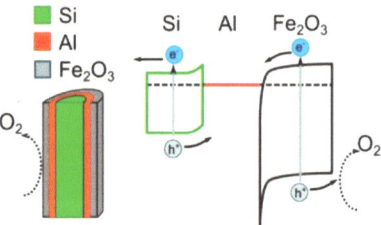

Figure 6. Schematic and energy diagram of Si–Al–Fe$_2$O$_3$ nanowire. Reprinted with permission from [58], copyright (2014) American Chemical Society.

Wang et al. prepared affordable α-Fe$_2$O$_3$ nanosheet photoanodes co-deposited with Ag nanoparticles and Co phosphate OEC (Figure 7). The α-Fe$_2$O$_3$ nanosheet was prepared by anodization of an Fe foil. Ag nanoparticles were separately synthesized via a modified Tollens' reaction [59]. The Ag nanoparticles were then deposited on the α-Fe$_2$O$_3$ nanosheet using pulse-current electrodeposition. Finally, cobalt phosphate was photo-assisted electro-deposited onto the nanosheet. A large photocurrent density of 4.68 mA cm^{-2} (1.23 V vs. reversible hydrogen electrode (RHE)) was achieved presumably because of the following three points: (1) plasmonic light harvesting by Ag nanoparticles; (2) a reduced charge recombination by cobalt phosphate; and (3) electrode surface stabilization, also by cobalt phosphate [60].

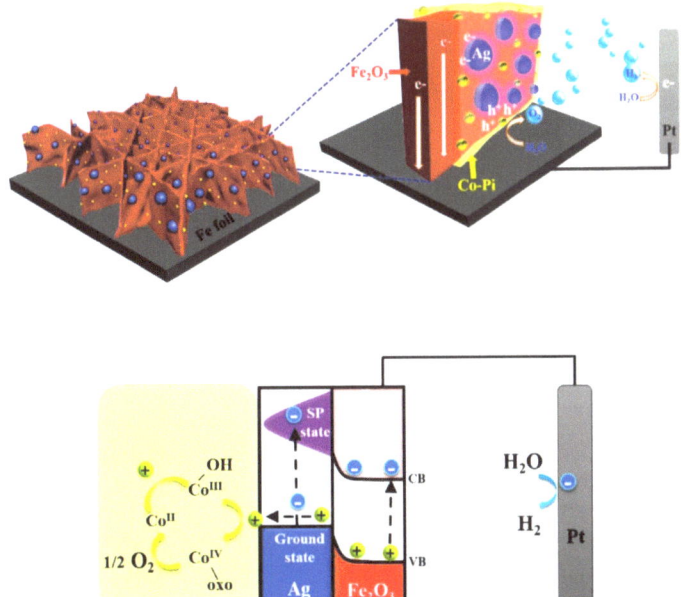

Figure 7. Schematic and energy diagram of an α-Fe$_2$O$_3$ nanosheet deposited with Ag nanoparticles. Reprinted with permission from [60], copyright (2016) Wiley-VCH Verlag GmbH and Co. KGaA, Weinheim.

Gap-plasmon resonance is known to occur when two or more plasmonic materials/parts are close enough and exhibits an extremely enhanced local field in the nanogap area, compared with individual plasmonic materials/nanostructures [61,62]. However, it is rarely used for enhancing photocatalytic water splitting. This is mainly because the fabrication processes generally become complicated and the area where a strong local field exists is quite limited [63]. Dutta et al. recently investigated gap-plasmon enhancement for water splitting using an α-Fe$_2$O$_3$ ultra-thin film. A 200 nm thick Au layer was first deposited on an Si substrate. Then, an α-Fe$_2$O$_3$ layer with a 15 nm thickness was deposited by pulsed laser deposition. Au nano-disks were fabricated on top of the α-Fe$_2$O$_3$ layer using e-beam lithography and lifted off with a polymer mask (Figure 8). Compared to bare α-Fe$_2$O$_3$ layer on a Si substrate, the α-Fe$_2$O$_3$ layer with Au nanogaps showed a two-fold increase in photocurrent density. The enhancement observed above the α-Fe$_2$O$_3$ bandgap was attributed to the enhanced light scattering by the Au nano-disks and back reflection from the Au mirror. On the other hand, a wavelength-dependent increase was observed below the bandgap of α-Fe$_2$O$_3$. This occurred presumably because of plasmon decay and subsequent hot hole generation. Further improvement of the photocurrent density can be expected by shifting the LSPR wavelength to the visible region above the α-Fe$_2$O$_3$ bandgap, which will enable an enhancement in inter-band excitation via plasmon energy transfer [64].

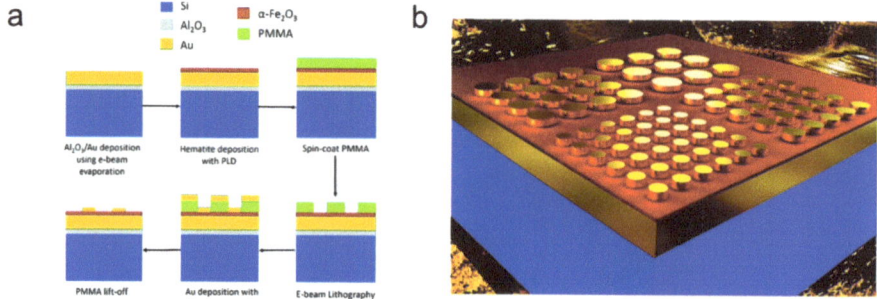

Figure 8. Schematic of the fabrication of an α-Fe$_2$O$_3$ layer with Au nanogaps. (**a**) Process flow and (**b**) 3-dimensional image. Reprinted with permission from [64], copyright (2019) Royal Society of Chemistry.

4.3. Other Semiconductors

Compared to water reduction by plasmonic hot electrons, fundamental research on water oxidation by plasmonic holes is relatively limited. One interesting report recently revealed that plasmonic holes tend to be concentrated at the Au–TiO$_2$ interface, which was identified as the main reaction site for water oxidation in the plasmonic system [65]. This interfacial effect of plasmonic holes was also validated by Au–SrTiO$_3$, Au–BaTiO$_3$, etc. [65]. Nb- or Rh-doped SrTiO$_3$ decorated with Au nanoparticles was studied as a potential component in a water splitting device [66,67]. Zhong et al. employed Nb-doped SrTiO$_3$ as a semiconductor component to investigate the cocatalyst effect on H$_2$ evolution in plasmon-induced water splitting. Since H$_2$ evolution is known to be disturbed by the Schottky barrier formation and reverse water splitting reactions, the Schottky barrier was removed by inserting an In–Ga alloy interlayer between the Pt board and the Nb–SrTiO$_3$ substrate. Reverse reactions were also eliminated by separating the chambers for H$_2$ and O$_2$ evolutions. As a result, the cocatalyst effect became effective and a 3-fold increase of H$_2$ evolution was achieved when a Rh thin layer was deposited on the Pt board as a cocatalyst [66]. Wang et al. designed a plasmon-based solid Z-scheme photocatalyst, which needed no redox mediator. Au, TiO$_2$, Rh-doped SrTiO$_3$, and Ru were employed as the plasmon source, semiconductors, and cocatalyst, respectively. Upon irradiation with visible light, plasmonic hot electrons generated on the Au surface were injected into the conduction band of TiO$_2$, and the remaining holes were used for O$_2$ generation at the interface of Au and TiO$_2$. (Figure 9). The hot electrons injected to the TiO$_2$ conduction band migrated to the valance band of Rh-doped SrTiO$_3$ through interconnecting Au nanoparticles, due to the large work function of Au compared to TiO$_2$. The designed hot electron migration largely reduced the plasmon-induced hot charge recombination, which is often the predominant phenomenon accompanying LSPR. Simultaneously, the excited electrons in Rh-doped SrTiO$_3$ were collected by Ru cocatalyst and used for H$_2$ generation, accomplishing complete water splitting [67].

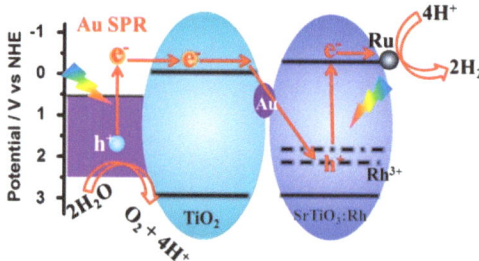

Figure 9. Energy diagram of the Au/TiO$_2$/Rh-doped SrTiO$_3$ nanocomposite. Reprinted with permission from [67], copyright (2017) Elsevier Inc.

Yang et al. made an attempt to use a non-metallic plasmonic component instead of conventional novel metals. SrTiO$_3$ nanotube arrays were first prepared by hydrothermal treatment of anodized TiO$_2$ nanotube arrays. Then, the surface of SrTiO$_3$ was thermochemically reduced to form a crystalline-core@amorphous-shell structure with abundant oxygen vacancies in the amorphous shell (Figure 10). Since the amorphous shell had abundant free charges, it was possible to have LSPR properties. The resonant wavelength was tunable and reversible by oxidation and reduction treatment and incident light angle adjusting, which provided flexibility for various applications in the targeted LSPR wavelength. The reduced core@shell sample showed much higher water splitting ability than the as-synthesized SrTiO$_3$ sample, proving that the use of non-metallic plasmonic semiconductors is a promising way to enhance catalytic reactions, especially water splitting [68].

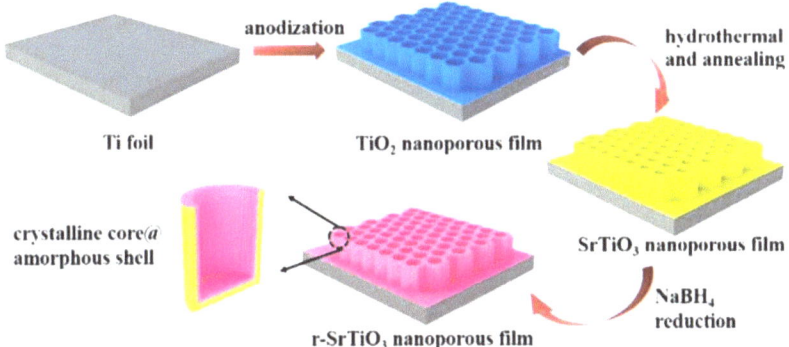

Figure 10. Schematic of the fabrication of SrTiO$_3$ nanotube arrays with a crystalline-core@amorphous-shell structure. Reprinted with permission from [68], copyright (2018) American Chemical Society.

WO$_3$ is also recognized as an applicable semiconductor for water splitting because of its moderate bandgap (2.6–2.8 eV), adequate valence band edge position (3.0 V vs. standard hydrogen electrode (SHE)), long minority carrier diffusion length (0.5–6 μm), and high chemical stability [69]. Hu et al. developed an electron-charging and reducing method for the preparation of a Au/WO$_3$ water splitting plasmonic photocatalyst. After WO$_3$ nanoplate arrays were formed on a FTO substrate by hydrothermal growth, electron-charging was carried out on the arrays by applying an electric bias in H$_2$SO$_4$ solution. The charged WO$_3$ was then immersed into HAuCl$_4$ solution to reductively deposit Au nanoparticles on WO$_3$. The Au/WO$_3$ nanocomposite photocatalyst exhibited a significant increase in oxygen evolution reaction (OER) photocurrent density. Moreover, it showed a negatively shifted OER onset potential and an improved Faraday efficiency for OER. The greatly enhanced performance observed was presumably due to the electrically good Au/WO$_3$ contact formed by the newly developed preparation method [70]. Several interesting attempts have also been reported using a variety of semiconductors such as CdS [71,72], ZnO [73,74], Cu$_2$O [75], and the Cu$_2$O–TiO$_2$ composite [76].

5. Summary and Outlook

Plasmonic enhancement for photocatalytic water splitting has been reported recently by many researchers. However, the mechanisms underlying the enhancement are still not fully understood, and so the enhancement factors reported are not as high as expected. In this paper, many combinations of plasmonic sources and semiconductors with various morphologies for plasmonic water splitting were reviewed. Noble metals such as Au and Ag have been widely used as typical plasmonic sources because of their high chemical stability, resonant wavelength in the visible range, and large field enhancement factor. On the other hand, Al and some transition metal nitrides are attracting increasing attention as they can be alternatives to the precious and expensive Au and Ag. As for semiconductors,

TiO$_2$ and α-Fe$_2$O$_3$ have been the most widely studied materials due to their appropriate redox potential and abundance. SrTiO$_3$, BaTiO$_3$, WO$_3$, CdS, ZnO, Cu$_2$O, etc. have also recently been employed as semiconductors because of their peculiar characteristics. It is noteworthy here that all cases reported thus far have shown a moderate improvement of water splitting efficiency by LSPR, and further studies to achieve much higher efficiency need to be carried out. Additionally, as many phenomena related to plasmonic enhancement of water splitting are well understood, the cost reduction and scale-up of plasmonic photocatalyst fabrication are vital steps to achieve practical application.

Author Contributions: G.K. prepared the original draft. A.M. reviewed and edited the draft.

Funding: This work was supported by JSPS KAKENHI (grant number 18K04701) and the Tatematsu Foundation.

Conflicts of Interest: The authors declare no conflicts of interest.

References

1. *BP Statistical Review of World Energy*, 68th ed.; 1 St James's Square: London, UK, 2019.
2. Hisatomi, T.; Kubota, J.; Domen, K. Recent advances in semiconductors for photocatalytic and photoelectrochemical water splitting. *Chem. Soc. Rev.* **2014**, *43*, 7520–7535. [CrossRef]
3. Emsley, J. *Nature's Building Blocks: An A-Z Guide to the Elements*; Oxford University Press: Oxford, UK, 2001; pp. 297–304, ISBN 978-0-19-850340-8.
4. Hou, W.; Cronin, S.B. A review of surface plasmon resonance-enhanced photocatalysis. *Adv. Funct. Mater.* **2013**, *23*, 1612–1619. [CrossRef]
5. Ma, X.-C.; Dai, Y.; Yu, L.; Huang, B.-B. Energy transfer in plasmonic photocatalytic composites. *Light Sci. Appl.* **2016**, *5*, e16017. [CrossRef] [PubMed]
6. Tatsuma, T.; Nishi, H.; Ishida, T. Plasmon-induced charge separation: Chemistry and wide applications. *Chem. Sci.* **2017**, *8*, 3325–3337. [CrossRef] [PubMed]
7. Wu, N. Plasmonic metal-semiconductor photocatalysts and photoelectrochemical cells: A review. *Nanoscale* **2018**, *10*, 2679–2696. [CrossRef] [PubMed]
8. Cushing, S.K.; Wu, N.Q. Plasmon-induced solar energy harvesting. *Electrochem. Soc. Interface* **2013**, *22*, 63–67. [CrossRef]
9. Shehzad, N.; Tahir, M.; Johari, K.; Murugesan, T.; Hussain, M. A critical review on TiO$_2$ based photocatalytic CO$_2$ reduction system: Strategies to improve efficiency. *J. CO2 Util.* **2018**, *26*, 98–122. [CrossRef]
10. Kumaravel, V.; Mathew, S.; Bartlett, H.; Pillai, S.C. Photocatalytic hyderogen production using metal doped TiO2: A review of recent advances. *Appl. Catal. B Environ.* **2019**, *244*, 1021–1064. [CrossRef]
11. Khan, M.E.; Cho, M.H. Surface plasmon-based nanomaterials as photocatalyst. In *Advanced Nanostructured Materials for Environmental Remediation. Environmental Chemistry for a Sustainable World*; Naushad, M., Rajendran, S., Gracia, F., Eds.; Springer: Cham, Switzerland, 2019; Volume 25, ISBN 978-3-030-04477-0. [CrossRef]
12. Chen, S.; Takata, T.; Domen, K. Particulate photocatalysts for overallwater splitting. *Nat. Rev. Mater.* **2017**, *2*, 17050. [CrossRef]
13. Hisatomi, T.; Domen, K. Reaction systems for solar hydrogen production via water splitting with particlulate semiconductor photocatalysts. *Nat. Catal.* **2019**, *2*, 387–399. [CrossRef]
14. Fujishima, A.; Honda, K. Electrochemical photolysis of water at a semiconductor electrode. *Nature* **1972**, *238*, 37–38. [CrossRef] [PubMed]
15. Khan, S.U.M.; Al-Shahry, M.; Ingler, W.B., Jr. Efficient photochemical water splitting by a chemically modified d-TiO$_2$. *Science* **2002**, *297*, 2243–2245. [CrossRef]
16. Ni, M.; Leung, M.K.H.; Leung, D.Y.C.; Sumathy, K. A review and recent developments in photocatalytic water-splitting using TiO$_2$ for hydrogen production. *Renew. Sust. Energy Rev.* **2007**, *11*, 401–425. [CrossRef]
17. Zaleska, A. Doped-TiO$_2$: A review. *Recent Pat. Eng.* **2008**, *2*, 157–164. [CrossRef]
18. Choi, W.; Termin, A.; Hoffmann, M.R. The role of metal ion dopants in quantum-sized TiO$_2$: Correlation between photoreactivity and charge carrier recombination dynamics. *J. Phys. Chem.* **1994**, *98*, 13669–13679. [CrossRef]

19. Basavarajappa, P.S.; Patil, S.B.; Ganganagappa, N.; Reddy, K.R.; Raghy, A.V.; Reddy, C.V. Recent progress in metal-doped TiO_2, non-metal doped/codoped TiO_2 and TiO_2 nanostructured hybrids for enhanced photocatalysis. *Int. J. Hydrog. Energy* **2019**, in press. [CrossRef]
20. Li, H.; Tu, W.; Zhou, Y.; Zou, Z. Z-scheme photocatalytic systems for promoting photocatalytic performance: Recent progress and future challenges. *Adv. Sci.* **2016**, *3*, 1500389. [CrossRef]
21. Xu, Q.; Zhang, L.; Yu, J.; Wageh, S. Al-Ghamdi, A.A.; Jaroniec, M. Direct Z-scheme photocatalysts: Principles, synthesis, and applications. *Mater. Today* **2018**, *21*, 1042–1063. [CrossRef]
22. Kudo, A.; Miseki, Y. Heterogeneous photocatalyst materials for water splitting. *Chem. Soc. Rev.* **2009**, *38*, 253–278. [CrossRef]
23. Fabian, D.M.; Hu, S.; Singh, N.; Houle, F.A.; Hisatomi, T.; Domen, K.; Osterloh, F.E.; Ardo, S. Particle suspension reactors and materials for solar-driven water splitting. *Energy Environ. Sci.* **2015**, *8*, 2825–2850. [CrossRef]
24. Wang, Y.; Suzuki, H.; Xie, J.; Tomita, O.; Martin, D.J.; Higashi, M.; Kong, D.; Abe, R.; Tang, J. Mimicking natural photosynthesis: Solar to renewable H2 fuel synthesis by Z-scheme water splitting systems. *Chem. Rev.* **2018**, *118*, 5201–5241. [CrossRef] [PubMed]
25. Awazu, K.; Fujimaki, M.; Rockstuhl, C.; Tominaga, J.; Murakami, H.; Ohki, Y.; Yoshida, N.; Watanabe, T. A plasmonic photocatalyst consisting of silver nanoparticles embedded in titanium dioxide. *J. Am. Chem. Soc.* **2008**, *130*, 1676–1680. [CrossRef]
26. Daniel, M.-C.; Astruc, D. Gold nanoparticles: Assembly, supramolecular chemistry, quantum-size-related properties, and applications toward biology, catalysis, and nanotechnology. *Chem. Rev.* **2004**, *104*, 293–346. [CrossRef]
27. Zhang, X.; Chen, Y.L.; Liu, R.-S.; Tsai, D.P. Plasmonic photocatalysis. *Rep. Prog. Phys.* **2013**, *76*, 046401. [CrossRef]
28. Kawamura, G.; Arai, T.; Muto, H.; Matsuda, A. Charge behavior in a plasmonic photocatalyst composed of Au and TiO_2. *Catal. Sci. Technol.* **2018**, *8*, 1813–1818. [CrossRef]
29. Linic, S.; Christopher, P.; Ingram, D.B. Plasmonic-metal nanostructures for efficient conversion of solar to chemical energy. *Nat. Mater.* **2011**, *10*, 911–921. [CrossRef]
30. Rosseler, O.; Shankar, M.V.; Du, M.K.L.; Schmidlin, L.; Keller, N.; Keller, V. Solar light photocatalytic hydrogen production from water over Pt and Au/TiO_2 (anatase/rutile) photocatalysts: Influence of noble metal and porogen promotion. *J. Catal.* **2010**, *269*, 179–190. [CrossRef]
31. Liu, Z.W.; Hou, W.B.; Pavaskar, P.; Aykol, M.; Cronin, S.B. Plasmon resonant enhancement of photocatalytic water splitting under visible illumination. *Nano Lett.* **2011**, *11*, 1111–1116. [CrossRef]
32. Silva, C.G.; Juarez, R.; Marino, T.; Molinari, R.; Garcia, H. Influence of excitation wavelength (UV or visible light) on the photocatalytic activity of titania containing gold nanoparticles for the generation of hydrogen or oxygen from water. *J. Am. Chem. Soc.* **2011**, *133*, 595–602. [CrossRef]
33. Ingram, D.B.; Linic, S. Water splitting on composite plasmonic-metal/semiconductor photoelectrodes: Evidence for selective plasmon-induced formation of charge carriers near the semiconductor surface. *J. Am. Chem. Soc.* **2011**, *133*, 5202–5205. [CrossRef]
34. Zhan, Z.; An, J.; Zhang, H.; Varghese, H.; Zheng, L. Three-dimensional plasmonic photoanodes based on Au-embedded TiO2 structures for enhanced visible-light water splitting. *ACS Appl. Mater. Interfaces* **2014**, *6*, 1139–1144. [CrossRef]
35. Zhang, J.; Jin, X.; Morales-Guzman, P.I.; Yu, X.; Liu, H.; Zhang, H.; Razzari, L.; Claverie, J.P. Engineering the absorption and field enhancement properties of $Au-TiO_2$ nanohybrids via whispering gallery mode resonances for photocatalytic water splitting. *ACS Nano* **2016**, *10*, 4496–4503. [CrossRef]
36. Nikoobakht, B.; El-Sayed, M.A. Preparation and growth mechanism of gold nanorods (NRs) using seed-mediated growth method. *Chem. Mater.* **2003**, *15*, 1957–1962. [CrossRef]
37. Gole, A.; Murphy, C.J. Seed-mediated synthesis of gold nanorods: Role of the size and nature of the seed. *Chem. Mater.* **2004**, *16*, 3633–3640. [CrossRef]
38. Kawamura, G.; Yang, Y.; Nogami, M. End-to-end assembly of CTAB-stabilized gold nanorods by citrate anions. *J. Phys. Chem. C* **2008**, *112*, 10632–10636. [CrossRef]
39. Lee, J.; Mubeen, S.; Ji, X.; Stucky, G.D.; Moskovits, M. Plasmonic photoanodes for solar water splitting with visible light. *Nano Lett.* **2012**, *12*, 5014–5019. [CrossRef]

40. Mubeen, S.; Lee, J.; Singh, N.; Kramer, S.; Stucky, G.D.; Moskovits, M. An autonomous photosynthetic device in which all charge carriers derive from surface plasmons. *Nat. Nanotechnol.* **2013**, *8*, 247–251. [CrossRef]
41. Zhong, Y.; Ueno, K.; Mori, Y.; Shi, X.; Oshikiri, T.; Murakoshi, K.; Inoue, H.; Misawa, H. Plasmon-assisted water splitting using two sides of the same SrTiO$_3$ single-crystal substrate: Conversion of visible light to chemical energy. *Angew. Chem. Int. Ed.* **2014**, *53*, 10350–10354. [CrossRef]
42. Takakura, R.; Oshikiri, T.; Ueno, K.; Shi, X.; Kondo, T.; Masuda, H.; Misawa, H. Water splitting using a three-dimensional plasmonic photoanode with titanium dioxide nano-tunnels. *Green Chem.* **2017**, *19*, 2398–2404. [CrossRef]
43. Shi, L.; Li, Z.; Dao, T.D.; Nagao, T.; Yang, Y. A synergistic interaction between isolated Au nanoparticles and oxygen vacancies in an amorphous black TiO$_2$ nanoporous film: Toward enhanced photoelectrochemical water splitting. *J. Mater. Chem. A* **2018**, *6*, 12978–12984. [CrossRef]
44. Naldoni, A.; Guler, U.; Wang, Z.; Marelli, M.; Malara, F.; Meng, X.; Besteiro, L.V.; Govorov, A.O.; Kildishev, A.V.; Boltasseva, A.; et al. Broadband hot-electron collection for solar water splitting with plasmonic titanium nitride. *Adv. Opt. Mater.* **2017**, *5*, 1601031. [CrossRef]
45. Sivula, K.; Formal, F.L.; Gratzel, M. Solar water splitting: Progress using hematite (α-Fe$_2$O$_3$) photoelectrodes. *Chem. Sus. Chem.* **2011**, *4*, 432–449. [CrossRef] [PubMed]
46. Chernomordik, B.D.; Russell, H.B.; Cvelbar, U.; Jasinski, J.B.; Kumar, V.; Deutsch, T.; Sunkara, M.K. Photoelectrochemical activity of as-grown, α-Fe$_2$O$_3$ nanowire array electrodes for water splitting. *Nanotechnology* **2012**, *23*, 194009. [CrossRef] [PubMed]
47. Li, J.; Cushing, S.K.; Zheng, P.; Meng, F.; Chu, D.; Wu, N. Plasmon-induced photonic and energy-transfer enhancement of solar water splitting by a hematite nanorod array. *Nat. Commun.* **2013**, *4*, 2651. [CrossRef]
48. Xi, L.; Lange, K.M. Surface modification of hematite photoanodes for improvement of photoelectrochemical performance. *Catalysts* **2018**, *8*, 497. [CrossRef]
49. Bassi, P.S.; Sritharan, T.; Wong, L.H. Recent progress in iron oxide based photoanodes for solar water splitting. *J. Phys. D Appl. Phys.* **2018**, *51*, 473002. [CrossRef]
50. Formal, F.L.; Gratzel, M.; Sivula, K. Controlling photoactivity in ultrathin hematite films for solar water splitting. *Adv. Funct. Mater.* **2010**, *10*, 1099–1107. [CrossRef]
51. Formal, F.L.; Tetreault, N.; Cornuz, M.; Moehl, T.; Gratzel, M.; Sivula, K. Passivating surface states on water splitting hematite photoanodes with alumina overlayers. *Chem. Sci.* **2011**, *2*, 737–743. [CrossRef]
52. Meng, F.; Li, J.; Cushing, S.K.; Bright, J.; Zhi, M.; Rowley, J.D.; Hong, Z.; Manivannan, A.; Bristow, A.D.; Wu, N. Photocatalytic water oxidation by hematite/reduced graphene oxide composites. *ACS Catal.* **2013**, *3*, 746–751. [CrossRef]
53. Li, D.; Shi, J.; Li, C. Transition-metal based electrocatalysts as cocatalysts for photoelectrochemical water splitting: A mini review. *Small* **2018**, *14*, 1704179. [CrossRef]
54. Dias, P.; Andrade, L.; Mendes, A. Hematite-based photoelectrode for solar water splitting with very high photovoltage. *Nano Energy* **2017**, *38*, 218–231. [CrossRef]
55. Chong, R.; Wang, B.; Su, C.; Li, D.; Mao, L.; Cheng, Z.; Zhang, L. Dual-functional CoAl layered double hydroxide decorated α-Fe$_2$O$_3$ as an efficient and stable photoanode for photoelectrochemical water oxidation in neutral electrolyte. *J. Mater. Chem. A* **2017**, *5*, 8583–8590. [CrossRef]
56. Liu, G.; Zhao, Y.; Yao, R.; Li, N.; Wang, M.; Ren, H.; Li, J.; Zhao, C. Realizing high performance solar water oxidation for Ti-doped hematite nanoarrays by synergistic decoration with ultrathin cobalt-iron phosphate nanolayers. *Chem. Eng. J.* **2019**, *355*, 49–57. [CrossRef]
57. Chong, R.; Du, Y.; Chang, Z.; Jia, Y.; Qiao, Y.; Liu, S.; Liu, Y.; Zhou, Y.; Li, D. 2D Co-incorporated hydroxyapatite nanoarchitecture as a potential efficient wxygen evolution cocatalyst for boosting photoelectrochemical water splitting on Fe$_2$O$_3$ photoanode. *Appl. Catal. B Environ.* **2019**, *250*, 224–233. [CrossRef]
58. Ramadurgam, S.; Lin, T.-G.; Yang, C. Aluminum plasmonics for enhanced visible light absorption and high efficiency water splitting in core-multishell nanowire photoelectrodes with ultrathin hematite shells. *Nano Lett.* **2014**, *14*, 4517–4522. [CrossRef] [PubMed]
59. Panacek, A.; Kvitek, L.; Prucek, R.; Kolar, M.; Vecerova, R.; Pizurova, N.; Sharma, V.K.; Nevecna, T.; Zboril, R. Silver colloid nanoparticles: Synthesis, characterization, and their antibacterial activity. *J. Phys. Chem. B* **2006**, *110*, 16248–16253. [CrossRef]
60. Peerakiatkhajohn, P.; Yun, J.-H.; Chen, H.; Lyu, M.; Butburee, T.; Wang, L. Stable hematite nanosheet photoanodes for enhanced photoelectrochemical water splitting. *Adv. Mater.* **2016**, *28*, 6405–6410. [CrossRef]

61. Mertens, J.; Eiden, A.L.; Sigle, D.O.; Huang, F.; Lombardo, A.; Sun, Z.; Sundaram, R.S.; Colli, A.; Tserkezis, C.; Aizpurua, J.; et al. Controlling busnanometer gaps in plasmonic dimers using graphene. *Nano Lett.* **2013**, *13*, 5033–5038. [CrossRef]
62. Thacker, V.V.; Herrmann, L.O.; Sigle, D.O.; Zhang, T.; Liedl, T.; Baumberg, J.J.; Keyse, U.F. DNA origami based assembly of gold nanoparticle dimers for surface-enhanced Raman scattering. *Nat. Commun.* **2014**, *5*, 3448. [CrossRef]
63. Sigle, D.O.; Zhang, L.; Ithyrria, S.; Dubertret, B.; Baumberg, J.J. Ultrathin CdSe in plasmonic nanogaps for enhanced photocatalytic water splitting. *J. Phys. Chem. Lett.* **2015**, *6*, 1099–1103. [CrossRef]
64. Dutta, A.; Naldoni, A.; Malara, F.; Govorov, A.O.; Shalaev, V.M.; Boltasseva, A. Gap-plasmon enhanced water splitting with ultrathin hematite films: The role of plasmonic-based light trapping and hot electrons. *Faraday Discuss.* **2019**, *214*, 283–295. [CrossRef] [PubMed]
65. Wang, S.; Gao, Y.; Miao, S.; Liu, T.; Mu, L.; Li, R.; Fan, F.; Li, C. Positioning the water oxidation reaction sites in plasmonic photocatalysts. *J. Am. Chem. Soc.* **2017**, *139*, 11771–11778. [CrossRef] [PubMed]
66. Zhong, Y.; Ueno, K.; Mori, Y.; Oshikiri, T.; Misawa, H. Cocatalyst effects on hydrogen evolution in a plasmon-induced water-splitting system. *J. Phys. Chem. C* **2015**, *119*, 8889–8897. [CrossRef]
67. Wang, S.; Gao, Y.; Qi, Y.; Li, A.; Fan, F.; Li, C. Achieving overall water splitting on plasmon-based solid Z-scheme photocatalysts free of redox mediators. *J. Catal.* **2017**, *354*, 250–257. [CrossRef]
68. Shi, L.; Zhou, W.; Li, Z.; Koul, S.; Kushima, A.; Yang, Y. Periodically ordered nanoporous perovskite photoelectrode for efficient photoelectrochemical water splitting. *ACS Nano* **2018**, *12*, 6335–6342. [CrossRef] [PubMed]
69. Solarska, R.; Bienkowski, K.; Zoladek, S.; Majcher, A.; Stefaniuk, T.; Kulesza, P.J.; Augustynski, J. Enhanced water splitting at thin film tungsten trioxide photoanodes bearing plasmonic gold-polyoxometalate particles. *Angew. Chem. Int. Ed.* **2014**, *53*, 14196–14200. [CrossRef]
70. Hu, D.; Diao, P.; Xu, D.; Wu, Q. Gold/WO3 nanocomposite photoanodes for plasmonic solar water splitting. *Nano Res.* **2016**, *9*, 1735–1751. [CrossRef]
71. Duan, H.; Xuan, Y. Enhancement of light absorption of cadmium sulfide nanoparticle at specific wave band by plasmon resonance shifts. *Phys. E Low Dimens. Syst. Nanostruct.* **2011**, *43*, 1475–1480. [CrossRef]
72. Torimoto, T.; Horibe, H.; Kameyama, T.; Okazaki, K.; Ikeda, S.; Matsumura, M.; Ishikawa, A.; Ishihara, H. Plasmon-enhanced photocatalytic activity of cadmium sulfide nanoparticle immobilized on silica-coated gold particles. *J. Phys. Chem. Lett.* **2011**, *2*, 2057–2062. [CrossRef]
73. Thiyagarajan, P.; Ahn, H.J.; Lee, J.S.; Yoon, J.C.; Jiang, J.H. Hierarchical metal/semiconductor nanostructure for efficient water splitting. *Small* **2013**, *9*, 2341–2347. [CrossRef]
74. Wu, M.; Chen, W.J.; Shen, Y.H.; Huang, F.Z.; Li, C.H.; Li, S.K. In situ growth of matchlike ZnO/Au plasmonic heterostructure for enhanced photoelectrochemical water splitting. *ACS Appl. Mater. Interfaces* **2014**, *6*, 15052–15060. [CrossRef]
75. Wang, B.; Li, R.; Zhang, Z.; Zhang, W.; Yan, X.; Wu, X.; Cheng, G.; Zheng, R. Novel Au/Cu$_2$O multi-shelled porous heterostructures for enhanced efficiency of photoelectrochemical water splitting. *J. Mater. Chem. A* **2017**, *5*, 14415–14421. [CrossRef]
76. Sinatra, L.; LaGrow, A.P.; Peng, W.; Kirmani, A.R.; Amassian, A.; Idriss, H.; Bakr, O.M. A Au/Cu$_2$O-TiO$_2$ system for photo-catalytic hydrogen production. A pn-junction effect or a simple case of in situ reduction? *J. Catal.* **2015**, *322*, 109–117. [CrossRef]

© 2019 by the authors. Licensee MDPI, Basel, Switzerland. This article is an open access article distributed under the terms and conditions of the Creative Commons Attribution (CC BY) license (http://creativecommons.org/licenses/by/4.0/).

Review

Plasmonic Photocatalysts for Microbiological Applications

Maya Endo-Kimura and Ewa Kowalska *

Institute for Catalysis, Hokkaido University, N21 W10, Sapporo 001-0021, Japan; m_endo@cat.hokudai.ac.jp
* Correspondence: kowalska@cat.hokudai.ac.jp

Received: 26 June 2020; Accepted: 22 July 2020; Published: 23 July 2020

Abstract: Wide-bandgap semiconductors modified with nanostructures of noble metals for photocatalytic activity under vis irradiation due to localized surface plasmon resonance (LSPR), known as plasmonic photocatalysts, have been intensively investigated over the last decade. Most literature reports discuss the properties and activities of plasmonic photocatalysts for the decomposition of organic compounds and solar energy conversion. Although noble metals, especially silver and copper, have been known since ancient times as excellent antimicrobial agents, there are only limited studies on plasmonic photocatalysts for the inactivation of microorganisms (considering vis-excitation). Accordingly, this review has discussed the available literature reports on microbiological applications of plasmonic photocatalysis, including antibacterial, antiviral and antifungal properties, and also a novel study on other microbiological purposes, such as cancer treatment and drug delivery. Although some reports indicate high antimicrobial properties of these photocatalysts and their potential for medical/pharmaceutical applications, there is still a lack of comprehensive studies on the mechanism of their interactions with microbiological samples. Moreover, contradictory data have also been published, and thus more study is necessary for the final conclusions on the key-factor properties and the mechanisms of inactivation of microorganisms and the treatment of cancer cells.

Keywords: plasmonic photocatalyst; vis-responsive material; antimicrobial effect; antifungal properties; antiviral effect; disinfection; bacteriocyte; noble metal; LSPR; environmental purification

1. Introduction

Clean water and sanitation, one of Sustainable Development Goals (adopted in United Nations summit in 2015), are highly necessary to achieve a better and sustainable society. Accordingly, the efficient technologies of water and wastewater treatment, also being accessible to everyone, have been considered as an urgent issue. It has been assumed that, in the world, one in three people do not have access to clean water for drinking and two out of five cannot wash hands properly, due to a lack of basic washing facilities [1]. Moreover, nearly 1000 children die every day due to diarrheal diseases, connected with contaminated water [1]. The coronavirus disease 2019/2020 (COVID-19) has unfortunately clarified the importance of sanitation, hygiene and adequate access to clean water for preventing the spear of infectious disease. In developing countries with insufficient water purification facilities, people must use water contaminated with feces, heavy metals, inorganic and organic pollutants, and pathogenic microorganisms. Waterborne pathogens might cause acute (e.g., diarrhea) and chronic (e.g., infectious hepatitis, cancer) health effects. Furthermore, not only slight symptoms (e.g., diarrhea), but also serious ones have been reported, such as cholera, hemolytic-uremic syndrome by *Escherichia coli* (*E. coli*) O157, and even cancer [2]. Whereas, even in developed countries, despite the fact that chemical contamination is well monitored and controlled [3–5], unpredicted outbreaks of pathogenic microorganisms have also occurred. It should be pointed out that, not only imported infectious diseases (e.g., typhus (caused by *Salmonella enterica* serovar Typhi, *Salmonella enterica* serovar Paratyphi A or *Rickettsia prowazekii*) and polio (Poliovirus)), but also illnesses caused by indigenous microorganisms via fecal-oral route (e.g., contaminated drinking water, or pools and public spas), have been observed, including a cryptosporidiosis by *Cryptosporidium* species (sp.), which is resistant to chlorination, and a Legionnaires' disease by *Legionella* sp.

In association with the development of society, water demand is increasing for both domestic and industrial applications, but unfortunately water sources are limited. Therefore, it is necessary to reuse water without additional loads for the environment. In developed countries, the treatment of (waste) water is carried out by combinations of various methods, including precipitation, filtration (e.g., membrane technologies), adsorption, microbial purification (by activated sludge), oxidation (ozonization, chlorination, advanced oxidation technologies (AOPs)), depending on the kind and levels of contamination (water and municipal or industrial wastewater). Chlorination is commonly used for various types of wastewater and water, mainly as a final step of water disinfection, because chlorine can completely inactivate some microorganisms. However, it might also negatively influence aquatic organisms, due to the ability to bind to nitrogen and organic compounds, resulting in the formation of highly toxic by-products (e.g., trihalomethane). In addition, chlorine disinfection has a low effect on some pathogenic protozoa (e.g., *Cryptosporidium* sp.) and viruses (e.g., norovirus). Similarly, UV-irradiation and ozonization both possess advantages and disadvantages (i.e., high efficiency against microorganisms, including viruses (UV and ozone), but low sensitivity to some viruses, such as adenovirus, no residual effect (UV), own toxicity, and the formation of toxic by-products (ozone)). Therefore, clean, inexpensive, and environmentally friendly methods for water purification and wastewater treatment are highly needed. Moreover, point-of-use water treatment methods are strongly desired in many areas that are inaccessible to municipal water treatment technologies.

Among oxidation methods, AOPs have been indicated as highly efficient and are recommended, due to the in situ generation of powerful reactive oxygen species (ROS), including hydroxyl radicals (HO$^\bullet$). However, many of them involve high investments and operating costs. Heterogenous photocatalysis on semiconductor oxides also belong to AOPs, since under irradiation the semiconductor is excited, and generates charge carriers (i.e., electrons in the conduction band (CB) and holes in the valence band (VB)) might form ROS in the presence of oxygen and water. There is only one main limitation of heterogenous photocatalysis as AOPs, i.e., low activity under solar radiation because the most active photocatalysts have wide bandgap, and thus must be excited with UV, being only ca. 3–4% of solar spectrum. Accordingly, vis-responsive materials have been intensively investigated for efficient photocatalysis under real solar conditions, including plasmonic photocatalysts (i.e., wide-bandgap semiconductor modified with plasmonic NPs, such as gold, silver, platinum, and copper). Although, plasmonic photocatalysis is a new topic, various reports on plasmonic photocatalysts have already been published, including some on antimicrobial activity. Accordingly, this review summarizes and discusses the available literature on plasmonic photocatalysis for microbial inactivation. Additionally, some other biological applications of plasmonic photocatalysts are presented.

2. Plasmonic Photocatalysis

Titanium(IV) oxide (titania, TiO_2) is one of the most studied semiconductor photocatalysts, due to various advantages, such as high photocatalytic activity, stability, abundance, inexpensiveness and low toxicity (except for the toxicity of nanomaterials [6]). It should be pointed out that photocatalysis has been considered as one of the best methods for environmental purification, since additional chemical compounds, such as strong oxidants (ozone, hydrogen peroxide (H_2O_2), chlorine [7–14]) are not introduced into the environment [15,16], and the energy consumption is much lower than that in other AOPs (e.g., wet air oxidation [17], supercritical water oxidation [18], or H_2O_2/UV-C [19]). However, titania has wide bandgap of ca. 3.0–3.2 eV (depending on the crystalline form), and thus it must be excited by UV-light irradiation (absorption edge at ca. 385–410 nm). The principle of heterogeneous photocatalysis might be presented by the band-structure model. In brief, irradiation with higher or equal energy than the bandgap excites electrons from VB to CB, generating positive holes in VB. Generated electrons (e^-) and holes (h^+) reduce and oxidize adsorbed substances (e.g., organic compounds, water), respectively. When electrons and holes react with water and oxygen, ROS are formed (e.g., hydroxyl radical (HO$^\bullet$), superoxide anion radical ($O_2^{\bullet-}$) and H_2O_2). On the other hand, charge carriers (e^-/h^+) might recombine either in the bulk or on the surface of titania, which is typical for all semiconductors, resulting in a lower than expected efficiency of photocatalytic reactions. Indeed, the quantum yields of photocatalytic reactions are usually much lower than 100% (e.g., 4% has been reported for the generation of hydroxyl radicals [20]). In addition, titania cannot absorb a large fraction of solar light, which contains only about 3–5% of UV, due to its wide bandgap. Therefore, in order to improve the photocatalytic activity of titania, a number of studies have been performed, focusing on the utilization of visible light and the suppression of charge carrier recombination (e.g., surface modification, doping (nitrogen, sulfur, carbon and self-doping) and preparation of coupled nanostructures [21–30]). Although titania doping has resulted in the appearance of vis activity, dopants could be recombination centers, and thus decreasing photocatalytic activity under UV irradiation.

Modification with noble metal nanoparticles (NMNPs) (e.g., Ag, Au, Pt, Pd) seems to be the most promising, since it is well known that NM (noble metal) works as an electron sink under UV irradiation, inhibiting the e^-/h^+ recombination [31–33]. When NMNPs are in a contact with semiconductors, a Schottky barrier is established, hindering the charge carriers' recombinations [34]. On the other hand, under vis irradiation, titania is activated by the plasmonic properties of NM (plasmonic photocatalysis) [35,36]. NM-modified titania photocatalysts show localized surface plasmon resonance (LSPR), and obtained vis response depends on the kind of metal (e.g., LSPR for small spherical NPs of Ag, Cu, and Au at ca. 420, 600, and 550 nm, respectively), their morphology, size, and environment (refractive index of medium). For example, Au nanospheres of several nanometers show LSPR at ca. 520 nm and bathochromic shift has been observed with an increase in the particle size [37]. A similar shift has been obtained for nanorods with an increase in their aspect ratio [37]. The most important findings by action spectrum analysis (action spectra resembling absorption spectra) confirm that plasmon resonance is responsible for vis-activity of wide-bandgap semiconductors [35,38,39].

Three main mechanisms of plasmonic photocatalysis under vis irradiation have been proposed (i.e., (i) charge transfer; (ii) energy transfer; (iii) plasmonic heating), as follows:

(i) Under irradiation with LSPR wavelength, NMNPs absorb the photons and "hot" electrons are transferred to the CB of the semiconductor. The oxygen on the semiconductor surface is reduced by the electron and the resultant electron-deficient NMNP might oxidize substances to recover to its original metallic state. For example, Tian et al. have shown an electron transfer from Au to titania and from electron donor to Au by the observation of anodic and cathodic photocurrents in Au/TiO$_2$/indium tin oxide (ITO) and TiO$_2$/Au/ITO, respectively [35,40,41].

(ii) Plasmon resonance energy transfer (PRET) might occur for energy overlapping between plasmonic metals and semiconductors (LSPR band of metals and the band gap absorption of semiconductors) (e.g., Au NPs (LSPR of ca. 2.2 eV) and Cu$_2$O (ca. 2.2 eV) [42]). Accordingly, although energy transfer is not expected for Au-modified titania, due to the wide band gap of titania (ca. 3–3.2 eV), PRET has been claimed for Au-modified nitrogen-doped titania, due to the generation of new energy levels (inside bandgap) [43]. Additionally, PRET has already been proposed for application in biomolecular imaging, e.g., cytochrome c in living cells [44].

(iii) Plasmonic heating has also been considered as one of the key factors for photocatalytic activities (e.g., oxidation of HCHO [45], degradation of methylene blue [46], reduction of 4-nitrobenzenethiol [47] and CO$_2$ reduction [48]). According to Chen et al., the plasmon-induced heating on Au NPs might activate organic molecules to induce their oxidation [45]. Although other studies have claimed that energy release by plasmonic heating is insufficient for chemical bond cleavage, it seems that local heating must be important for microorganism killing, as microorganisms are highly sensitive to any environmental changes.

As described above, plasmonic photocatalysts induce redox reactions of organic compounds on the surface of NMNP and titania. However, those photocatalysts do not possess as strong oxidation abilities as that of titania under UV irradiation (2–3 orders in magnitude lower photocatalytic activity under vis than that under UV have been reported [33,38,39]), due to the low energy of visible light and fast charge carrier recombination (i.e., back electron transfer to NMNPs (e.g., NM→TiO_2→NM) [49]). Therefore, various studies on the improvement of the photocatalytic activities of plasmonic photocatalysts have been proposed (e.g., modification of NMNPs' morphology [50], heterojunction with other semiconductors/complexes [33,51,52], deposition of second NMNPs [53–55], the addition of adsorbents [56–58], and morphology modifications of semiconductor [49,59–61]).

3. Heterogeneous Photocatalysis for Microbiological Purposes

The first report on microbial inactivation by heterogeneous photocatalysis was published by Matsunaga et al. In 1985 [62]. It was found that irradiated titania and platinum (Pt)-loaded titania could sterilize yeast, Gram-positive and Gram-negative bacteria, and algae, due to the photoelectrochemical oxidation of coenzyme A (CoA). Accordingly, it has been proposed that the oxidation of CoA has inhibited respiratory activity, inducing the death of cells [62–64]. The bacterial killing mechanisms by titania photocatalysts has been summarized comprehensively by Markowska-Szczupak et al. [65], pointing to three main pathways (i.e., peroxidation of cell membrane phospholipids [66,67], direct DNA damage [64], and the oxidation of CoA [62]). All these mechanisms are attributed to ROS formation from oxygen and water by reduction/oxidation reactions on irradiated titania. It has been reported that mainly HO$^\bullet$ radicals, the most active ROS, participate in bactericidal effect [4,68], despite their short half-lifetime. Moreover, other ROS (e.g., $O_2^{\bullet-}$ and H_2O_2) and direct redox reactions by charge carriers (surface e^-/h^+) have also been proposed for bacterial inactivation [61,69]. In addition, Du et al. have found that proteins are an initial target of radicals (HO$^\bullet$), preceding lipids and DNA [70]. On the other hand, some bacteria possess their protection system against oxidation stress (i.e., forming ROS scavenger enzymes, such as superoxide dismutase (SOD) and catalase (CAT)). SOD enzymes might decompose ROS to H_2O_2 and O_2, and not only in the cytoplasm, but also on the surface of cells [71,72]. Generated H_2O_2 is further decomposed to O_2 and H_2O by CAT. Importantly, some bacteria secreting CAT might be released outside of cells [73,74], which also might contribute to their resistance against extracellular ROS. For example, it has been found that the levels of SOD and CAT increase during first 30 min of titania irradiation with UV/vis, and then decrease (after 60 min) [75]. According to these changes, the two-step mechanism of bacteria response (Gram-negative *E. coli* and Gram-positive *Staphylococcus epidermidis*) has been proposed, where the changes in the enzyme activity during the photocatalytic disinfection might indicate that the defense capacity has been overwhelmed by the rapidly created ROS at the initial stage. The correlation between high content of generated HO$^\bullet$ radicals during 90 min of irradiation and the loss of enzymatic activity suggests that oxidative stress might act as an important step, in which the photocatalyst induces bacterial death.

Moreover, heat shock proteins protect cells under an oxidative stressed environment [76], and DNA repair enzymes might repair damaged DNA [77]. In addition to enzymes, cell morphology also influences microbial inactivation by photocatalysis. It has been proposed that an inactivation effect of titania photocatalyst against microorganisms might be put in the following order: Gram-negative bacteria > Gram-positive bacteria >> fungi (yeast) > fungi (mold) [24], possibly due to the complexity of cells. Gram-negative bacteria have a thin (ca. 10 nm) peptidoglycan layer, whereas the cell walls in Gram-positive bacteria are much thicker (ca. 10–100 nm). Moreover, fungal cells (i.e., eukaryotic organisms), have rigid cell walls, containing chitin, proteins, polysaccharide polymers and lipids [78]. Furthermore, they have nuclear membranes and multicellular structures (except for unicellular organisms).

It should be pointed out that the intrinsic and surface properties of photocatalysts (e.g., bandgap energy, specific surface area, crystallite/particle size, aggregation, surface charge, defects distribution and impurities content), are also crucial for the overall antimicrobial effect [49,54,75,79]. For example, rutile (polymorphic form of titania) has shown much higher activity than anatase (usually the most active titania form) under natural indoor light against mold fungi, probably because of its narrower bandgap, and thus ability to absorb more photons [79]. Moreover, it has been found that sporulation and mycotoxin generation have been highly inhibited by photocatalytic reactions on titania photocatalysts, as exemplarily shown in Figure 1. Interestingly, some fungi could be stimulated by titania presence (i.e., although the growth of *Pseudallescheria boydii* and *Aspergillus versicolor* has been highly inhibited by titania and light, the growth of *Stachybotrys chartarum* has been accelerated by titania). It should be remembered that different fungi species have different water and nutrient demand. Accordingly, adsorbed water and some impurities on the titania surface might either stimulate or inhibit the fungal growth [79].

Figure 1. Images of three-day growth of *Aspergillus melleus*: (**a**) in the dark and (**b**) under irradiation in the presence of commercial titania photocatalyst P25 (after homogenization); adapted with permission (after formatting) from [79]. Copyright (2015) Elsevier.

Various applications of titania photocatalysts for microbial inactivation have already been proposed, including commercialized ones (e.g., the surfaces of equipment/walls covered with titania for efficient sterilization in hospitals). Titania modified with glucose has also been proposed for water disinfection (e.g., by placing concrete plates covered with titania in aqueous reservoirs or disinfecting water in TiO_2/concrete-covered containers used for storage of drinking water or fish farms), as exemplified for a home aquarium in Figure 2a [75]. Other microbiological applications of titania photocatalysts have also been proposed, including for biomedical purposes (e.g., for decomposition of human breast cancer cells (Figure 2b) [80]).

3.1. Plasmonic Photocatalysts for Inactivation of Microorganisms

Plasmonic photocatalysts are surely promising candidates as antimicrobial agents. Although there are only several studies on microbial inactivation on plasmonic photocatalysts (i.e., under vis irradiation (not considering activities of NMNP-modified titania tested under UV and in the dark)), high activity is expected, coming from: (i) intrinsic activity of noble metals; (ii) enhanced activity under UV irradiation (inhibition of e^-/h^+ recombination); (iii) activity under vis irradiation, resulting from the plasmonic activation of wide-bandgap semiconductors. The exemplary effects of the intrinsic properties of NM are described below.

Although the safety of silver for human and animals has been confirmed, it is also known that a slight amount of silver shows remarkable antimicrobial efficiency. Generally, the antibacterial activity of silver is more effective for Gram-negative bacteria than Gram-positive ones, presumably due to the differences in their biological structures. Although, the antimicrobial effect of silver has been well-studied, the mechanism of its action is still under discussion. Therefore, different antimicrobial pathways of Ag NPs have been proposed in the literature, as follows:

(1) Adsorption of Ag cations on a negatively charged bacterial cell wall, followed by the collapse of the cell wall and the plasma membrane, and a release of substances outside of the cell, leading to cell lysis and death [81,82];

(2) Interaction with thiol groups of transport and respiratory enzymes that are fatal to the survival of cells, causing the uncoupling of respiration from the synthesis of ATP [83];

(3) Collapse of membrane potential and a leakage of protons, as a result of the destabilization of plasma membrane and de-energization of bacterial cells [84].

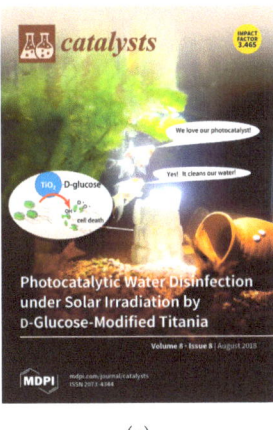
(a)

(b)

Figure 2. Examples of possible microbiological applications of titania photocatalysts: (a) modified titania with D-glucose for disinfection of water tanks (journal cover image; https://res.mdpi.com/data/covers/catalysts/big_cover-catalysts-v8-i8.png) and (b) decomposition of breast adenocarcinoma MCF7 cells (ATCC® HTB-22™, breast adenocarcinoma cell line from human) on commercial titania (P25) irradiated with UVA/vis (graphical abstract); adapted with permission from [75] and [80], respectively. Copyright 2018 and 2020, respectively, Creative Commons Attribution.

Accordingly, it has been considered that the release of Ag cations from carrier (free Ag cations in solution/suspension) is important for bactericidal activity [85–89]. For example, it has been reported that the activity of Ag/TiO$_2$ in the dark is higher than that in the presence of light, due to the instability of Ag on TIO$_2$ without irradiation (i.e., Ag release) [85]. In order to obtain the continuous release of Ag ions from Ag NPs (to keep long-term antimicrobial activities), Ag NPs have been deposited on some supporting materials (e.g., silica [90], clay [91], and zeolite [57]). Silver NPs also possess antimicrobial activities against fungi, algae, and viruses. The proposed mechanism of antivirus action is based on the interaction between Ag NPs and glycoproteins from the virus surface, which induces an inhibition of virus binding to target cells, as well as the possible inhibition of virus replication when viral cells are penetrated by Ag NPs [92,93].

Copper (Cu) is also considered as an excellent purifier, due to its low price and high activity. The proposed bactericidal mechanisms are based on the adsorption of Cu ions on the surface of bacteria and, subsequently: (1) the structure of surface proteins is denatured [94], and/or (2) adsorbed Cu ions induce oxidative stress in bactericidal process [95]. The accumulation of Cu ions inside bacteria has been reported as the main mechanism of bacteria inactivation by Cu-modified blotting paper (i.e., accumulation of Cu ions from direct contact with Cu NPs in the paper [96]). Furthermore, in the case of copper oxides, it has been proposed that Cu$_2$O has higher bactericidal activity than CuO and Ag [94]. On the other hand, Cu NPs exhibit lower toxicity than Ag NPs against some species of fungi [97].

Recently, the bactericidal activity of gold NPs has been proposed, however, it is still under discussion. There are many contradictory reports on the activity of Au NPs (i.e., Au NPs exhibit negligible activity for both Gram-negative and Gram-positive bacteria [98–100], possibly due to a lack of ability to surround bacterial cells [101], and oppositely, high activity [100,102–104]). For size dependence, Zheng et al. have shown that only Au nanoclusters (sub-nanometer size) exhibit bactericidal activity, but NPs (>2 nm) do not [100]. On the contrary, Badwaik et al. have demonstrated that, although 25-nm Au NPs show low activity, larger NPs (60 and 120-nm Au NPs) are highly active [103]. It has been proposed that the bactericidal mechanisms of Au NPs is based on: (i) the ability to change the membrane potential; (ii) inhibition of ATP synthase activities to decrease the ATP level; (iii) inhibition of the subunit of ribosome for tRNA binding and, importantly, all these mechanisms are independent on ROS generation [104]. Moreover, Au NPs possess antifungal activity, depending on the particle size of Au NPs (i.e., activity increase with a decrease in the particle size, due to an increase in the specific surface area of Au NPs, enhancing the interaction between Au and the binding sites of the plasma membrane proteins, and thus resulting in the inhibition of H^+-ATPase-mediated proton pumping [105]). Au NPs also show antiviral activity (e.g., inhibition of viral attachment, entry, and cell-to-cell spread [106]).

Therefore, due to the high antimicrobial activity of NMNPs, as described above, various studies on silver- [57,69,85,89,107–109] and copper (and copper oxide)-modified titania have been reported [52,110–112]. Moreover, Au-modified titania has also been investigated, due to high photocatalytic activity (ROS generation), and the ability to disrupt the electron transport in the cells [59]. The proposed mechanisms of microbial inactivation and exemplary applications of plasmonic photocatalysts with antimicrobial properties (summarized in Table 1 for property-governed activity) are briefly discussed in the following paragraphs.

It is widely known that silver-modified titania shows high antimicrobial activity, due to the intrinsic activity of silver in the dark and enhanced activity of titania under irradiation. The release of Ag^+, silver adsorption onto bacteria, and ROS generation are important factors for bactericidal action. For example, Castro et al. have suggested that increased Ag^+ content on the surface of TiO_2 by vis pre-irradiation (oxidation of Ag^0 to Ag^+ and Ag^{2+}) enhances bacteriostatic activity in the dark [109]. Importantly, not only the inactivation of bacteria, but also the decomposition of bacterial cells is essential as the complete removal of possible allergen [113]. Accordingly, the destruction of bacteria cells (e.g., protoplast formation; Figure 3e) and initiation of their complete decomposition (mineralization), estimated by continuous CO_2 evolution (Figure 4), has been observed on Ag/TiO_2 photocatalysts only under irradiation (negligible CO_2 evolution in the dark due to bacteria breathing), as exemplarily shown in Figures 4 and 5 [114,115].

The importance of ROS generation for the inactivation and decomposition of bacterial cells has been proposed by many researchers (e.g., vis irradiation (λ > 420 nm) of $Ag/AgBr/TiO_2$ nanotube array causes the oxidative attack of E. coli by $HO^\bullet$, O_2^-, h^+, and Br^0 from the exterior to the interior, resulting in cell death as the primary mechanism of photo-electrocatalytic (PEC) inactivation (Figure 5) [116]).

Silver-modified TiO_2 samples show high antibacterial activity under visible light irradiation (much higher than that in the dark) with complete decomposition of E. coli cells [114,115]. The surface properties of photocatalysts are crucial for the antimicrobial effect and, generally, a decrease in particle sizes results in an increase in activity, due to a larger interface between microorganisms and photocatalysts. Moreover, morphology-controlled titania NPs (e.g., faceted anatase with octahedral shape-octahedral anatase particles (OAP)), which are highly active for the decomposition of organic compounds, also show high antimicrobial activity against bacteria (E. coli) and fungi (Candida albicans) [117]. Interestingly, it has been found that an Ag-modified OAP sample has superior bactericidal activity under visible light than that under UV–vis irradiation (Figure 6), possibly because of Ag oxidation with simultaneous release of Ag cations (an electron transfer from Ag to the CB of TiO_2 under LSPR excitation; in contrast to a reverse electron transfer under UV irradiation (i.e., from TiO_2 to Ag)), and thus free Ag cations might penetrate the cell membrane, resulting in the death of bacteria [117].

Similarly, Ye et al. have shown that the flower-like hierarchical TiO$_2$/Ag composites with slightly higher activity than spherical ones decompose bacterial cells (both *E. coli* and *Staphylococcus aureus*), due to the synergistic effect of the generated ROS and release of Ag ions [118]. On the other hand, the stability of Ag/TiON (on polyester) photocatalyst has been shown in repetitive experiments for bacterial inactivation under indoor light, suggesting that leaching of Ag into the environment does not happened [119]. Interestingly, van Grieken et al. have shown that substantial lixiviation of Ag (Ag$^+$) occurs in the dark, increasing the bactericidal activity, but UV irradiation stabilizes Ag deposits [85]. Liga et al. have described that increased virus (bacteriophage MS2) adsorption onto silver sites and leaching of Ag$^+$ contribute to virus inactivation [89]. In addition, enhanced ROS generation by Ag-modified titania photocatalysts has correlated with the microbial inactivation under UV–vis irradiation [69,87,120]. Swetha et al. have proposed that ROS in large quantity, generated under UV irradiation of nano titania in aqueous medium, might attack the cell membrane, peroxidized the lipid layer with the release of proteins, K$^+$ ions, nucleic acids and β-D-galactosidase [121]. However, it should be pointed out that different mechanisms of photocatalyst activation, and thus different mechanisms of microorganism inactivation should be considered under vis and UV irradiation, i.e., LSPR excitation with possible "hot" electron transfer from Ag to TiO$_2$ (resulting in more positively charged silver with high possibility of its leakage) and TiO$_2$ excitation with electron transfer to Ag (resulting in less positively charged Ag with lower possibility of leakage), respectively. Interesting data have been found by action spectra analyses (activity depending on the irradiation wavelengths) for titania modified with mono- and bi-metallic Ag and Au NPs, where high stability of Au@Ag/TiO$_2$ photocatalysts under vis irradiation resulted from the possible stabilization of silver by gold (electron transfer from Au to electron-deficient Ag) [122,123].

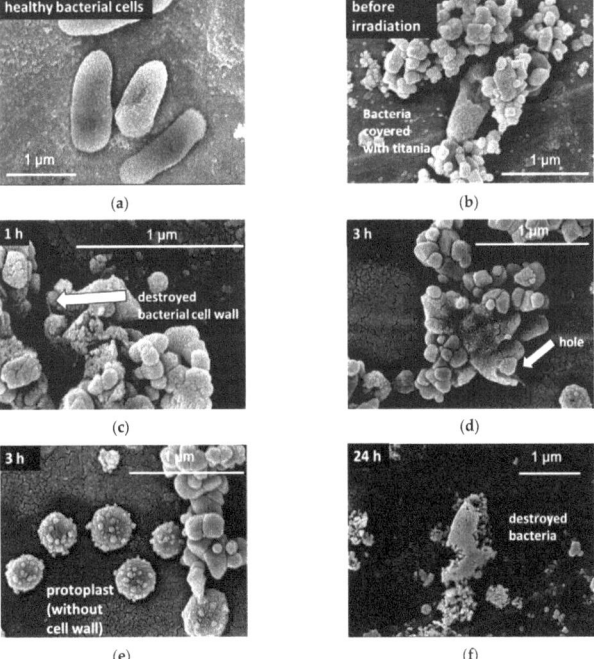

Figure 3. SEM images of the decomposition of *E. coli* under vis (λ > 420 nm) irradiation of Ag/TiO$_2$ photocatalyst: (**a**) healthy bacteria cells, (**b**) bacteria covered with titania before photocatalytic reaction, (**c**,**f**) destroyed bacteria cells after: (**c**) 1 h, (**d**,**e**) 3 hand (**f**) 24 hof irradiation; adapted with permission from [115]. Copyright 2018, Creative Commons Attribution.

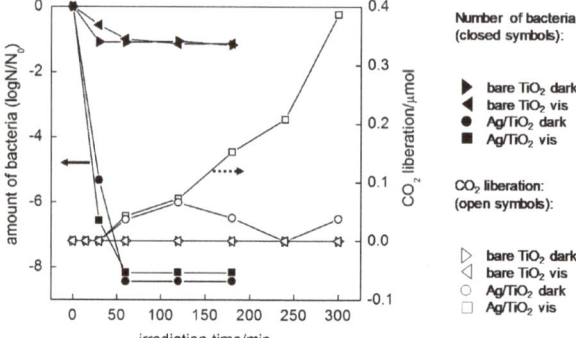

Figure 4. *E. coli* decomposition presented as a decrease in bacteria number (closed symbols) and CO_2 evolution (open symbols), tested in the suspension of titania under vis irradiation ($\lambda > 450$ nm) and in the dark; adapted with permission (after formatting) from [114]. Copyright 2015 Elsevier.

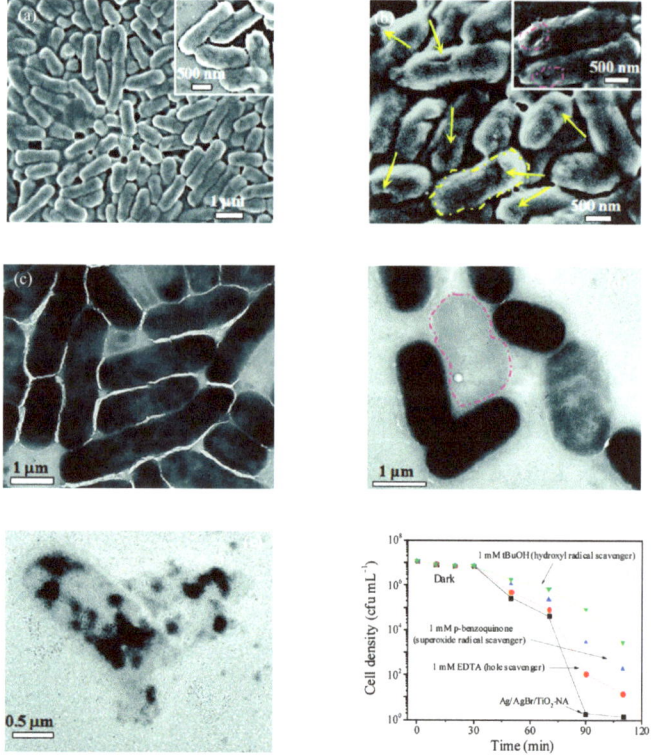

Figure 5. *E. coli* inactivation by PEC on Ag/AgBr/TiO_2^- nanotube electrode: (**a,b**) ESEM images of *E. coli*: (**a**) untreated and (**b**) after PEC inactivation for 80 min; (**c,e**) TEM images of E. coli: (**c**) untreated and after PEC inactivation for: (**d**) 40 min and (**e**) 80 min; (**f**) inhibition of *E. coli* inactivation by scavengers. Reprinted with permission from [116]. Copyright (2012) American Chemical Society.

Recently, the fungicidal activity of Ag-modified titania has also been studied intensively. Although fungal cells are more complex than bacterial cells and quite resistant to antimicrobial agents, plasmonic photocatalysts have been proposed as effective fungicides [117,124–126]. For example, the antifungal activity (*C. albicans*) of Ag-modified faceted titania (OAP) is much higher than that by OAP modified with other NMNPs (Cu, Au, and Pt), both under visible light and in the dark (ca. 50% higher activity under vis during 1 h) [117]. Moreover, it has been reported that the fungicidal activity of Ag–TiO_2 depends on the fungal strain and Ag content [124], and thus the higher content of Ag–TiO_2 or prolonged exposure time might be necessary for specific fungal species [125]. Interestingly, although mono Ag or Cu-modified titania does not show enhanced activity under visible light, bi-metallic titania (Ag/Pt and Cu/Ag) promote vis activity, due to the oxidation of organic compounds by superoxide anions [126]. Similarly, a synergistic effect has been observed for titania modified with Ag@CuO in respect to mono-modified titania (Ag/TiO_2 and CuO/TiO_2) against mold fungi (*A. melleus* and *P. chrysogenum*) [127]. Interestingly, only the sample with molar ratio of Ag to CuO of 1:3 shows synergism, and only bi-modified samples exhibit higher activity in the dark than that under vis irradiation.

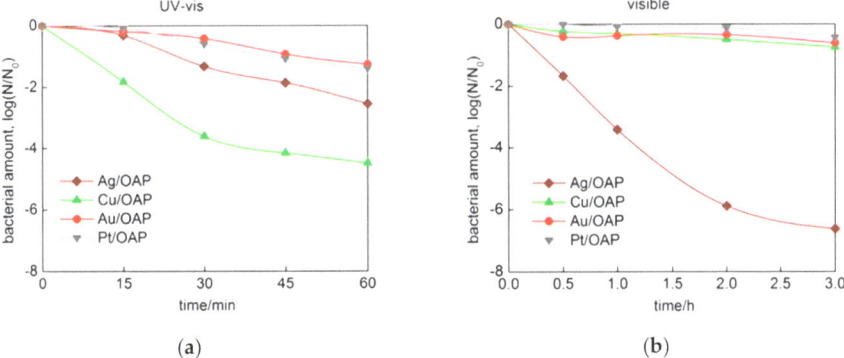

Figure 6. Antibacterial activity of OAP samples modified with NMNPs under: (**a**) UV/vis irradiation and (**b**) vis irradiation; adapted with permission (after formatting) from [117]. Copyright 2017 Creative Commons Attribution.

Table 1. Key-factors of antimicrobial properties of NM-modified titania photocatalysts.

Sample	Preparation Method	Important Properties	Experimental Procedure	Best Performance	Ref.
Ag/TiO_2	hydrothermal synthesis	Ag oxidation	UV or vis-preirradiated samples under vis irradiation with *Escherichia coli*	non-treated and vis-preirradiated samples under vis	[109]
Ag/TiO_2	photodeposition	Ag size	catalysts suspension with bacteriophage MS2 under vis, UV and dark	small Ag in dark and large Ag under vis	[114]
Ag/OAP Cu/OAP	hydrothermal synthesis (OAP) and Ag or Cu by photodeposition	Ag oxidation, Ag release	samples under vis, UV–vis or dark with *E. coli* or *Candida albicans*	Ag/OAP under vis for both *E. coli* and *C. albicans*	[117]
Ag/TiO_2 (spherical or flower-like hierarchical)	template induction (spherical) and solvothermal (hierarchical)	hierarchical structure, Ag release, ROS	minimal inhibitory concentration, growth curves of bacteria and zone of inhibition against *E. coli* and *S. aureus* under Xe arc	Ag-modified flower-like hierarchical titania	[118]
Ag/TiO_2	photochemical reduction of silver nitrate	Ag size and content, Ag release	catalysts suspension with bacteriophage MS2 under UV–A	high content of Ag on TiO_2 and small size of Ag	[89]
Cu/TiO_2	impregnation method	Cu_2O/CuO ratio	suspension of *E. coli*, *S. aureus* or Qβ bacteriophage on Cu_xO/TiO_2 under vis	Cu_xO clusters; Cu_2O:CuO = 1.3:1 on titania	[52]
Ag@CuO/TiO_2	[60]Co panoramic gamma irradiation	Ag/CuO ratio	colony growth; suspension of fungal (*A. melleus* and *P. chrisogenum*) spores with sample under vis	Ag: CuO = 1:3 on titania	[126]
Au/TiO_2 nanotube	magnetron sputtering	size and distribution of Au	*E. coli* or *S. aureus* attached on Au/TiO_2 in dark	annealed Au/TiO_2 (large and isolated Au)	[59]
Au/TiO_2	photodeposition	sizes of Au and titania	inhibition of spore- generation (*A. melleus* and *P. chrisogenum*) under vis	small Au	[115]

Copper, especially copper (I) oxide (although it could not be considered as plasmonic, usually a mixed-oxidation state of copper has been reported), has been well known as an antimicrobial agent since ancient times. It has been applied to improve the photo-induced antimicrobial activity of titania. The proposed mechanisms include: (i) the structure of surface proteins being denatured [94]; (ii) the adsorbed copper ions inducing oxidative stress in the bactericidal processes [95], and the accumulation of copper ions inside bacteria [128]. It has been found that the optimal balance between Cu_2O and CuO in the Cu_xO/TiO_2 composite photocatalyst is important to achieve good antibacterial performance under visible light irradiation and dark conditions, and furthermore that Cu_2O/TiO_2 is more active than CuO/TiO_2 and $CuNPs/TiO_2$ [52]. The mechanistic study by Rtimi et al. In the presence of scavengers (dimethyl sulfoxide and superoxide dismutase) has shown that the VB holes in TiO_2 and the toxicity of the Cu ion are responsible for *E. coli* inactivation under actinic light [111]. In the case of two kinds of faceted anatase photocatalysts (i.e., OAP and DAP (decahedral anatase particle)), the modification with NM (Ag, Cu, Au and Pt) resulted in enhanced antibacterial activity (in comparison to the activity of bare samples) under vis irradiation (λ > 420 nm) against *E. coli* only for Ag- and Cu-modified samples [129]. It has been proposed that the surface oxidation states of NM deposits have been responsible for this behavior (i.e., mainly +1 for Ag and Cu, and 0 for Au and Pt), resulting in the facile adsorption of Ag/Cu-TiO_2 on the bacteria surface. Moreover, these results indirectly support the mechanism of plasmonic activation of titania by the charge transfer mechanism, resulting in the formation of more positively charged Ag/Cu under vis irradiation, and thus an increase in the overall bactericidal activity [122,123,129]. The change in surface oxidation states of NM has been confirmed in another study by XPS analysis for mono- and bi-metal (Au/Ag)-modified titania under monochromatic irradiation [122,123].

In addition to Ag and Cu, Au-modified titania has also been considered as an antimicrobial agent. Moreover, not only the plasmonic photocatalysis (ROS generation) under irradiation [61,115,130,131], but also electron transfer between Au and bacteria (extracellular electron transfer) [59,132] might cause bacterial death. On the other hand, another report insists that Au-modified titania does not possess any bactericidal property [120]. Although silver-modified samples usually show much higher activity than other plasmonic photocatalysts, gold-modified samples prove to be the most active against mold fungi, especially for the inhibition of the sporulation, as shown in Figure 7 [54,115,122].

3.2. Biomedical Applications of Plasmonic Materials

Considering the use of titania for other biological applications, cancer therapy should be noticed, as already presented for bare titania in Figure 2b. Titania has various advantages (e.g., effective ROS generation and non-toxicity), but also disadvantages (e.g., usage of UV light (low penetration through skin, thus, limited to superficial cancer) and no-specification to the target. To overcome these disadvantages, titania has been modified in various ways to obtain visible–near infrared (NIR) response and specification for cancer cells. For example, hydrogenated black titania [133–135], green titania [136], titania modified with NM nanocomposites [137–139], and core/shell photocatalysts with NaYF4: Yb,Tm up-conversion NPs as a core and titania as a shell [140], show NIR absorption and have been proposed for cancer therapy. Among them, plasmonic photocatalysts seem to be the most promising antitumor agents, due to the enhanced photocatalytic property and vis/NIR plasmonic absorption. For example, in the case of Au nanorods and nanoshells, LSPR might be tuned to the NIR region, enabling us to perform in vivo imaging and therapy through the selective localized photothermal heating of cancer cells [141].

Xu et al. [142] have found that Au/TiO_2 nanocomposites, synthesized by the deposition–precipitation (DP) method, might kill the carcinoma cells efficiently, and the amount of Au on the surface of TiO_2 strongly affects the photocatalytic inactivation efficiency (i.e., 2 wt% of Au being the most active). Seo et al. have demonstrated that Ag/AgBr/TiO_2 NPs induce the killing of mammalian cancer cell lines in vitro under visible light illumination (>450 nm) and, moreover, reduced tumor volume in vivo (Figure 8) [139].

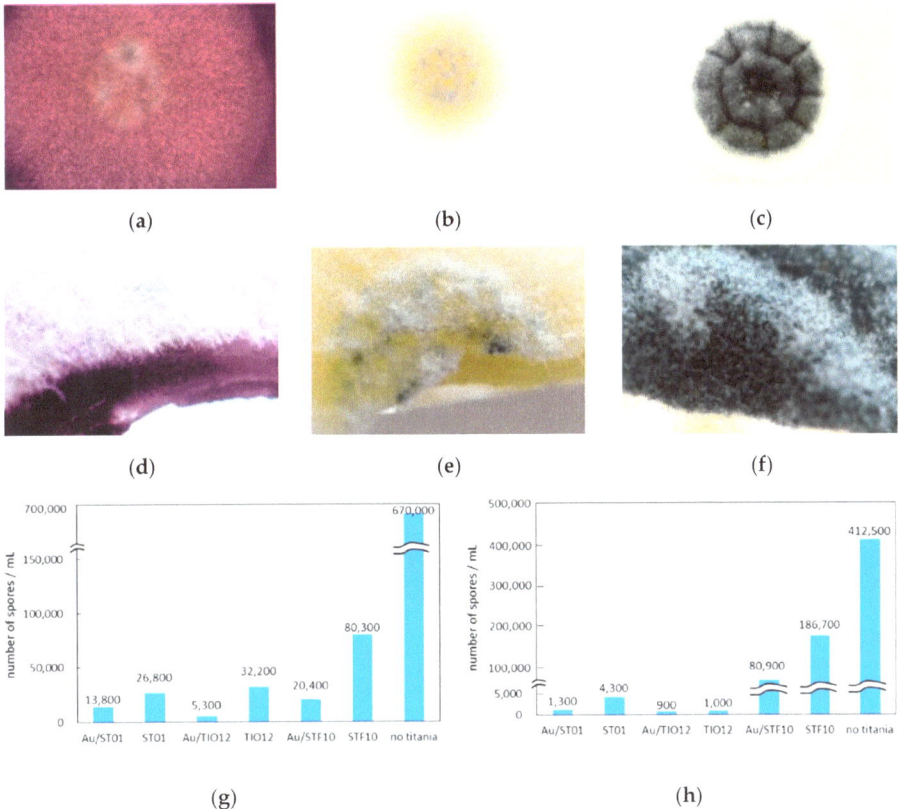

Figure 7. (**a**,**f**) Representative photographs (top view: (**a**–**c**), and cross section: (**d**–**f**)) of *P. chrysogenum* cultivated for four days under fluorescent-light irradiation on: (**a**,**d**) Au/TiO$_2$, (**b**,**e**) TiO$_2$, (**c**,**f**) without photocatalyst. (**g**,**h**) Number of spores after five days of growth under vis irradiation for: (**g**) *P. chrysogenum* and (**h**) *A. melleus*; ST01, TIO12 and STF10—different samples of commercial titania—and Au/ST01, Au/TIO12 and AU/STF10—respective titania modified with gold NPs; adapted with permission from [115]. Copyright 2018, Creative Commons Attribution.

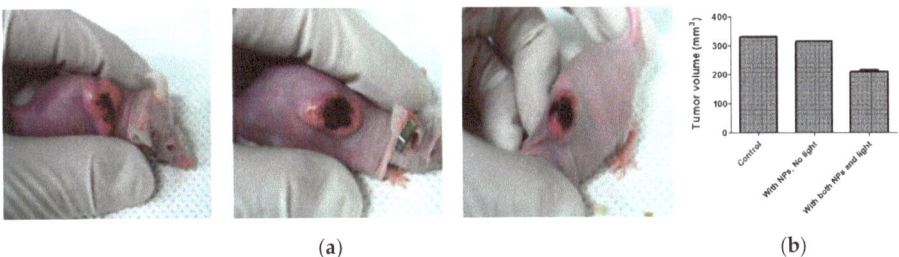

Figure 8. (**a**) Photographs of A431-heterograft tumor on mice in the absence of Ag/AgBr/TiO$_2$ NPs and irradiation for 10 min (left), in the presence of NPs without irradiation (middle), and in the presence of both NPs and irradiation for 10 min (right); (**b**) The measured tumor volumes (mm^3) for the three types of samples. Reprinted with permission (after formatting) from [139]. Copyright (2011) Elsevier.

Although plasmonic photocatalysts might cause an anticancer effect, they are deficient in the specificity. Therefore, similar to drug delivery systems, titania has been modified with various components. For example, some receptors are overexpressed on the surface of cancer cells (e.g., epidermal growth factor receptor (EGFR) [143], interleukin-13R2 receptor domain (IL13R2R) [144] and folic acid-folate receptor [145]). Accordingly, in order to add the binding ability to cancer cells to titania (or other materials), an 11 amino acid peptide fragment of EGF [143], anti-IL13R2R antibody [144], and folic acid [145] are first conjugated with titania. Accordingly, the bound titania to receptor is uptaken by endocytosis, incorporating into cytoplasm, and then probably into nucleus. Next, ROS are generated under irradiation and attack proteins, lipids, nucleic acid, and so on, inducing cell death by apoptosis or necrosis. Recently, it has been found that nuclear-targeted AuNPs (AuNPs with three ligands, methoxypolyethylene glycol thiol (PEG), RGD (RGDRGDRGDRGDPGC) peptides, and nuclear localization signal (NLS, CGGGPKKKRKVGG) peptides), inhibit cancer cell migration by increasing their nuclear stiffness, which greatly reduces the AuNP dosage, resulting in the suppression of the metastasis [146]. Moreover, Ali et al. have proposed that integrin-targeted Au nanorods (Au nanorods with Arg–Gly–Asp (RGD) peptide), activated by NIR light, causes changes in cell morphology and a decrease in cell migration [138], since integrin plays critical roles in controlling the organization of cytoskeletons (strongly related to cell migration).

On the other hand, plasmonic photocatalysts might be used as drug-delivery carriers for anti-tumor drugs [137]. For example, gold nanorods (GNR)/TiO$_2$ has been used as the carrier of gambogic acid (GA), a potent anticancer agent with poor solubility in aqueous solutions, providing the stable dispersion and enhanced intracellular GA delivery. Moreover, GNR/TiO$_2$ shows high photothermal conversion efficiency, and thus irradiation with low-intense laser at 808 nm has enhanced the anticancer effect of the GA-loaded GNR/TiO$_2$, resulting in the improvement of the therapeutic efficacy of GA (Figure 9).

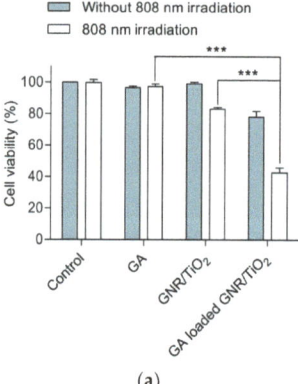 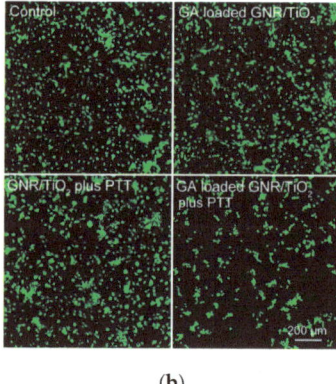

(a) (b)

Figure 9. Enhanced cytotoxic effect of GA by using GNR/TiO2 nanostructure-mediated photothermal therapy. U-87 MG cells were incubated with GNR/TiO2 or GA-loaded GNR/TiO2 nanostructures for 24 h, followed by 808 nm NIR irradiation (5.3 W cm^{-2}) for 2 min. After incubation for 24 h, cell viability was determined using an ATP assay (a) and calcein AM staining; (b) Live cells were stained with green fluorescence by calcein AM. The data shown represent the mean ± S.E.M., *** $p < 0.001$. Reprinted with permission (after formatting) from [137]. Copyright (2017) The Royal Society of Chemistry.

4. Conclusions

Plasmonic photocatalysts have proved to be efficient agents against various microorganisms. Their activity is usually much higher under vis irradiation than in the dark (intrinsic properties of noble metals), resulting from the generation of ROS and the direct redox reactions between microorganisms and photocatalysts. The properties of both photocatalysts and target microorganisms are crucial for the overall antimicrobial effect. Generally, smaller NPs (and thus larger specific surface area) and more simple organisms have shown fast and efficient microbial inactivation.

Unfortunately, some contradictory results have been published. It might be expected that some experiments have been performed in the presence of UV irradiation (e.g., indoor light, artificial or natural solar light), and thus different mechanisms of microorganism inactivation should be considered, i.e., direct excitation of wide-bandgap semiconductor (under UV) and plasmonic excitation (under vis). Considering "hot" electron transfer under plasmonic excitation, the opposite electron transfer under vis and UV is expected, i.e., from NM to semiconductor and from semiconductor to NM, respectively. Therefore, the change in the oxidation state of less noble metals (e.g., Ag and Cu) might result in the facile adsorption of the photocatalyst on the surface of microorganisms, but also a release of metal cations into solution/suspension. For the mechanisms study, it seems that the tests under sole vis (also under sole UV for the comparison) and in the dark are the most recommended (e.g., by action spectra analysis).

The other microbiological applications of plasmonic photocatalysts are also quite promising, especially for cancer treatment and anticancer therapy, but more studies on the improved selectivity against only danger cells are highly necessary.

Author Contributions: Conceptualization, E.K.; writing—original draft preparation, M.E.-K.; writing—review and editing, E.K. and M.E.-K.; supervision, E.K. All authors have read and agreed to the published version of the manuscript.

Funding: This research was funded by a Grand Challenges Explorations Grant (GCE RB, OPP1060234) from Bill & Melinda Gates Foundation, and "Yugo-Sohatsu Kenkyu" for an Integrated Research Consortium on Chemical Sciences (IRCCS) project from the Ministry of Education and Culture, Sport, Science and Technology-Japan (MEXT). The APC was funded by E.K.

Acknowledgments: Authors thank Agata Markowska-Szczupak from Western Pomeranian University of Technology in Szczecin, for fruitful discussion, motivation, and support.

Conflicts of Interest: The authors declare no conflict of interest.

References

1. United Nations. Sustainable Development Goals; 6. Clean Water and Sanitation, (n.d.). Available online: https://www.un.org/sustainabledevelopment/water-and-sanitation/ (accessed on 1 March 2020).
2. WHO. *Guidelines for Drinking-Water Quality*, 4th ed.; WHO: Geneva, Switzerland, 2011.
3. Thomas, V.G.; Guitart, R. Limitations of European Union policy and law for regulating use of lead shot and sinkers: Comparisons with North American regulation. *Environ. Policy Gov.* **2010**, *20*, 57–72. [CrossRef]
4. Van Grinsven, H.J.M.; Ten Berge, H.F.M.; Dalgaard, T.; Fraters, B.; Durand, P.; Hart, A.; Hofman, G.; Jacobsen, B.H.; Lalor, S.T.J.; Lesschen, J.P.; et al. Management, regulation and environmental impacts of nitrogen fertilization in northwesternEurope under the Nitrates Directive; A benchmark study. *Biogeosciences* **2012**, *9*, 5143–5160. [CrossRef]
5. Beulke, S.; van Beinum, W.; Suddaby, L. Interpretation of aged sorption studies for pesticides and their use. In European Union regulatory leaching assessments. *Integr. Environ. Assess. Manag.* **2015**, *11*, 276–286. [CrossRef] [PubMed]
6. Pichat, P. A Brief Survey of the Potential Health Risks of TiO_2. *J. Adv. Oxid. Technol.* **2010**, *13*, 238–246.
7. Arslan, I.; Balcioglu, I.A.; Tuhkanen, T. Advanced Oxidation of Synthetic Dyehouse Effluent by O_3, H_2O_2/O_3 and H_2O_2/UV Processes. *Environ. Technol.* **1999**, *20*, 921–931. [CrossRef]
8. Kowalska, E.; Janczarek, M.; Hupka, J.; Grynkiewicz, M. H_2O_2/UV enhanced degradation of pesticides in wastewater. *Water Sci. Technol.* **2004**, *49*, 261–266. [CrossRef]
9. Barbeni, M.; Minero, C.; Pelizzetti, E.; Borgarello, E.; Serpone, N. Chemical degradation of chlorophenols with Fenton's reagent ($Fe^{2+} + H_2O_2$). *Chemosphere* **1987**, *16*, 2225–2237. [CrossRef]
10. Karci, A.; Arslan-Alaton, I.; Olmez-Hanci, T.; Bekbolet, M. Transformation of 2,4-dichlorophenol by H_2O_2/UV-C, Fenton and photo-Fenton processes: Oxidation products and toxicity evolution. *J. Photochem. Photobiol. A Chem.* **2012**, *230*, 65–73. [CrossRef]
11. Mozia, S.; Tomaszewska, M.; Morawski, A.W. Application of an ozonation-adsorption-ultrafiltration system for surface water treatment. *Desalination* **2006**, *190*, 308–314. [CrossRef]

12. King, C.H.; Shotts, E.B.; Wooley, R.E.; Porter, K.G. Survival of coliforms and bacterial pathogens within protozoa during chlorination. *Appl. Environ. Microbiol.* **1988**, *54*, 3023–3033. [CrossRef]
13. Mokrini, A.; Ousse, D.; Esplugas, S. Oxidation of aromatic compounds with UV radiation/ozoneihydrogen peroxide. *Water Sci. Technol.* **1997**, *35*, 95–102. [CrossRef]
14. Tomova, D.; Iliev, V.; Rakovsky, S.; Anachkov, M.; Eliyas, A.; Puma, G.L. Photocatalytic oxidation of 2,4,6 trinitrotoluene in the presence of ozone under irradiation with UV and visible light. *J. Photochem. Photobiol. A Chem.* **2012**, *231*, 1–8. [CrossRef]
15. Hoffmann, M.R.; Martin, S.T.; Choi, W.; Bahnemann, D.W. Environmental applications of photocatalysis. *Chem. Rev.* **1995**, *95*, 69–96. [CrossRef]
16. Bahnemann, D.W.; Cunningham, J.; Al-Sayyed, G.; Srijaranai, S. Photocatalytic treatment of waters. In *Aquatic and Surface Photochemistry*; Lewis Publ.: Boca Raton, FL, USA, 1994; pp. 261–314.
17. Duffy, J.E.; Anderson, M.A.; Hill, C.G.; Zeltner, W.A. Photocatalytic oxidation as a secondary treatment method following wet air oxidation. *Ind. Eng. Chem. Res.* **2000**, *39*, 3698–3706. [CrossRef]
18. Sánchez-Oneto, J.; Mancini, F.; Portela, J.R.; Nebot, E.; Cansell, F.; Martínez de la Ossa, E.J. Kinetic model for oxygen concentration dependence in the supercritical water oxidation of an industrial wastewater. *Chem. Eng. J.* **2008**, *144*, 361–367. [CrossRef]
19. Legrini, O.; Oliveros, E.; Braun, A.M. Photochemical processes for water treatment. *Chem. Rev* **1993**, *93*, 671–698. [CrossRef]
20. Sun, L.; Bolton, J.R. Determination of the quantum yield for the photochemical generation of hydroxyl adicals in TiO_2 suspensions. *J. Phys. Chem.* **1996**, *100*, 4127–4134. [CrossRef]
21. Pelaez, M.; Nolan, N.T.; Pillai, S.C.; Seery, M.K.; Falaras, P.; Kontos, A.G.; Dunlop, P.S.M.; Hamilton, J.W.J.; Byrne, J.A.; O'Shea, K.; et al. A review on the visible light active titanium dioxide photocatalysts for environmental applications. *Appl. Catal. B Environ.* **2012**, *125*, 331–349. [CrossRef]
22. Zaleska, A. Doped-TiO_2: A review. *Recent Pat. Eng.* **2008**, *2*, 157–164. [CrossRef]
23. Asahi, R.; Morikawa, T.; Ohwaki, T.; Aoki, K.; Taga, Y. Visible-light photocatalysis in nitrogen-doped. titanium oxides. *Science* **2011**, *293*, 269–271. [CrossRef]
24. Mitoraj, D.; Jańczyk, A.; Strus, M.; Kisch, H.; Stochel, G.; Heczko, P.B.; Macyk, W. Visible light inactivation of bacteria and fungi by modified titanium dioxide. *Photochem. Photobiol. Sci.* **2007**, *6*, 642–648. [CrossRef] [PubMed]
25. Sakthivel, S.; Janczarek, M.; Kisch, H. Visible light activity and photoelectrochemical properties of nitrogen-doped TiO_2. *J. Phys. Chem. B* **2004**, *108*, 19384–19387. [CrossRef]
26. Bledowski, M.; Wang, L.; Ramakrishnan, A.; Khavryuchenko, O.V.; Khavryuchenko, V.D.; Ricci, P.C.; Strunk, J.; Cremer, T.; Kolbeck, C.; Beranek, R. Visible-light photocurrent response of TiO_2-polyheptazine hybrids: Evidence for interfacial charge-transfer absorption. *Phys. Chem. Chem. Phys.* **2011**, *13*, 21511–21519. [CrossRef] [PubMed]
27. Dozzi, M.V.; Selli, E. Doping TiO_2 with p-block elements: Effects on photocatalytic activity. *J. Photochem. Photobiol. C Photochem. Rev.* **2013**, *14*, 13–28. [CrossRef]
28. Georgieva, J.; Sotiropoulos, S.; Armyanov, S.; Philippidis, N.; Poulios, I. Photoelectrocatalytic activity of bi-layer TiO_2/WO_3 coatings for the degradation of 4-chlorophenol: Effect of morphology and catalyst loading. *J. Appl. Electrochem.* **2011**, *41*, 173–181. [CrossRef]
29. Li, G.; Nie, X.; Gao, Y.; An, T. Can environmental pharmaceuticals be photocatalytically degraded and completely mineralized in water using $g-C_3N_4/TiO_2$ under visible light irradiation? -Implications of persistent toxic intermediates. *Appl. Catal. B Environ.* **2016**, *180*, 726–732. [CrossRef]
30. Serpone, N.; Maruthamuthu, P.; Pichat, P.; Pelizzetti, E.; Hidaka, H. Exploiting the interparticle electron transfer process in the photocatalysed oxidation of phenol, 2-chlorophenol and pentachlorophenol: Chemical evidence for electron and hole transfer between coupled semiconductors. *J. Photochem. Photobiol. A Chem.* **1995**, *85*, 247–255. [CrossRef]
31. Kraeutler, B.; Bard, A.J. Heterogeneous photocatalytic preparation of supported catalysts. Photodeposition of platinum on TiO_2 powder and other substrates. *J. Am. Chem. Soc.* **1978**, *100*, 4317–4318. [CrossRef]
32. Kowalska, E.; Remita, H.; Colbeau-Justin, C.; Hupka, J.; Belloni, J. Modification of titanium dioxide with platinum ions and clusters: Application in photocatalysis. *J. Phys. Chem. C* **2008**, *112*, 1124–1131. [CrossRef]

33. Kowalska, E.; Yoshiiri, K.; Wei, Z.; Zheng, S.; Kastl, E.; Remita, H.; Ohtani, B.; Rau, S. Hybrid photocatalysts composed of titania modified with plasmonic nanoparticles and ruthenium complexes for decomposition of organic compounds. *Appl. Catal. B Environ.* **2015**, *178*, 133–143. [CrossRef]
34. Sze, S.M.; Ng, K.K. *Physics of Semiconductor Devices*, 3rd ed.; John Wiley & Sons: Hoboken, NJ, USA, 2007.
35. Tian, Y.; Notsu, H.; Tatsuma, T. Visible-light-induced patterning of Au- and Ag-TiO_2 nanocomposite film surfaces on the basis of plasmon photoelectrochemistry. *Photochem. Photobiol. Sci.* **2005**, *4*, 598–601. [CrossRef] [PubMed]
36. Verbruggen, S.W.; Keulemans, M.; Goris, B.; Blommaerts, N.; Bals, S.; Martens, J.A.; Lenaerts, S. Plasmonic "rainbow" photocatalyst with broadband solar light response for environmental applications. *Appl. Catal. B Environ.* **2016**, *188*, 147–153. [CrossRef]
37. Jain, P.K.; Lee, K.S.; El-Sayed, I.H.; El-Sayed, M.A. Calculated absorption and scattering properties of gold nanoparticles of different size, shape, and composition: Applications in biological imaging and biomedicine. *J. Phys. Chem. B* **2006**, *110*, 7238–7248. [CrossRef]
38. Kowalska, E.; Abe, R.; Ohtani, B. Visible light-induced photocatalytic reaction of gold-modified titanium(IV) oxide particles: Action spectrum analysis. *Chem. Commun.* **2009**, *14*, 241–243. [CrossRef] [PubMed]
39. Kowalska, E.; Prieto-Mahaney, O.O.; Abe, R.; Ohtani, B. Visible-light-induced photocatalysis through surface plasmon excitation of gold on titania surfaces. *Phys. Chem. Chem. Phys.* **2010**, *12*, 2344–2355. [CrossRef]
40. Sakai, N.; Fujiwara, Y.; Takahashi, Y.; Tatsuma, T. Plasmon-resonance-based generation of cathodic photocurrent at electrodeposited gold nanoparticles coated with TiO_2 films. *ChemPhysChem* **2009**, *10*, 766–769. [CrossRef]
41. Sakai, N.; Fujiwara, Y.; Arai, M.; Yu, K.; Tatsuma, T. Electrodeposition of gold nanoparticles on ITO: Control of morphology and plasmon resonance-based absorption and scattering. *J. Electroanal. Chem.* **2009**, *628*, 7–15. [CrossRef]
42. Cushing, S.K.; Li, J.; Meng, F.; Senty, T.R.; Suri, S.; Zhi, M.; Li, M.; Bristow, A.D.; Wu, N. Photocatalytic activity enhanced by plasmonic resonant energy transfer from metal to semiconductor. *J. Am. Chem. Soc.* **2012**, *134*, 15033–15041. [CrossRef]
43. Bouhadoun, S.; Guillard, C.; Dapozze, F.; Singh, S.; Amans, D.; Bouclé, J.; Herlin-Boime, N. One step synthesis of N-doped and Au-loaded TiO_2 nanoparticles by laser pyrolysis: Application in photocatalysis. *Appl. Catal. B Environ.* **2015**, *174*, 367–375. [CrossRef]
44. Choi, Y.; Kang, T.; Lee, L.P. Plasmon resonance energy transfer (PRET)-based molecular imaging of cytochrome C in living cells. *Nano Lett.* **2009**, *9*, 85–90. [CrossRef]
45. Chen, X.; Zhu, H.Y.; Zhao, J.C.; Zheng, Z.F.; Gao, X.P. Visible-light-driven oxidation of organic contaminants in air with gold nanoparticle catalysts on oxide supports. *Angew. Chem.* **2008**, *47*, 5353–5356. [CrossRef] [PubMed]
46. Bora, T.; Zoepfl, D.; Dutta, J. Importance of plasmonic heating on visible light driven photocatalysis of gold nanoparticle decorated zinc oxide nanorods. *Sci. Rep.* **2016**, *6*, 1–10. [CrossRef] [PubMed]
47. Golubev, A.A.; Khlebtsov, B.N.; Rodriguez, R.D.; Chen, Y.; Zahn, D.R.T. Plasmonic heating plays a dominant role in the plasmon-induced photocatalytic reduction of 4-nitrobenzenethiol. *J. Phys. Chem. C* **2018**, *122*, 5657–5663. [CrossRef]
48. Wang, C.; Ranasingha, O.; Natesakhawat, S.; Ohodnicki, P.R.; Andio, M.; Lewis, J.P.; Matranga, C. Visible light plasmonic heating of Au–ZnO for the catalytic reduction of CO_2. *Nanoscale* **2013**, *5*, 6968–6974. [CrossRef] [PubMed]
49. Wei, Z.; Janczarek, M.; Endo, M.; Balcytis, A.; Nitta, A.; Mendez Medrano, M.G.; Colbeau-Justin, C.; Juodkazis, S.; Ohtani, B.; Kowalska, E. Noble metal-modified facetted anatase titania photocatalysts: Octahedron versus decahedron. *Appl. Catal. B Environ.* **2018**, *237*, 574–587. [CrossRef]
50. Kowalska, E.; Rau, S.; Ohtani, B. Plasmonic titania photocatalysts active under UV and visible-light irradiation: Influence of gold amount, size, and shape. *J. Nanotechnol.* **2012**. [CrossRef]
51. Huang, L.; Peng, F.; Wang, H.; Yu, H.; Li, Z. Preparation and characterization of Cu_2O/TiO_2 nano-nano heterostructure photocatalysts. *Catal. Commun.* **2009**, *10*, 1839–1843. [CrossRef]
52. Qiu, X.; Miyauchi, M.; Sunada, K.; Minoshima, M.; Liu, M.; Lu, Y.; Li, D.; Shimodaira, Y.; Hosogi, Y.; Kuroda, Y.; et al. Hybrid Cu_xO/TiO_2 nanocomposites as risk-reduction materials in indoor environments. *ACS Nano.* **2012**, *6*, 1609–1618. [CrossRef]

53. Zielińska-Jurek, A.; Kowalska, E.; Sobczak, J.W.; Lisowski, W.; Ohtani, B.; Zaleska, A. Preparation and characterization of monometallic (Au) and bimetallic (Ag/Au) modified-titania photocatalysts activated by visible light. *Appl. Catal. B Environ.* **2011**, *101*, 504–514. [CrossRef]
54. Kowalska, E.; Wei, Z.S.; Karabiyik, B.; Janczarek, M.; Endo, M.; Wang, K.L.; Rokicka, P.; Markowska-Szczupak, A.; Ohtani, B. Development of plasmonic photocatalysts for environmental application. *Adv. Sci. Technol.* **2014**, *93*, 174–183. [CrossRef]
55. Méndez-Medrano, M.G.; Kowalska, E.; Lehoux, A.; Herissan, A.; Ohtani, B.; Bahena, D.; Briois, V.; Colbeau-Justin, C.; Rodríguez-López, J.L.; Remita, H. Surface modification of TiO_2 with Ag nanoparticles and CuO nanoclusters for application in photocatalysis. *J. Phys. Chem. C* **2016**, *120*, 5143–5154. [CrossRef]
56. Mo, A.; Liao, J.; Xu, W.; Xian, S.; Li, Y.; Bai, S. Preparation and antibacterial effect of silver-hydroxyapatite/titania nanocomposite thin film on titanium. *Appl. Surf. Sci.* **2008**, *255*, 435–438. [CrossRef]
57. Padervand, M.; Reza Elahifard, M.; Vatan Meidanshahi, R.; Ghasemi, S.; Haghighi, S.; Reza Gholami, M. nvestigation of the antibacterial and photocatalytic properties of the zeolitic nanosized $AgBr/TiO_2$ composites. *Mater. Sci. Semicond. Process.* **2012**, *15*, 73–79. [CrossRef]
58. Jansson, I.; Yoshiiri, K.; Hori, H.; Garcia-Garcia, F.J.; Rojas, S.; Sanchez, B.; Ohtani, B.; Suerez, S. Cerebral aneurysm surgery based on preoperative computerized tomography angiography. *J. Neurosurg.* **2016**, *521*, 208–219.
59. Li, J.; Zhou, H.; Qian, S.; Liu, Z.; Feng, J.; Jin, P.; Liu, X. Plasmonic gold nanoparticles modified titania nanotubes for antibacterial application. *Appl. Phys. Lett.* **2014**, *104*, 261110. [CrossRef]
60. Bian, Z.; Tachikawa, T.; Zhang, P.; Fujitsuka, M.; Majima, T. Au/TiO_2 superstructure-based plasmonic photocatalysts exhibiting efficient charge separation and unprecedented activity. *J. Am. Chem. Soc.* **2014**, *136*, 458–465. [CrossRef]
61. Guo, L.; Shan, C.; Liang, J.; Ni, J.; Tong, M. Bactericidal mechanisms of Au@TNBs under visible light irradiation. *Colloids Surf. B Biointerfaces* **2015**, *128*, 211–218. [CrossRef]
62. Matsunaga, T.; Tomoda, R.; Nakajima, T.; Wake, H. Photoelectrochemical sterilization of microbial cells by semiconductor powders. *FEMS Microbiol. Lett.* **1985**, *29*, 211–214. [CrossRef]
63. Matsunaga, T.; Tomoda, R.; Nakajima, T.; Nakamura, N.; Komine, T. Continuous-sterilization system that uses photosemiconductor powders. *Appl. Environ. Microbiol.* **1988**, *54*, 1330–1333. [CrossRef]
64. Huang, Z.; Maness, P.-C.; Blake, D.M.; Wolfrum, E.J.; Smolinski, S.L.; Jacoby, W.A. Bactericidal mode of titanium dioxide photocatalysis. *J. Photochem. Photobiol. A Chem.* **2000**, *130*, 163–170. [CrossRef]
65. Markowska-Szczupak, A.; Ulfig, K.; Morawski, A.W. The application of titanium dioxide for deactivation of bioparticulates: An overview. *Catal. Today* **2011**, *169*, 249–257. [CrossRef]
66. Kiwi, J.; Nadtochenko, V. New evidence for TiO_2 photocatalysis during bilayer lipid peroxidation. *J. Phys. Chem. B.* **2004**, *108*, 17675–17684. [CrossRef]
67. Kiwi, J.; Nadtochenko, V. Evidence for the mechanism of photocatalytic degradation of the bacterial wall membrane at the TiO_2 interface by ATR-FTIR and laser kinetic spectroscopy. *Langmuir* **2005**, *21*, 4631–4641. [CrossRef] [PubMed]
68. Cho, M.; Chung, H.; Choi, W.; Yoon, J. Linear correlation between inactivation of *E. coli* and OH radical concentration in TiO_2 photocatalytic disinfection. *Water Res.* **2004**, *38*, 1069–1077. [CrossRef]
69. Wang, X.; Tang, Y.; Chen, Z.; Lim, T.T. Highly stable heterostructured $Ag-AgBr/TiO_2$ composite: A bifunctional visible-light active photocatalyst for destruction of ibuprofen and bacteria. *J. Mater. Chem.* **2012**, *22*, 23149–23158. [CrossRef]
70. Du, J.; Gebicki, J.M. Proteins are major initial cell targets of hydroxyl free radicals. *Int. J. Biochem. Cell Biol.* **2004**, *36*, 2334–2343. [CrossRef]
71. Battistoni, A. Role of prokaryotic Cu, Zn superoxide dismutase in pathogenesis. *Biochem. Soc. Trans.* **2003**, *31*, 1326–1329. [CrossRef]
72. Wintjens, R.; Noël, C.; May, A.C.; Gerbod, D.; Dufernez, F.; Capron, M.; Viscogliosi, E.; Rooman, M. Specificity and Phenetic Relationships of Iron- and Manganese-containing Superoxide Dismutases on the Basis of Structure and Sequence Comparisons. *J. Biol. Chem.* **2004**, *279*, 9248–9254. [CrossRef]
73. Harris, A.G.; Hinds, F.E.; Beckhouse, A.G.; Kolesnikow, T.; Hazell, S.L. Resistance to hydrogen peroxide in Helicobacter pylori: Role of catalase (KatA) and Fur, and functional analysis of a novel gene product designated "KatA-associated protein", KapA (HP0874). *Microbiology* **2002**, *148*, 3813–3825. [CrossRef]

74. Harris, A.G.; Hazell, S.L. Localisation of *Helicobacter pylori* catalase in both the periplasm and cytoplasm, and its dependence on the twin-arginine target protein, KapA, for activity. *FEMS Microbiol. Lett.* **2003**, *229*, 283–289. [CrossRef]
75. Markowska-Szczupak, A.; Rokicka, P.; Wang, K.; Endo, M.; Morawski, A.W.; Kowalska, E. Photocatalytic water disinfection under solar radiation by D-glucose modified titania. *Catalysts* **2018**, *8*, 316. [CrossRef]
76. Jakob, U.; Muse, W.; Eser, M.; Bardwell, J.C.A. Chaperone activity with a redox switch. *Cell* **1999**, *96*, 341–352. [CrossRef]
77. Volkert, M.R.; Landini, P. Transcriptional responses to DNA damage. *Curr. Opin. Microbiol.* **2001**, *4*, 178–185. [CrossRef]
78. Chaffin, W.L.; López-Ribot, J.L.; Casanova, M.; Gozalbo, D.; Martínez, J.P. Cell wall and secreted proteins of *Candida albicans*: Identification, Function, and Expression. *Microbiol. Mol. Biol. Rev.* **1998**, *62*, 130–180. [CrossRef] [PubMed]
79. Markowska-Szczupak, A.; Wang, K.; Rokicka, P.; Endo, M.; Wei, Z.; Ohtani, B.; Morawski, A.W.; Kowalska, E. The effect of anatase and rutile isolated from titania P25 on pathogenic fungi. *J. Photochem. Photobiol. B* **2015**, *151*, 54–62. [CrossRef] [PubMed]
80. Markowska-Szczupak, A.; Wei, Z.; Kowalska, E. The influence of the light-activated titania P25 on human breast cancer cells. *Catalysts* **2020**, *10*, 238. [CrossRef]
81. Chwalibog, A.; Sawosz, E.; Hotowy, A.; Szeliga, J.; Mitura, S.; Mitura, K.; Grodzik, M.; Orlowski, P.; Sokolowska, A. Visualization of interaction between inorganic nanoparticles and bacteria or fungi. *Int. J. Nanomed.* **2010**, *5*, 1085–1094. [CrossRef]
82. Sambhy, V.; MacBride, M.M.; Peterson, B.R.; Sen, A. Silver bromide nanoparticle/polymer composites: Dual action tunable antimicrobial materials. *J. Am. Chem. Soc.* **2006**, *128*, 9798–9808. [CrossRef]
83. Holt, K.B.; Bard, A.J. Interaction of silver(I) ions with the respiratory chain of *Escherichia coli*: Anm electrochemical and scanning electrochemical microscopy study of the antimicrobial mechanism of micromolar Ag. *Biochemistry* **2005**, *44*, 13214–13223. [CrossRef]
84. Lok, C.-N.; Ho, C.-M.; Chen, R.; He, Q.-Y.; Yu, W.-Y.; Sun, H.; Tam, P.K.-H.; Chiu, J.-F.; Che, C.-M. Proteomic analysis of the mode of antibacterial action of silver nanoparticles. *J. Proteome Res.* **2006**, *5*, 916–924. [CrossRef]
85. van Grieken, R.; Marugán, J.; Sordo, C.; Martínez, P.; Pablos, C. Photocatalytic inactivation of bacteria in water using suspended and immobilized silver-TiO_2. *Appl. Catal. B Environ.* **2009**, *93*, 112–118. [CrossRef]
86. Akhavan, O. Lasting antibacterial activities of Ag-TiO_2/Ag/a-TiO_2 nanocomposite thin film photocatalysts under solar light irradiation. *J. Colloid Interface Sci.* **2009**, *336*, 117–124. [CrossRef] [PubMed]
87. Li, M.; Noriega-Trevino, M.E.; Nino-Martinez, N.; Marambio-Jones, C.; Wang, J.; Damoiseaux, R.; Ruiz, F.; Hoek, E.M.V. Synergistic bactericidal activity of Ag-TiO_2 nanoparticles in both light and dark conditions. *Environ. Sci. Technol.* **2011**, *45*, 8989–8995. [CrossRef] [PubMed]
88. Lin, W.C.; Chen, C.N.; Tseng, T.T.; Wei, M.H.; Hsieh, J.H.; Tseng, W.J. Micellar layer-by-layer synthesis of TiO_2/Ag hybrid particles for bactericidal and photocatalytic activities. *J. Eur. Ceram. Soc.* **2010**, *30*, 2849–2857. [CrossRef]
89. Liga, M.V.; Bryant, E.L.; Colvin, V.L.; Li, Q. Virus inactivation by silver doped titanium dioxide nanoparticles for drinking water treatment. *Water Res.* **2011**, *45*, 535–544. [CrossRef]
90. Tian, Y.; Qi, J.; Zhang, W.; Cai, Q.; Jiang, X. Facile, one-pot synthesis, and antibacterial activity of mesoporous silica nanoparticles decorated with well-dispersed silver nanoparticles. *Appl. Mater. Interfaces* **2014**, *6*, 12038–12045. [CrossRef]
91. Su, H.L.; Chou, C.C.; Hung, D.J.; Lin, S.H.; Pao, I.C.; Lin, J.H.; Huang, F.L.; Dong, R.X.; Lin, J.J. The disruption of bacterial membrane integrity through ROS generation induced by nanohybrids of silver and clay. *Biomaterials* **2009**, *30*, 5979–5987. [CrossRef]
92. Galdiero, S.; Falanga, A.; Vitiello, M.; Cantisani, M.; Marra, V.; Galdiero, M. Silver nanoparticles as potential antiviral agents. *Molecules* **2011**, *16*, 8894–8918. [CrossRef]
93. Speshock, J.L.; Murdock, R.C.; Braydich-Stolle, L.K.; Schrand, A.M.; Hussain, S.M. Interaction of silver nanoparticles with HIV-1. *J. Nanobiotechnol.* **2010**, *8*, 19–27. [CrossRef]
94. Sunada, K.; Minoshima, M.; Hashimoto, K. Highly efficient antiviral and antibacterial activities of solid-state cuprous compounds. *J. Hazard. Mater.* **2012**, *235*, 265–270. [CrossRef]

95. Deng, C.H.; Gong, J.L.; Zeng, G.M.; Zhang, P.; Song, B.; Zhang, X.G.; Liu, H.Y.; Huan, S.Y. Graphene sponge decorated with copper nanoparticles as a novel bactericidal filter for inactivation of *Escherichia coli*. *Chemosphere* **2017**, *184*, 347–357. [CrossRef] [PubMed]
96. Dankovich, T.A.; Gray, D.G. Bactericidal paper impregnated with silver nanoparticles for emergency water disinfection. *Environ. Sci. Technol.* **2011**, *45*, 1992–1998. [CrossRef]
97. Jafari, A.; Pourakbar, L.; Farhadi, K.; Mohamadgolizad, L.; Goosta, Y. Biological synthesis of silver nanoparticles and evaluation of antibacterial and antifungal properties of silver and copper nanoparticles. *Turk. J. Biol.* **2015**, *39*, 556–561. [CrossRef]
98. Amin, R.M.; Mohamed, M.B.; Ramadan, M.A.; Verwanger, T.; Krammer, B. Rapid and sensitive microplate assay for screening the effect of silver and gold nanoparticles on bacteria. *Nanomedicine* **2009**, *4*, 637–643. [CrossRef] [PubMed]
99. Zhang, W.; Li, Y.; Niu, J.; Chen, Y. Photogeneration of reactive oxygen species on uncoated silver, gold, nickel, and silicon nanoparticles and their antibacterial effects. *Langmuir* **2013**, *29*, 4647–4651. [CrossRef]
100. Zheng, K.; Setyawati, M.I.; Leong, D.T.; Xie, J. Antimicrobial gold nanoclusters. *ACS Nano* **2017**, *11*, 6904–6910. [CrossRef]
101. Ratte, H.T. Bioaccumulation and toxicity of silver compounds: A review. *Environ. Toxicol. Chem.* **1999**, *18*, 89–108. [CrossRef]
102. Rai, A.; Prabhune, A.; Perry, C.C. Antibiotic mediated synthesis of gold nanoparticles with potent antimicrobial activity and their application in antimicrobial coatings. *J. Mater. Chem.* **2010**, *20*, 6789–6798. [CrossRef]
103. Badwaik, V.D.; Vangala, L.M.; Pender, D.S.; Willis, C.B.; Aguilar, Z.P.; Gonzalez, M.S.; Paripelly, R.; Dakshinamurthy, R. Size-dependent antimicrobial properties of sugarencapsulated gold nanoparticles synthesized by a green method. *Nanoscale Res. Lett.* **2012**, *7*, 1–11. [CrossRef]
104. Cui, Y.; Zhao, Y.; Tian, Y.; Zhang, W.; Lü, X.; Jiang, X. The molecular mechanism of action of bactericidal gold nanoparticles on *Escherichia coli*. *Biomaterials* **2012**, *33*, 2327–2333. [CrossRef]
105. Ahmad, T.; Wani, I.A.; Lone, I.H.; Ganguly, A.; Manzoor, N.; Ahmad, A.; Ahmed, J.; Al-Shihri, A.S. Antifungal activity of gold nanoparticles prepared by solvothermal method. *Mater. Res. Bull.* **2013**, *48*, 12–20. [CrossRef]
106. Baram-Pinto, D.; Shukla, S.; Gedanken, A.; Sarid, R. Inhibition of HSV-1 attachment, entry, and cell-to-cell spread by functionalized multivalent gold nanoparticles. *Small* **2010**, *6*, 1044–1050. [CrossRef] [PubMed]
107. Kubacka, A.; Ferrer, M.; Martínez-Arias, A.; Fernández-García, M. Ag promotion of TiO_2-anatase disinfection capability: Study of *Escherichia coli* inactivation. *Appl. Catal. B Environ.* **2008**, *84*, 87–93. [CrossRef]
108. Mai, L.; Wang, D.; Zhang, S.; Xie, Y.; Huang, C.; Zhang, Z. Synthesis and bactericidal ability of Ag/TiO_2 composite films deposited on titanium plate. *Appl. Surf. Sci.* **2010**, *257*, 974–978. [CrossRef]
109. Castro, C.A.; Osorio, P.; Sienkiewicz, A.; Pulgarin, C.; Centeno, A.; Giraldo, S.A. Photocatalytic production of $^1O_2$ and OH mediated by silver oxidation during the photoinactivation of Escherichia coli with TiO_2. *J. Hazard. Mater.* **2012**, *211*, 172–181. [CrossRef]
110. Baghriche, O.; Rtimi, S.; Pulgarin, C.; Sanjines, R.; Kiwi, J. Innovative TiO_2/Cu nanosurfaces inactivating bacteria in the minute range under low-intensity actinic light. *ACS Appl. Mater. Interfaces* **2012**, *4*, 5234–5240. [CrossRef]
111. Rtimi, S.; Giannakis, S.; Sanjines, R.; Pulgarin, C.; Bensimon, M.; Kiwi, J. Insight on the photocatalytic bacterial inactivation by co-sputtered TiO_2-Cu in aerobic and anaerobic conditions. *Appl. Catal. B Environ.* **2016**, *182*, 277–285. [CrossRef]
112. Liu, L.; Yang, W.; Li, Q.; Gao, S.; Shang, J.K. Synthesis of Cu_2O nanospheres decorated with TiO_2 nanoislands, their enhanced photoactivity and stability under visible light illumination, and their post-illumination catalytic memory. *ACS Appl. Mater. Interfaces* **2014**, *6*, 5629–5639. [CrossRef]
113. Stentzel, S.; Teufelberger, A.; Nordengrün, M.; Kolata, J.; Schmidt, F.; van Crombruggen, K.; Michalik, S.; Kumpfmüller, J.; Tischer, S.; Schweder, T.; et al. Staphylococcal serine protease–like proteins are pacemakers of allergic airway reactions to *Staphylococcus aureus*. *J. Allergy Clin. Immunol.* **2017**, *139*, 492–500. [CrossRef]
114. Kowalska, E.; Wei, Z.; Karabiyik, B.; Janczarek, M.; Endo, M.; Markowska-Szczupak, A.; Remita, H.; Ohtani, B. Silver modified titania with enhanced photocatalytic and antimicrobial properties under UV and visible light irradiation. *Cata. Today* **2015**, *252*, 136–142. [CrossRef]
115. Endo, M.; Wei, Z.; Wang, K.; Karabiyik, B.; Yoshiiri, K.; Rokicka, P.; Ohtani, B.; Markowska-Szczupak, A.; Kowalska, E. Noble metal-modified titania with visible-light activity for the decomposition of microorganisms. *Beilstein J. Nanotechnol.* **2018**, *9*, 829–841. [CrossRef]

116. Hou, Y.; Li, X.; Zhao, Q.; Chen, G.; Raston, C.L. Role of hydroxyl radicals and mechanism of escherichia coli inactivation on Ag/AgBr/TiO$_2$ nanotube array electrode under visible light irradiation. *Environ. Sci. Technol.* **2012**, *46*, 4042–4050. [CrossRef] [PubMed]
117. Wei, Z.; Endo, M.; Wang, K.; Charbit, E.; Markowska-Szczupak, A.; Ohtani, B.; Kowalska, E. Noble metal-modified octahedral anatase titania particles with enhanced activity for decomposition of chemical and microbiological pollutants. *Chem. Eng. J.* **2017**, *318*, 121–134. [CrossRef] [PubMed]
118. Ye, J.; Cheng, H.; Li, H.; Yang, Y.; Zhang, S.; Rauf, A.; Zhao, Q.; Ning, G. Highly synergistic antimicrobial activity of spherical and flower-like hierarchical titanium dioxide/silver composites. *J. Colloid Interface Sci.* **2017**, *504*, 448–456. [CrossRef] [PubMed]
119. Rtimi, S.; Baghriche, O.; Sanjines, R.; Pulgarin, C.; Bensimon, M.; Kiwi, J. TiON and TiON-Ag sputtered surfaces leading to bacterial inactivation under indoor actinic light. *J. Photochem. Photobiol. A Chem.* **2013**, *256*, 52–63. [CrossRef]
120. Veréb, G.; Manczinger, L.; Bozsó, G.; Sienkiewicz, A.; Forró, L.; Mogyorósi, K.; Hernádi, K.; Dombi, A. Comparison of the photocatalytic efficiencies of bare and doped rutile and anatase TiO$_2$ photocatalysts under visible light for phenol degradation and *E. coli* inactivation. *Appl. Catal. B Environ.* **2013**, *129*, 566–574. [CrossRef]
121. Swetha, S.; Kumari Singh, M.; Minchitha, K.U.; Geetha Balakrishna, R. Elucidation of cell killing mechanism by comparative analysis of photoreactions on different types of bacteria. *Photochem. Photobiol.* **2012**, *88*, 414–422. [CrossRef]
122. Endo-Kimura, M. A Mechanistic Study on Noble Metal-Modified Titania Photocatalysts for Inactivation of Bacteria and Fungi. Ph.D. Thesis, Hokkaido University, Sapporo, Japan, 2019.
123. Endo-Kimura, M.; Wang, K.; Wei, Z.; Juodkazis, S.; Markowska-Szczupak, A.; Ohtani, B.; Kowalska, E. Surface-Plasmon-Driven Inactivation of Microorganisms on Mono- and Bi-Metal (Au/Ag)-Modified Titania: A Mechanistic Study. (under preparation).
124. Zielińska, A.; Kowalska, E.; Sobczak, J.W.; Łacka, I.; Gazda, M.; Ohtani, B.; Hupka, J.; Zaleska, A. Silver-doped TiO$_2$ prepared by microemulsion method: Surface properties, bio- and photoactivity. *Sep. Purif. Technol.* **2010**, *72*, 309–318. [CrossRef]
125. Kowal, K.; Wysocka-Król, K.; Kopaczyńska, M.; Dworniczek, E.; Franiczek, R.; Wawrzyńska, M.; Vargová, M.; Zahoran, M.; Rakovský, E.; Kuš, P.; et al. In situ photoexcitation of silver-doped titania nanopowders for activity against bacteria and yeasts. *J. Colloid Interface Sci.* **2011**, *362*, 50–57. [CrossRef]
126. Wysocka, I.; Markowska-Szczupak, A.; Szweda, P.; Ryl, J.; Endo-Kimura, M.; Kowalska, E.; Nowaczyk, G.; Zielińska-Jurek, A. Gas-phase removal of indoor volatile organic compounds and airborne microorganisms over mono- and bimetal-modified (Pt, Cu, Ag) titanium(IV) oxide nanocomposites. *Indoor Air.* **2019**, *29*, 979–992. [CrossRef] [PubMed]
127. Mendez-Medrano, M.G.; Kowalska, E.; Endo, M.; Wang, K.; Bahena, D.; Rodriguez-Lopez, J.L.; Remita, H. Inhibition of fungal growth using modified TiO$_2$ with core@shell structure of Ag@CuO clusters. *ACS Appl. Bio. Mater.* **2019**, *2*, 5626–5633. [CrossRef]
128. Dankovich, T.A.; Smith, J.A. Incorporation of copper nanoparticles into paper for point-of-use water purification. *Water Res.* **2014**, *63*, 245–251. [CrossRef] [PubMed]
129. Endo, M.; Janczarek, M.; Wei, Z.; Wang, K.; Mrkowska-Szczupak, M.; Ohtani, B.; Kowalska, E. Bactericidal properties of plasmonic photocatalysts composed of noble metal nanoparticles on faceted anatase titania. *J. Nanosci. Nanotechnol.* **2019**, *19*, 442–452. [CrossRef]
130. He, W.; Huang, H.; Yan, J.; Zhu, J. Photocatalytic and antibacterial properties of Au-TiO$_2$ nanocomposite on monolayer graphene: From experiment to theory. *J. Appl. Phys.* **2013**, *114*, 204701. [CrossRef]
131. Noimark, S.; Page, K.; Bear, J.C.; Sotelo-Vazquez, C.; Quesada-Cabrera, R.; Lu, Y.; Allan, E.; Darr, J.A.; Parkin, I.P. Functionalised gold and titania nanoparticles and surfaces for use as antimicrobial coatings. *Faraday Discuss.* **2014**, *175*, 273–287. [CrossRef]
132. Wang, G.; Feng, H.; Gao, A.; Hao, Q.; Jin, W.; Peng, X.; Li, W.; Wu, G.; Chu, P.K. Extracellular Electron Transfer from Aerobic Bacteria to Au-Loaded TiO$_2$ Semiconductor without Light: A New Bacteria-Killing Mechanism Other than Localized Surface Plasmon Resonance or Microbial Fuel Cells. *ACS Appl. Mater. Interfaces* **2016**, *8*, 24509–24516. [CrossRef]

133. Ren, W.; Yan, Y.; Zeng, L.; Shi, Z.; Gong, A.; Schaaf, P.; Wang, D.; Zhao, J.; Zou, B.; Yu, H.; et al. A near infrared light triggered hydrogenated black TiO_2 for cancer photothermal therapy. *Adv. Healthc. Mater.* **2015**, *4*, 1526–1536. [CrossRef]
134. Mou, J.; Lin, T.; Huang, F.; Chen, H.; Shi, J. Black titania-based theranostic nanoplatform for single NIR laser induced dual-modal imaging-guided PTT/PDT. *Biomaterials* **2016**, *84*, 13–24. [CrossRef]
135. Han, X.; Huang, J.; Jing, X.; Yang, D.; Lin, H.; Wang, Z.; Li, P.; Chen, Y. Oxygen-deficient black titania for synergistic/enhanced sonodynamic and photoinduced cancer therapy at near infrared-II biowindow. *ACS Nano* **2018**, *12*, 4545–4555. [CrossRef]
136. Mou, J.; Lin, T.; Huang, F.; Shi, J.; Chen, H. A new green titania with enhanced NIR absorption for mitochondria-targeted cancer therapy. *Theranostics* **2017**, *7*, 1531–1542. [CrossRef] [PubMed]
137. Wan, H.Y.; Chen, J.L.; Yu, X.Y.; Zhu, X.M. Titania-coated gold nanorods as an effective carrier for gambogic acid. *RSC Adv.* **2017**, *7*, 49518–49525. [CrossRef]
138. Ali, M.R.K.; Wu, Y.; Tang, Y.; Xiao, H.; Chen, K.; Han, T.; Fang, N.; Wu, R.; El-Sayed, M.A. Targeting cancer cell integrins using gold nanorods in photothermal therapy inhibits migration through affecting cytoskeletal proteins. *Proc. Natl. Acad. Sci. USA* **2017**, *114*, E5655–E5663. [CrossRef] [PubMed]
139. Seo, J.H.; Jeon, W.I.; Dembereldorj, U.; Lee, S.Y.; Joo, S.W. Cytotoxicity of serum protein-adsorbed visible-light photocatalytic Ag/AgBr/TiO_2 nanoparticles. *J. Hazard. Mater.* **2011**, *198*, 347–355. [CrossRef]
140. Lucky, S.S.; Muhammad Idris, N.; Li, Z.; Huang, K.; Soo, K.C.; Zhang, Y. Titania coated upconversion nanoparticles for near-infrared light triggered photodynamic therapy. *ACS Nano* **2015**, *9*, 191–205. [CrossRef] [PubMed]
141. Jain, P.K.; Huang, X.; El-Sayed, I.H.; El-Sayed, M.A. Noble metals on the nanoscale: Optical and photothermal properties and some applications in imaging, sensing, biology, and medicine. *Acc. Chem. Res.* **2008**, *41*, 1578–1586. [CrossRef] [PubMed]
142. Xu, J.; Sun, Y.; Zhao, Y.; Huang, J.; Chen, C.; Jiang, Z. Photocatalytic inactivation effect of gold-doped TiO_2 (Au/ TiO_2) nanocomposites on human colon carcinoma LoVo cells. *Int. J. Photoenergy* **2007**, *21*. [CrossRef]
143. Yuan, Y.; Chen, S.; Paunesku, T.; Gleber, S.C.; Liu, W.C.; Doty, C.B.; Mak, R.; Deng, J.; Jin, Q.; Lai, B.; et al. Epidermal growth factor receptor targeted nuclear delivery and high-resolution whole cell x-ray imaging of Fe_3O_4@TiO_2 nanoparticles in cancer cells. *ACS Nano* **2013**, *7*, 10502–10517. [CrossRef]
144. Rozhkova, E.A.; Ulasov, I.; Lai, B.; Dimitrijevic, N.M.; Lesniak, M.S.; Rajh, T. A High-performance nanobio photocatalyst for targeted brain cancer therapy. *Nano Lett.* **2009**, *9*, 3337–3342. [CrossRef]
145. Tong, L.; Zhao, Y.; Huff, T.B.; Hansen, M.N.; Wei, A.; Cheng, J.X. Gold Nanorods Mediate Tumor Cell Death by Compromising Membrane Integrity. *Adv. Mater.* **2007**, *19*, 3136–3141. [CrossRef]
146. Ali, M.R.K.; Wu, Y.; Ghosh, D.; Do, B.H.; Chen, K.; Dawson, M.R.; Fang, N.; Sulchek, T.A.; El-Sayed, M.A. Nuclear membrane-targeted gold nanoparticles inhibit cancer cell migration and invasion. *ACS Nano* **2017**, *11*, 3716–3726. [CrossRef] [PubMed]

© 2020 by the authors. Licensee MDPI, Basel, Switzerland. This article is an open access article distributed under the terms and conditions of the Creative Commons Attribution (CC BY) license (http://creativecommons.org/licenses/by/4.0/).

Review

Photocatalytic Reversible Reactions Driven by Localized Surface Plasmon Resonance

Zheng Gong, Jialong Ji and Jingang Wang *

Computational Center for Property and Modification on Nanomaterials, College of Sciences, Liaoning Shihua University, Fushun 113001, China; gongz640@nenu.edu.cn (Z.G.); jjlzs5201314@163.com (J.J.)
* Correspondence: Jingang_wang@lnpu.edu.cn; Tel.: +86-18040036755

Received: 14 January 2019; Accepted: 18 February 2019; Published: 20 February 2019

Abstract: In this study, we review photocatalytic reversible surface catalytic reactions driven by localized surface plasmon resonance. Firstly, we briefly introduce the synthesis of 4,4′-dimercaptoazobenzene (DMAB) from 4-nitrobenzenethiol (4NBT) using surface-enhanced Raman scattering (SERS) technology. Furthermore, we study the photosynthetic and degradation processes of 4NBT to DMAB reduction, as well as factors associated with them, such as laser wavelength, reaction time, substrate, and pH. Last but not least, we reveal the competitive relationship between photosynthetic and degradation pathways for this reduction reaction by SERS technology on the substrate of Au film over a nanosphere.

Keywords: plasmon; 4-nitrobenzenethiol; 4,4′-dimercaptoazobenzene; Au film

1. Introduction

Surface-enhanced Raman scattering (SERS) and tip-enhanced Raman spectroscopy (TERS) are widely used in the fields of physics, chemistry, biology, medicine, and materials, and they are highly sensitive molecular-detection tools that play a significant role in research on the electronic structure and spectral properties of molecules and plasmon-driven chemical reactions [1–9]. Recently, we studied the physical mechanism of plasmon-enhanced resonance Raman and fluorescence spectra [10] to understand the applications and principles of photoinduced charge transfer [11]. Furthermore, SERS made significant contributions to the study of heterostructures in two-dimensional materials [12], and significant advances were made in the application of nonlinear optical microscopy [13]. At the same time, it is a significantly important tool for probing chemical signals about molecules adsorbed onto metal substrates, such as Au, Ag, and Cu. It is also a promising technology for detecting plasmon-driven catalytic mechanisms. This technique is extremely sensitive to detecting high field-enhancement regions on a plasmon substrate. Many research results show that an electromagnetic field is an important part of the SERS enhancement mechanism, which is generated by the amplification and localization of incident electromagnetic waves.

A lot of research shows that conversion of 4-nitrobenzenethiol (4NBT) and *p*-aminothiophenol (PATP) to 4,4′-dimercaptoazobenzene (DMAB) via reduction and oxidation reactions was achieved with SERS technology [2–11,14–18]. Theoretical and experimental studies showed that SERS synthesis of DMAB from PATP is achieved via the selective catalytic coupling of Ag nanoparticles (Ag NPs). It was concluded that, in the field of electromagnetic fields, the six strong generated vibration modes are enhanced by the plasmon; this is direct evidence of the reaction of PATP to DMAB on Ag NPs via selective catalytic coupling [19,20]. Therefore, DMAB is an important topic in the study of plasmon-assisted catalytic reactions [21–24].

Authors researched the effects of photoreduction-field enhancement from 4NBT to DMAB driven by plasmon on Au, Ag, and Cu [25–27]. The molecular structure of 4NBT and DMAB are shown in

Figure 1 [28]. Studies showed that there was a higher active SERS in a diluted HNO$_3$ solution for Cu substrates by using a 632.8-nm He/Ne laser. However, it could not be ruled out that Cu may also be an effective photoelectron emitter under a wavelength condition of 632.8 nm. The SERS spectrum on Cu-adsorbed 4NBT could be significantly converted into an SERS spectrum of another substance, such as 4-aminobenzenethiol (4ABT). Simultaneously, the authors studied the time-, substrate-, and wavelength-dependent surface catalysis reduction reaction of 4NBT to DMAB under the assisting condition of plasmon on Au, Ag, and Cu [29]. In Figure 2, atomic force microscopy (AFM) images of Au, Ag, and Cu films are displayed. Studies showed that this kind of reduction reaction of 4NBT to DMAB was significantly dependent on wavelength, substrates, and reaction time. When the wavelength was 632.8 nm, the reaction of 4NBT to DMAB did not occur on the Cu film substrate when the reaction time exceeded two hours. In contrast, 4NBT was quickly converted to DMAB when the wavelength was 514.5 nm. Thus, it was rational to control this reduction reaction by adjusting some of the factors mentioned above. Plasmon hot electrons provide high kinetic energy to overcome the reaction barrier for dissociation, and the electrons for photodissociation.

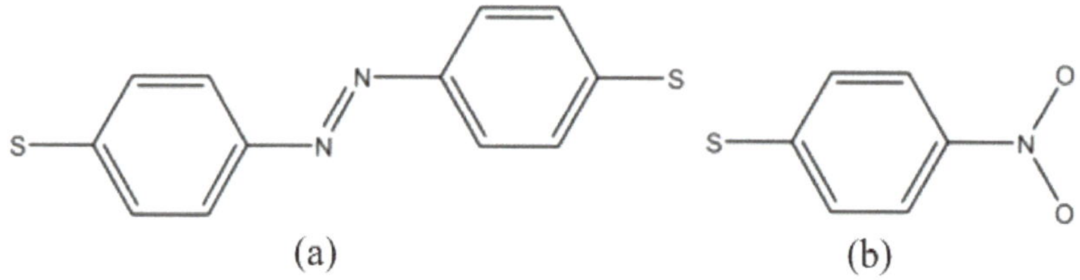

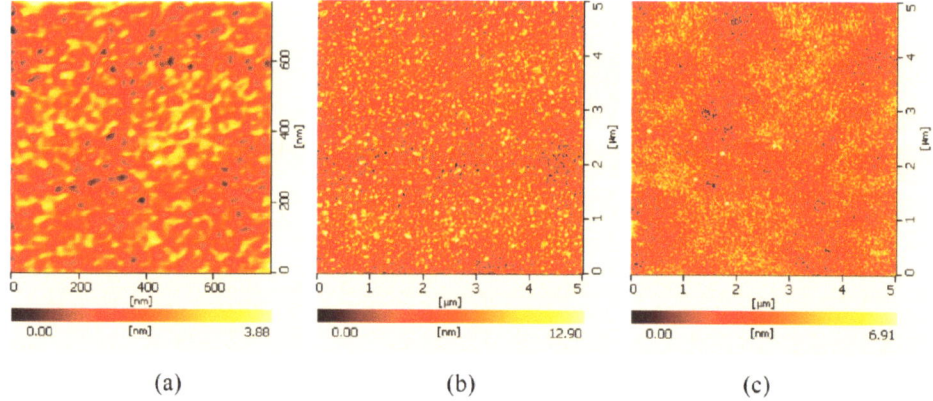

Figure 1. Chemical structures of (**a**) 4,4′-dimercaptoazobenzene (DMAB), and (**b**) 4-nitrobenzenethiol (4NBT) [28].

Figure 2. Atomic force microscopy (AFM) images of the (**a**) Au, (**b**) Ag, and (**c**) Cu films [29].

2. Plasmon-Driven Surface Catalysis Reduction Reaction of 4NBT to DMAB

In this section, we focus on the time- and wavelength-dependent plasmon-driven surface catalysis reduction reaction of 4NBT to DMAB on an Au film. Sun et al. conducted some research on Au films and determined whether the rate and yield of the reaction were related to the wavelength of the used laser and the reaction time. In Figure 3a,b, SERS spectra of 4NBT are shown at the junction of an Ag NP on the Au film with the incident wavelength of 514.5 and 632.8 nm, respectively. Studies showed that, when the incident laser had a wavelength of 632.8 nm and the reaction time reached 2.5 h, 4NBT did not convert to DMAB, revealed by the Raman intensity of $v_s(NO_2)$ on 4NBT. When the wavelength of the incident laser was 632.8 nm, the corresponding local electric-field distribution map of an Ag

NP–Au film was as displayed in Figure 4. However, when the incident laser wavelength was 514.5 nm, the Raman signal was hardly detected after 2.5 h, indicating that the Au film was not a reasonable substrate under the condition of a 514.5-nm laser.

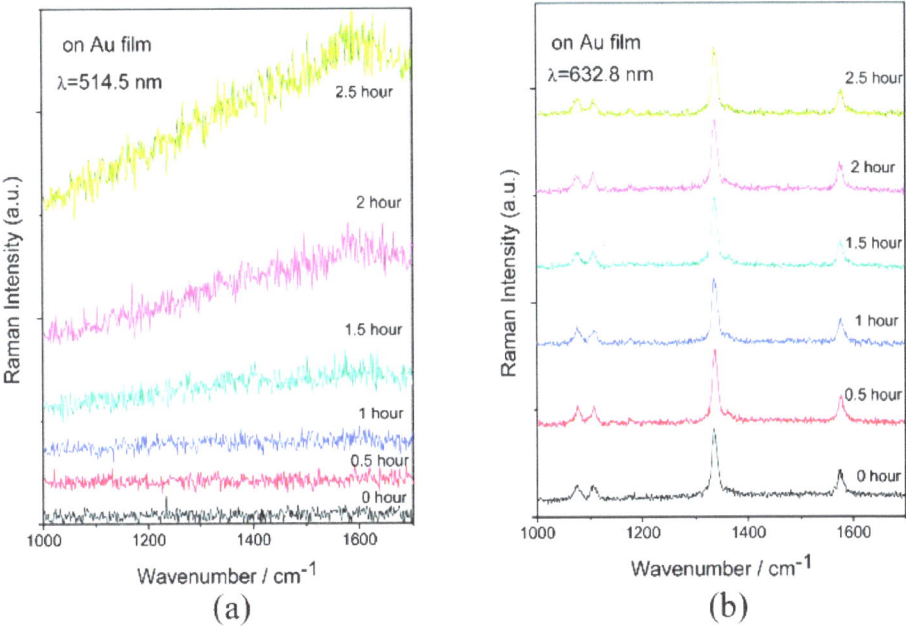

Figure 3. Surface-enhanced Raman scattering (SERS) spectra of 4NBT at the junction of an Ag nanoparticle (NP) on the Au film at incident wavelengths of (**a**) 514.5 nm and (**b**) 632.8 nm [29].

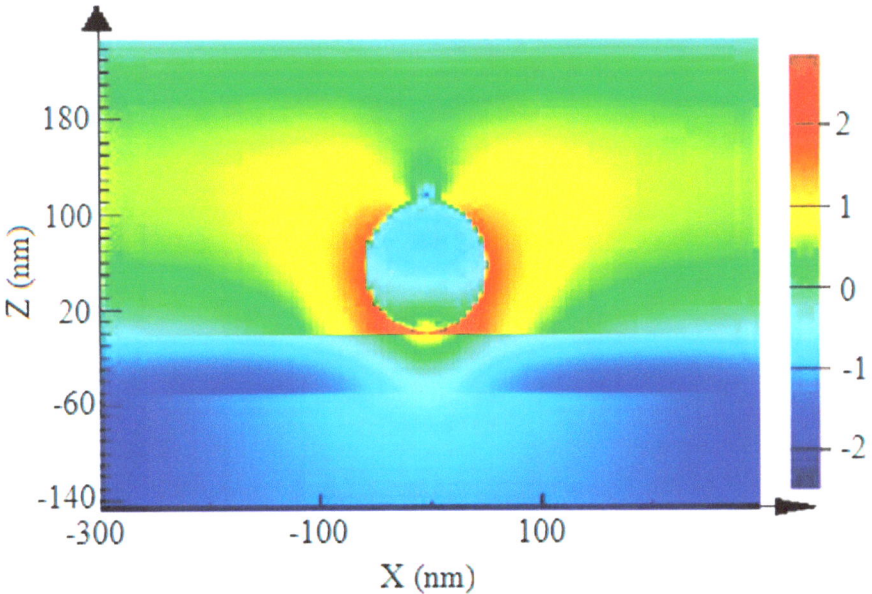

Figure 4. Ag NP–Au film at an incident wavelength of 632.8 nm [29].

Here, the authors proposed a reaction mechanism for a surface catalytic reaction when the 4NBT is transformed into DMAB, as displayed in Figure 5. There are many ways to provide energy for plasmon-driven surface catalytic reactions, such as enhanced thermal surroundings, hot electric transfer, and increased scattering. Here, the authors mainly studied energy supply by hot electrons. This mechanism indicates that the conversion of 4NBT to DMAB requires four electrons, and the required hot electrons were generated by plasmon decay, which had high kinetic energy, and the generated energy could drive the surface catalytic reaction.

By detecting the Raman signals of different incident-laser wavelengths and reaction times, the relationship between the surface catalytic reduction reaction of 4NBT to DMAB and the length of the reaction time was shown. The wavelength of the incident laser was also revealed, thus finding the optimal incident-laser wavelength and reaction time. Therefore, these two factors can be adjusted to promote or inhibit the surface catalytic reduction of 4NBT to DMAB on Au substrates. At the same time, plasmon resonance plays a key role in the surface catalytic reaction, and the energy that drives the reaction is generated by plasmon decay.

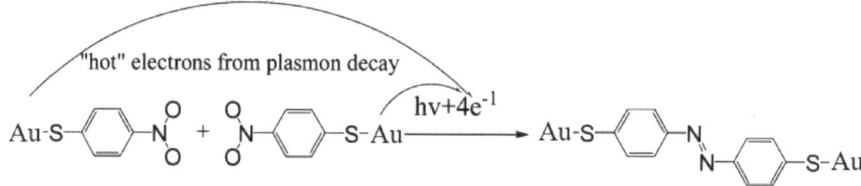

Figure 5. Mechanism of plasmon-assisted surface-catalyzed reaction [29].

In a previous study, control of the degradation reaction of DMAB was achieved, the chemical bond of DMAB was selectively broken using hot electrons as plasmon scissors, and the degradation process was related to various factors, such as pH. The authors controlled the dissociation products by pH, attaching hydrogen ions under acidic conditions to produce PATP. Under alkaline conditions, 4NBT was produced by attaching oxygen ions. In the next section, we present an in-depth discussion on what was more dominant in the synthesis and degradation process under different environmental conditions, as well as what had a higher conversion rate and a faster response rate.

3. Competition between Reaction and Degradation Pathways in Plasmon-Driven Photochemistry of 4NBT to DMAB

Plasmon materials are important for driving photochemical reactions. Previous studies showed that plasmon induced many photochemical reactions, but the corresponding reaction mechanism was not clear. In this section, the plasmon excitation field effect of plasmon excitation to promote the conversion of 4NBT to DMAB was investigated. For this reduction reaction, the increase in DMAB and decrease in 4NBT were related to time [30–32].

Plasmon nanostructures significantly contribute to driving photochemical reactions due to the powerful ability of plasmon nanostructures to focus and amplify various spectra [33–36]. Therefore, plasmon materials often act as a photocatalytic platform for inducing a lot of photochemical reactions. In previous studies, photoreactions were carried out, such as dissociation of H_2 [37,38], water decomposition [39,40], and reduction of CO_2 to hydrocarbon compounds [41,42]. This proved that plasmon-driven photoreactions have great potential. In this section, the authors used a substrate of Au film over a nanosphere (AuFON) to measure the yield and reaction rate of 4NBT photoreactions. In general, more field enhancement resulted in higher reaction yields. Unfortunately, this conjecture was not proven with the current technology. Studies showed that, for molecular degradation reactions, the reaction yield is highly dependent on electric-field strength.

To calculate the enhancement factor (EF) of each sampling area of the AuFON substrate, the specific relationship could be expressed by the following equation (Brooks, J.L., et al.):

$$EF = \frac{I_{SERS}/N_{surf}}{I_{NRS}/N_{vol}},\qquad(1)$$

where I_{SERS} is the Raman intensity of 4NBT when wavenumber position is 1074 cm^{-1} on the AuFON substrate, and the corresponding Raman intensity in the solution is I_{NRS}. N_{surf} is the number of molecules that are adsorbed onto the substrate of AuFON, and N_{vol} is the number of molecules in the sample collected by normal Raman scattering spectroscopy. The Raman cross-section of cyclohexane and 4NBT was collected in previous experiments [43,44].

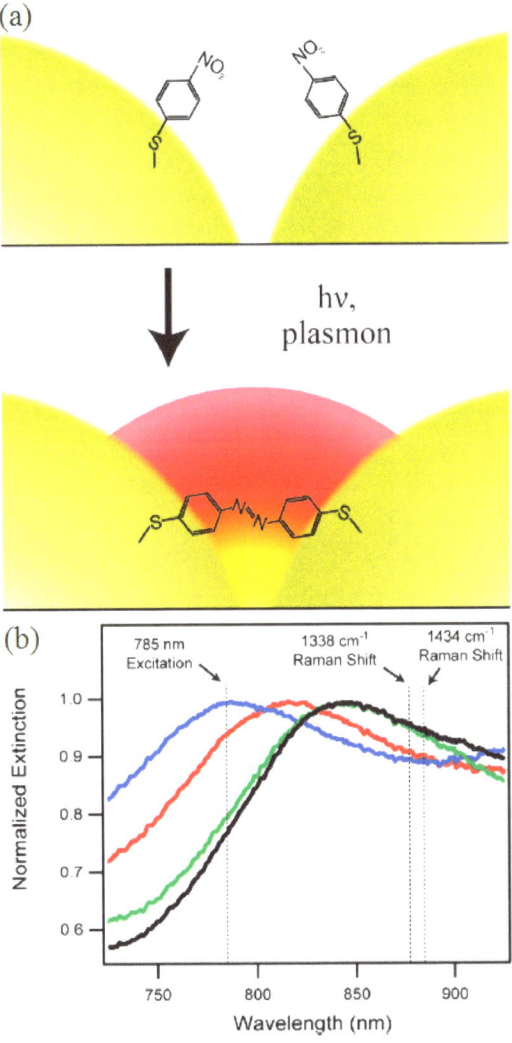

Figure 6. (a) Depiction of plasmon-driven conversion of 4NBT to DMAB; (b) localized surface plasmon (SP) measurements of SERS substrates. Different spots on the surface have varying plasmon-resonance wavelengths, affecting the SERS enhancement factor (EF) and electric-field enhancement [45].

The AuFON substrate used here has many advantages because it was spherical, its random variation changed the detected SERS EF, and the resulting value was the enhanced average in the selected region. The authors schematically illustrated the reaction of field-enhancement effects for plasmon-driven photoreaction on 4NBT transfer to DMAB, as shown in Figure 6a. It is shown in Figure 6b that defects in the spherical fill might cause a change in the local surface plasmon resonance (LSPR) of the detection region, thereby increasing/decreasing extinction and widening at the position of the excitation wavelength. In addition, Figure 6b displays the wavelength of the 785-nm laser and the scattering wavelengths of the two important vibration modes detected during the photoreactions [45]. With the red-shift of laser wavelength, the Raman peaks at 1338 and 1433 cm^{-1} also changed, which revealed the relationship of surface catalytic wavelength reactions.

In Figure 7 [45], we can see the time-resolved Raman spectroscopy of 4NBT plasmon-driven photocatalytic reactions. At the beginning, the vibration pattern appeared at 1338 cm^{-1}, as shown in the red part of the figure, indicating that 4NBT corresponds to NO_2 symmetric stretching vibration [46]. During laser irradiation, new Raman peaks appeared at 1140, 1388, and 1434 cm^{-1}, as shown in the green portion of the figure. This demonstrated DMAB generation, but it could be seen from the spectrum that no intermediate was produced. As reaction time increased, the DMAB peak increased and the 4NBT peak decreased. Among them, the Raman peak appearing at a 1434 cm^{-1} was selected as symbolic of the growth kinetics of the reaction product because the vibration mode of such a peak was found to be a symmetric azo bond in the DMAB spectrum [47]. The authors could calculate the yield and average kinetic energy of the 4NBT-to-DMAB reduction reactions. Here, the authors estimated plasmonic-field enhancement by using SERS enhancement factors (SERS EFs) [48,49]. With the increase in radiation time, the Raman peak at 1338 cm^{-1} was significantly reduced, while the Raman peak at 1433 cm^{-1} increased, which demonstrated surface catalytic reactions.

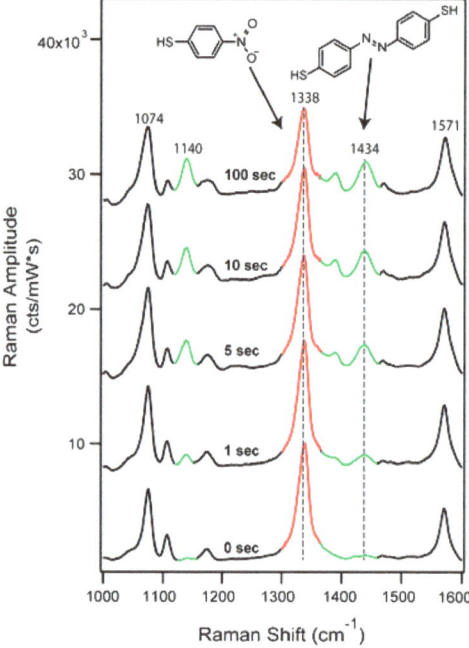

Figure 7. Time-resolved SERS measurements during the plasmon-driven conversion of 4NBT to DMAB. The reactant peak at 1338 cm^{-1} decays during SERS measurement, and the product peak at 1434 cm^{-1} grows during SERS measurement [45].

Figure 8 displays the variation in Raman peak amplitude during growth and decay, and plots the SERS EF function. It can be seen from the figure that the amplitude of the Raman peak during the growth process was positive, and the amplitude of the Raman peak corresponding to the attenuation process was negative. Figure 8 displays an amplitude plot of the Raman peaks of the reactants and products as a function of SERS EF. Studies showed that the change in peak amplitude of reactants and products was related to SERS EF. The Raman signal amplitude of the reactants and products increased with the increase of SERS EF when the molecule was bound to the Au surface.

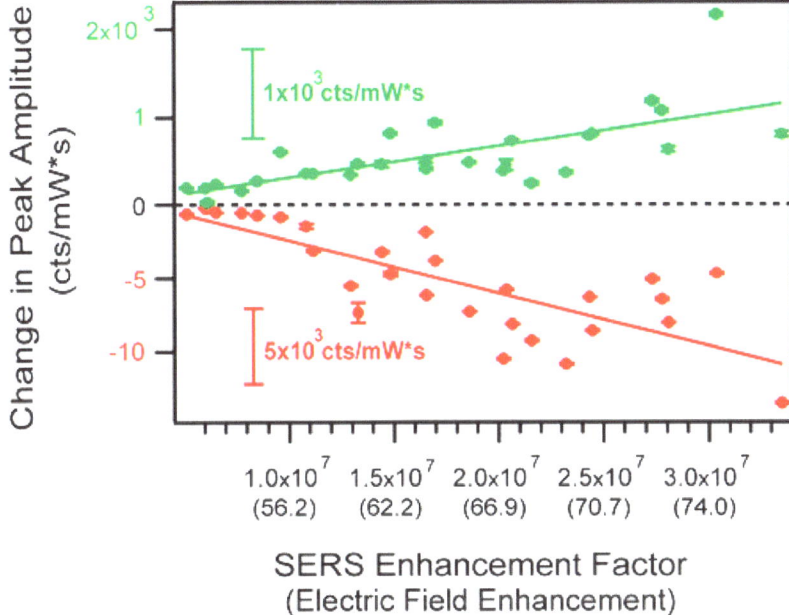

Figure 8. Total change in the reactant (red) and product (green) peak amplitudes as a function of SERS EF. The y-axes for reactant and product are different to clearly highlight the correlation with the SERS EF [45].

In order to further investigate the difference in reactant reduction and product increase in plasmon excitation reactions, reaction conversion efficiency and yield are proposed here. The reaction yield of the system is shown in Figure 9a, defining the ratio of the ending Raman amplitude of a product to a reactant. In fact, this was the percentage of the number of remaining 4NBT molecules on the surface when reactions were finished. There was no effect on the conversion efficiency of the SERS EF function. In this section, conversion efficiency is defined as the quotient of total product growth to total reactant reduction, where increase or decrease is expressed by the amplitude of the corresponding Raman peak. In contrast to the end-reaction yield, conversion efficiency concentrated on how intensely the reactant signal was lost, rather than recording the end amplitude. It can be observed from Figure 9b that, as the SERS EF increased, conversion efficiency gradually decreased. When the intensity of SERS EF was less than 10^7, conversion efficiency was between 20% and 100%; when the electric field was enhanced, conversion efficiency was generally low. Figure 9b displays that, when the intensity of the SERS EF was small, for example, less than 10^7, the conversion rate dropped significantly.

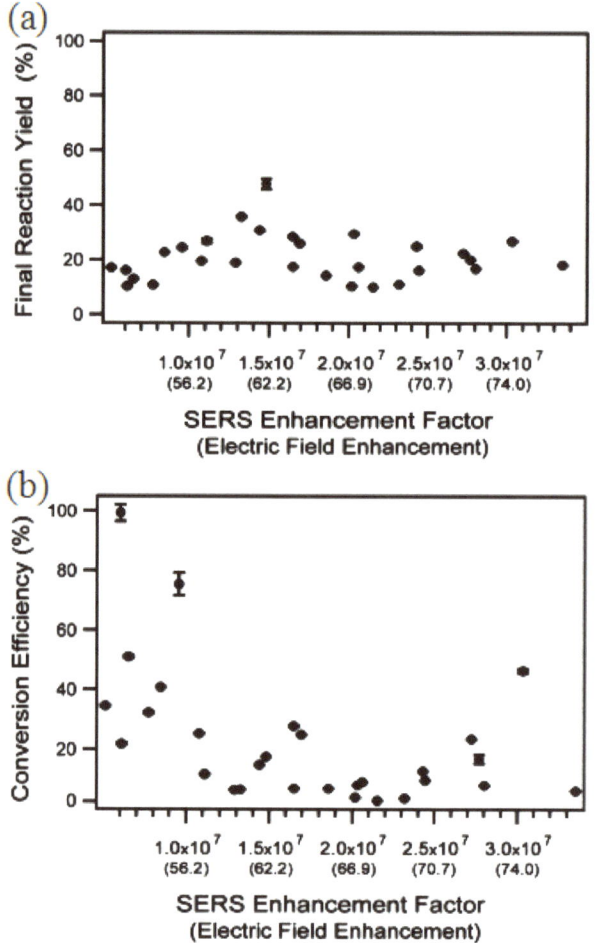

Figure 9. (a) Final reaction yield, which is the ratio between product final amplitudes and reactant amplitudes of photoreactions and dependence on SERS EF. As localized field enhancement increased, the reaction yield stayed between 10% and 40%; (b) dependence of conversion efficiency, defined as the ratio between the amplitude changes of the product and reactant amplitudes on the SERS EF. As field enhancement increased, the reactant peak (1338 cm^{-1}) decayed more significantly than the product peak formed (1434 cm^{-1}) [45].

Since the ability of the AuFON substrate was significantly improved, a large amount of the product could quickly be collected at the beginning; thus, the authors focused on average reaction kinetics. Figure 10 displays three independent kinetic measurements of the plasmon-induced reactions of 4NBT to DMAB, with different local field enhancements. Comparing Figure 10a–c, it can be seen that the resulting kinetics vary extensively in different areas of the substrate. Furthermore, in Figure 10a,b, product growth increased significantly within 10 s before the reactions, as indicated by the broken green line. In Figure 10c, we can see that the product was produced at a slower rate. It is, therefore, concluded that there was significant loss of reactant signal during product formation.

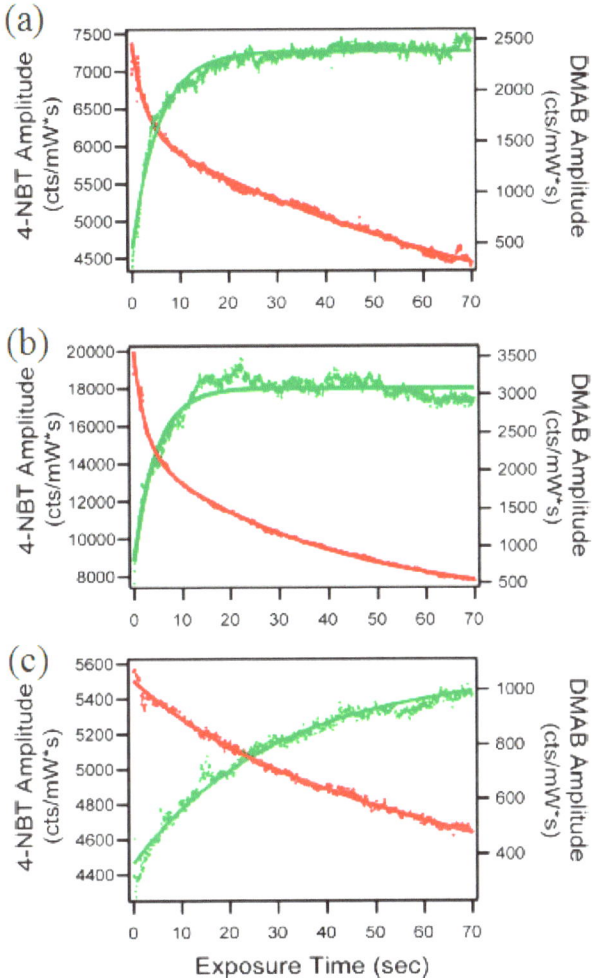

Figure 10. Reaction kinetics of 4NBT to DMAB plasmon-induced photoreactions. (**a**) The 4NBT reactant amplitude rapidly decreases with concurrent growth of product amplitude. At longer times, reactant peak experiences additional decay, while product amplitude remains constant; (**b**) similar to (a), but with a slight decrease in product amplitude at longer times; (**c**) in a weakly enhancing environment, reactions occur more slowly [45].

From Figure 10a,b, it can be found that 4NBT amplitude gradually decreased, and a strange phenomenon occurred after reactions proceeded for approximately 30 s. The 4NBT amplitude quickly decreased, but the DMAB amplitude was substantially unchanged. As demonstrated in Figure 10b, the authors detected minor decrease in the signal of DMAB amplitude, indicating a similar degradation-reaction process. However, it can be seen in Figure 10c that 4NBT signal amplitude did not significantly decrease, and exhibited a lower mean field enhancement. In this region, the rate of 4NBT reduction was very slow and consistent with DMAB product-formation kinetics.

Figure 11a displays the relationship between the photoreaction rate constant and the field enhancement. This was achieved by measuring product increase and reactant reduction. In general, the rate constant for reactant loss was generally higher compared to the rate constant of product increase, although the two constants were independent of the SERS EF. Thus, the rate-limiting stage of the photoconversion of the 4NBT molecule to DMAB reduction was independent of plasmon-field enhancement. In addition, the hydration layer was the root cause of the difference in photoreaction rate constants of the reactants and product. Studies showed that, in order for the reaction to better proceed, the nitro 4NBT group should be reduced, and protonation is needed to produce reaction intermediates [50–54]. In addition, the protons that came from the hydration layer on the Au surface were needed to limit the photoreaction rate [35–39].

Figure 11b displays the relationship between the rate constant of 4NBT degradation and photoreactions and field enhancement. Similarly, there was no significant relationship between field enhancement and rate constant for the scope of observed SERS EFs. Studies showed that, for other detection areas, the photoreaction rate constant was significantly higher than the degradation rate constant. This indicated that the reaction process was generally much quicker than degradation, which resulted in loss of light due to prolonged light exposure.

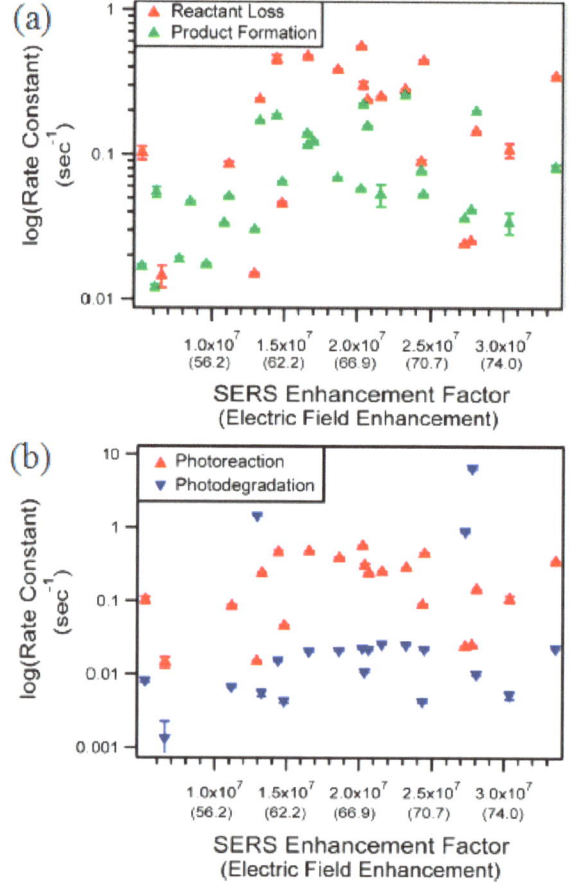

Figure 11. Rate constants for plasmon-induced processes. (a) Comparison of photoreaction rate constants with the loss of the 4NBT reactant (red), and formation of the DMAB product (green); (b) comparison of the photoreaction rate for 4NBT (red) and the photodegradation rate of 4NBT (blue). All observed ensemble-averaged rates are independent of field enhancement for the range of probed EFs [45].

4. Conclusions

In this review, we described plasmon catalyzed applications, such as plasmon-driven dissociations of hydrogen, nitrogen, water, and carbon dioxide. Photoreaction and degradation processes dominated under different environmental conditions. Therefore, in order to improve the rate of plasmon-driven processes for different chemical reactions, we should carefully consider a variety of photoinduced correlation rates.

In this work, we briefly introduced two ways to synthesize DMAB, and studied the photosynthetic reduction reaction and degradation process of 4NBT to DMAB, as well as the competitive relationship between reaction and degradation pathways. Note that the rate constant during competition is important because the degradation effect is usually more pronounced in a high field-enhancement region. The photosynthetic reaction was faster than the degradation process. However, in all regions, reactants were continuously reduced due to the progress of the degradation reaction, so a method of optimizing the process is to cause the photosynthetic reaction to occur before the degradation reaction, which effectively avoids the above problems.

In the future, plasmon–exciton interaction can promote efficiency and probability for plasmon-driven chemical reactions. Detailed information can be obtained from References [8,14,18,19,53–55].

Author Contributions: Resources, J.W.; data curation, Z.G. and J.J.; writing—original draft preparation, Z.G. and J.J.; writing—review and editing, Z.G. and J.W.; supervision, J.W.; project administration, J.W.; funding acquisition, J.W.

Funding: This work was supported by the National Nature Science Foundation of China (Grant No. 91436102, 11374353), the Fundamental Research Funds for the Central Universities, and the Talent Scientific Research Fund of LSHU (No. 6008).

Conflicts of Interest: The authors declare no conflicts of interest.

References

1. Sun, M.; Zhang, Z.; Wang, P.; Li, Q.; Ma, F.; Xu, H. Remotely excited Raman optical activity using chiral plasmon propagation in Ag nanowires. *Light Sci. Appl.* **2013**, *2*, e112. [CrossRef]
2. Quan, J.; Cao, E.; Mu, X.; Sun, M. Surface catalytic reaction driven by plasmonic waveguide. *Appl. Mater. Today* **2018**, *11*, 50–56. [CrossRef]
3. Fang, Y.; Zhang, Z.; Sun, M. High vacuum tip-enhanced Raman spectroscope based on a scanning tunneling microscope. *Rev. Sci. Instrum.* **2016**, *87*, 033104. [CrossRef] [PubMed]
4. Wang, J.; Qiao, W.; Mu, X. Au Tip-Enhanced Raman Spectroscopy for Catalysis. *Appl. Sci.* **2018**, *8*, 2026. [CrossRef]
5. Lin, W.; Ren, X.; Cui, L.; Zong, H.; Sun, M. Electro-optical tuning of plasmon-driven double reduction interface catalysis. *Appl. Mater. Today* **2018**, *11*, 189–192. [CrossRef]
6. Lin, W.; Xu, X.; Quan, J.; Sun, M. Propagating surface plasmon polaritons for remote excitation surface-enhanced Raman scattering spectroscopy. *Appl. Spectrosc. Rev.* **2018**, *53*, 771–782. [CrossRef]
7. Cao, E.; Lin, W.; Sun, M.; Liang, W.; Song, Y. Exciton-plasmon coupling interactions: From principle to applications. *Nanophotonics* **2018**, *7*, 145–167. [CrossRef]
8. Liu, W.; Lin, W.; Zhao, H.; Wang, P.; Sun, M. The nature of plasmon-exciton codriven surface catalytic reaction. *J. Raman Spectrosc.* **2017**, *49*, 383–387. [CrossRef]
9. Zhang, Z.; Sheng, S.; Wang, R.; Sun, M. Tip-Enhanced Raman Spectroscopy. *Anal. Chem.* **2016**, *88*, 9328–9346. [CrossRef]
10. Wang, X.; Cao, E.; Zong, H.; Sun, M. Plasmonic electrons enhanced resonance Raman scattering (EERRS) and electrons enhanced fluorescence (EEF) spectra. *Appl. Mater. Today* **2018**, *13*, 298–302. [CrossRef]
11. Sun, M.T.; Xu, H.X. A novel application of plasmonics: Plasmon-driven surface-catalyzed reactions. *Small* **2012**, *8*, 2777–2786. [CrossRef] [PubMed]
12. Wang, J.; Mu, X.; Wang, X.; Wang, N.; Ma, F.; Liang, W.; Sun, M. The thermal and thermoelectric properties of in-plane C-BN hybrid structures and graphene/h-BN van der Waals heterostructures. *Mater. Today Phys.* **2018**, *5*, 29–57. [CrossRef]

13. Li, R.; Zhang, Y.; Xu, X.; Zhou, Y.; Chen, M.; Sun, M. Optical characterizations of two-dimensional materials using nonlinear optical microscopies of CARS, TPEF, and SHG. *Nanophotonics* **2018**, *7*, 873–881. [CrossRef]
14. Lin, W.; Cao, Y.; Wang, P.; Sun, M. Unified treatment for plasmon-exciton co-driven reduction and oxidation reactions. *Langmuir* **2017**, *33*, 12102–12107. [CrossRef] [PubMed]
15. Ding, Q.; Li, R.; Chen, M.; Sun, M. Ag nanoparticles-TiO$_2$ film hybrid for plasmon-exciton co-driven surface catalytic reactions. *Appl. Mater. Today* **2017**, *9*, 251–258. [CrossRef]
16. Fang, Y.; Li, Y.; Xu, H.; Sun, M. Ascertaining p,p-dimercaptoazobenzene produced from p-aminothiophenol by selective catalytic coupling reaction on silver nanoparticles. *Langmuir* **2010**, *26*, 7737–7746. [CrossRef] [PubMed]
17. Li, P.; Ma, B.; Yang, L.; Liu, J. Hybrid single nanoreactor for in situ SERS monitoring of plasmon-driven and small Au nanoparticles catalyzed reactions. *Chem. Commun.* **2015**, *51*, 11394–11397. [CrossRef]
18. Cao, E.; Guo, X.; Zhang, L.; Shi, Y.; Lin, W.; Liu, X.; Fang, Y.; Zhou, L.; Sun, Y.; Song, Y.; et al. Electrooptical synergy on plasmon-exciton-codriven surface reduction reactions. *Adv. Mater. Interfaces* **2017**, *4*, 1700869. [CrossRef]
19. Lin, W.; Cao, E.; Zhang, L.; Xu, X.; Song, Y.; Liang, W.; Sun, M. Electrically enhanced hot hole driven oxidation catalysis at the interface of a plasmon-exciton hybrid. *Nanoscale* **2018**, *10*, 5482–5488. [CrossRef]
20. Sun, M.; Zhang, Z.; Zheng, H.; Xu, H. In-situ plasmon-driven chemical reactions revealed by high vacuum tip-enhanced Raman spectroscopy. *Sci. Rep.* **2013**, *2*, 647. [CrossRef]
21. Choi, H.K.; Shon, H.K.; Yu, H.; Lee, T.G.; Kim, Z.H. b2 Peaks in SERS Spectra of 4-Aminobenzenethiol: A Photochemical artifact or a real chemical enhancement? *J. Phys. Chem. Lett.* **2013**, *4*, 1079–1086. [CrossRef] [PubMed]
22. Sun, M.; Fang, Y.; Zhang, Z.; Xu, H. Activated vibrational modes and Fermi resonance in tip-enhanced Raman spectroscopy. *Phys. Rev. E* **2013**, *87*, 020401. [CrossRef] [PubMed]
23. Van Schrojenstein Lantman, E.M.; Deckert-Gaudig, T.; Mank, A.J.; Deckert, V.; Weckhuysen, B.M. Catalytic processes monitored at the nanoscale with tip-enhanced Raman spectroscopy. *Nat. Nanotechnol.* **2012**, *7*, 583–586. [CrossRef] [PubMed]
24. Xie, W.; Walkenfort, B.; Schlucker, S. Label-free SERS monitoring of chemical reactions catalyzed by small gold nanoparticles using 3D plasmonic superstructures. *J. Am. Chem. Soc.* **2013**, *135*, 1657–1660. [CrossRef] [PubMed]
25. Kim, K.; Choi, J.Y.; Shin, K.S. Photoreduction of 4-nitrobenzenethiol on Au by hot electrons plasmonically generated from Ag nanoparticles: Gap-mode surface-enhanced Raman scattering observation. *J. Phys. Chem. C* **2015**, *119*, 5187–5194. [CrossRef]
26. Lee, S.J.; Kim, K. Surface-induced photoreaction of 4-nitrobenzenethiol on silver: Influence of SERS-active sites. *Chem. Phys. Lett.* **2003**, *378*, 122–127. [CrossRef]
27. Shin, K.S.; Lee, H.S.; Joo, S.W.; Kim, K. Surface-induced photoreduction of 4-nitrobenzenethiol on Cu revealed by surface enhanced Raman scattering spectroscopy. *J. Phys. Chem. C* **2007**, *111*, 15223–15227. [CrossRef]
28. Mia, X.; Wang, Y.; Li, R.; Sun, M.; Zhang, Z.; Zheng, H. Multiple surface plasmon resonances enhanced nonlinear optical microscopy. *Nanophotonics* **2019**, *8*. [CrossRef]
29. Dong, B.; Fang, Y.; Chen, X.; Xu, H.; Sun, M. Substrate-, wavelength-, and time-dependent plasmon-assisted surface catalysis reactions of 4-nitrobenzenethiol dimerizing to p,p'-dimercaptoazobenzene on Au, Ag, and Cu films. *Langmuir* **2011**, *27*, 10677–10682. [CrossRef]
30. Kim, K.; Kim, K.L.; Shin, K.S. Photoreduction of 4,4'-dimercaptoazobenzene on Ag revealed by Raman scattering spectroscopy. *Langmuir* **2013**, *29*, 183–190. [CrossRef]
31. Zhang, Z.; Xu, P.; Yang, X.; Liang, W.; Sun, M. Surface plasmon-driven photocatalysis in ambient, aqueous and high-vacuum monitored by SERS and TERS. *J. Photochem. Photobiol. C Photochem. Rev.* **2016**, *27*, 100–112. [CrossRef]
32. Zhang, Q.; Wang, H. Facet-dependent catalytic activities of Au nanoparticles enclosed by high-index facets. *ACS Catal.* **2014**, *4*, 4027–4033. [CrossRef]
33. Brongersma, M.L.; Halas, N.J.; Nordlander, P. Plasmon-induced hot carrier science and technology. *Nat. Nanotechnol.* **2015**, *10*, 25–34. [CrossRef] [PubMed]
34. Xu, X.; Shi, Y.; Liu, X.; Sun, M. Femtosecond dynamics of monolayer MoS$_2$-Ag nanoparticles hybrid probed at 532 nm. *Chem. Phys. Lett.* **2018**, *692*, 208–213.

35. Hou, W.; Cronin, S.B. A review of surface plasmon resonance-enhanced photocatalysis. *Adv. Funct. Mater.* **2013**, *23*, 1612–1619. [CrossRef]
36. Lin, W.; Shi, Y.; Yang, X.; Li, J.; Cao, E.; Xu, X.; Pullerits, T.; Liang, W.; Sun, M. Physical mechanism on exciton-plasmon coupling revealed by femtosecond pump-probe transient absorption spectroscopy. *Mater. Today Phys.* **2017**, *3*, 33–40. [CrossRef]
37. Mukherjee, S.; Libisch, F.; Large, N.; Neumann, O.; Brown, L.V.; Cheng, J.; Lassiter, J.B.; Carter, E.A.; Nordlander, P.; Halas, N.J. Hot electrons do the impossible: Plasmon-induced dissociation of H_2 on Au. *Nano Lett.* **2013**, *13*, 240–247. [CrossRef] [PubMed]
38. Amenomiya, Y. Adsorption of hydrogen and H_2-D_2 exchange reaction on alumina. *J. Catal.* **1971**, *22*, 109–122. [CrossRef]
39. Mubeen, S.; Lee, J.; Singh, N.; Kramer, S.; Stucky, G.D.; Moskovits, M. An autonomous photosynthetic device in which all charge carriers derive from surface plasmons. *Nat. Nanotechnol.* **2013**, *8*, 247–251. [CrossRef]
40. Liu, Z.; Hou, W.; Pavaskar, P.; Aykol, M.; Cronin, S.B. Plasmon resonant enhancement of photocatalytic water splitting under visible illumination. *Nano Lett.* **2011**, *11*, 111–1116. [CrossRef]
41. Hou, W.; Hung, W.H.; Pavaskar, P.; Goeppert, A.; Aykol, M.; Cronin, S.B. Photocatalytic conversion of CO_2 to hydrocarbon fuels via plasmon-enhanced absorption and metallic interband transitions. *ACS Catal.* **2011**, *1*, 929–936. [CrossRef]
42. Kleinman, S.L.; Frontiera, R.R.; Henry, A.L.; Dieringer, J.A.; Van Duyne, R.P. Characterizing, and controlling chemistry with SERS hot spots. *Phys. Chem. Chem. Phys.* **2013**, *15*, 21–36. [CrossRef] [PubMed]
43. Valley, N.; Greeneltch, N.; Van Duyne, R.P.; Schatz, G.C. A look at the origin and magnitude of the chemical contribution to the enhancement mechanism of surface-enhanced Raman spectroscopy (SERS): Theory and experiment. *J. Phys. Chem. Lett.* **2013**, *4*, 2599–2604. [CrossRef]
44. Mccreery, R.L. *Raman Spectroscopy for Chemical Analysis*; Wiley-Interscience: New York, NY, USA, 2000.
45. Brooks, J.L.; Frontiera, R.R. Competition between reactions and degradation pathways in plasmon-driven photochemistry. *J. Phys. Chem. C* **2016**, *120*, 20869–20876. [CrossRef]
46. Skadtchenko, B.O.; Aroca, R. Surface-enhanced Raman scattering of p-nitrothiophenol. *Spectrochim. Acta Part A* **2001**, *57*, 1009–1016. [CrossRef]
47. Kim, K.; Shin, D.; Kim, K.L.; Shin, K.S. Surface-enhanced Raman scattering of 4,4′-dimercaptoazobenzene trapped in Au nanogaps. *Phys. Chem. Chem. Phys.* **2012**, *14*, 4095–4100. [CrossRef]
48. Hao, E.; Schatz, G.C. Electromagnetic fields around silver nanoparticles and dimers. *J. Chem. Phys.* **2004**, *120*, 357–366. [CrossRef]
49. Stranahan, S.M.; Willets, K.A. Super-resolution optical imaging of single-molecule SERS hot spots. *Nano Lett.* **2010**, *10*, 3777–3784. [CrossRef]
50. Xu, P.; Kang, L.; Mack, N.H.; Schanze, K.S.; Han, X.; Wang, H.L. Mechanistic understanding of surface plasmon assisted catalysis on a single particle: Cyclic redox of 4-aminothiophenol. *Sci. Rep.* **2013**, *3*, 2997. [CrossRef]
51. Dong, B.; Fang, Y.R.; Xia, L.; Xu, H.X.; Sun, M.T. Is 4-nitrobenzenethiol converted to p,p′-dimercaptoazobenzene or 4-aminothiophenol by surface photochemistry reaction? *J. Raman Spectrosc.* **2011**, *42*, 1205–1206. [CrossRef]
52. Wang, J.; Wang, X.; Mu, X. Plasmonic Photocatalysts Monitored by Tip-Enhanced Raman Spectroscopy. *Catalysts* **2019**, *9*, 109. [CrossRef]
53. Wang, J.; Lin, W.; Xu, X.; Ma, F.; Sun, M. Plasmon-Exciton Coupling Interaction for Surface Catalytic Reactions. *Chem. Rec.* **2018**, *18*, 481–490. [CrossRef] [PubMed]
54. Wang, J.; Feng, N.; Sun, Y.; Mu, X. Nanoplasmon–Semiconductor Hybrid for Interface Catalysis. *Catalysts* **2018**, *8*, 429. [CrossRef]
55. Cao, E.; Sun, M.; Song, Y.; Liang, W. Exciton-plasmon hybrids for surface catalysis detected by SERS. *Nanotechnology* **2018**, *29*, 372001. [CrossRef] [PubMed]

© 2019 by the authors. Licensee MDPI, Basel, Switzerland. This article is an open access article distributed under the terms and conditions of the Creative Commons Attribution (CC BY) license (http://creativecommons.org/licenses/by/4.0/).

Review

Plasmonic Photocatalysts Monitored by Tip-Enhanced Raman Spectroscopy

Jingang Wang [1,†], Xinxin Wang [2,†] and Xijiao Mu [2,*]

1 Computational Center for Property and Modification on Nanomaterials, College of Sciences, Liaoning Shihua University, Fushun 113001, China; jingang_wang@sau.edu.cn
2 School of Mathematics and Physics, Beijing Key Laboratory for Magneto-Photoelectrical Composite and Interface Science, University of Science and Technology Beijing, Beijing 100083, China; m13717927930@163.com
* Correspondence: shumuxijiao@163.com
† Contributed equally.

Received: 30 December 2018; Accepted: 21 January 2019; Published: 22 January 2019

Abstract: In this review, we first prove the resonance dissociation process by using time-dependent measurements of tip-enhanced resonance Raman spectroscopy (TERRS) under high vacuum conditions. Second, we show how to use thermal electrons to dissociate Malachite Green (MG) and the hot electrons in the nanogap of the high vacuum tip-enhanced Raman spectroscopy (TERS) device that are generated by plasma decay. Malachite Green is excited by resonance and adsorbed on the Ag and Au surfaces. Finally, we describe real-world and real-time observations of plasmon-induced general chemical reactions of individual molecules.

Keywords: plasmonic photocatalysis; hydrogen dissociation; hot electron

1. Introduction

A surface plasmon is a common oscillation that excites free electrons at the interface between metal and dielectric [1–3]. Surface plasmon resonance, at a specific wavelength, is excited by an incident laser, resulting in a large enhancement in the local electromagnetic field. The field of electromagnetic field enhancement for surface-enhanced Raman spectroscopy and tip-enhanced Raman spectroscopy is a physical enhancement mechanism. Generally speaking, metal nanostructures, based on local surface plasmon (LSP) and propagating surface plasmon (PSP), have revealed many fantastic properties. As demonstrated in Figure 1a, the LSP is a density oscillation that limits the charge on the surface of the Au or Ag nanostructures. However, PSPs are the collective excitation conduction of electrons, which propagates along the interface between metal and dielectric medium through surface electromagnetic waves (Figure 1b). This wave is limited to the vicinity of the interface.

Because of the addition of the catalyst, there is a decreased reaction barrier at the surface plasmon, causing surface catalysis to increase the rate of general chemical reactions. Detailed physical mechanism for plasmon-driven surface catalysis reactions can refer to recent review papers [2,3]. Low energy and high output have always been our goal for the catalytic reaction of plasma surfaces [2].

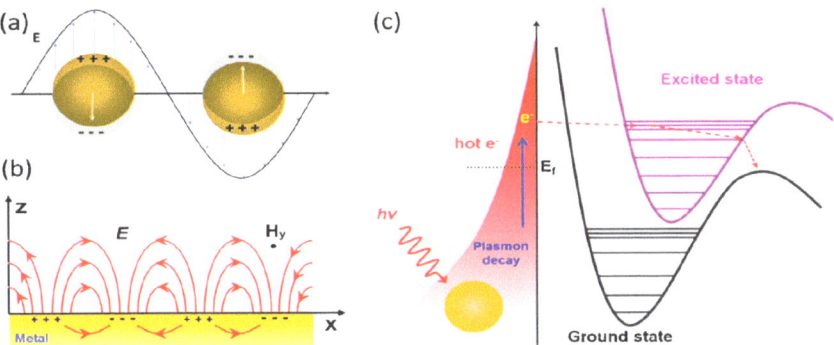

Figure 1. Sketches of the mechanism of local surface plasmons (LSPs), propagating surface plasmons (PSPs), and plasmon-driven catalytic reactions. With laser excitation on the Au or Ag nanostructures' surface, (**a**) LSPs are density oscillations of charge confined on the metal nanoparticles' surface; (**b**) PSPs are the electrons' collective excitation conduction that propagates near the interface; (**c**) hot electrons generated from plasmon decay can transfer to the excited state of molecule 1, and from there initiate the catalytic reaction to molecule 2. Adapted from [3].

The local surface plasmon, excited on the surface of Ag/Au nanoparticles, could be non-radioactively decayed into hot electrons (holes); these hot electrons have very high energy levels. These hot electrons can be dispersed into excited states of absorbing molecules and then initiate a chemical reaction by reducing the activation energy. As demonstrated in Figure 1c, thermal electrons are first produced by non-radioactive decay of plasma on gold or silver surfaces. The hot electrons, at high energy levels, have enough energy to transfer to the excited state of molecule 1, with a short-lived negative ion being produced. Plasmonic catalysis has been demonstrated in a variety of chemical reactions, for instance, dissociation of H_2 [4], and water splitting, in which the required activation energy is provided by plasma-induced thermo electrons.

Through our previous research, we understand the past advances of plasmon-driven surface catalytic reactions, in atmosphere, aqueous, and high-vacuum environments, monitored by surface-enhanced Raman spectroscopy (SERS) and tip-enhanced Raman spectroscopy (TERS). Next, we have introduced the high-vacuum TERS (HV-TERS) and TERS techniques, which are a perfect excuse for learning plasmonic catalysis. The catalytic process can be easily controlled by time- and intensity-dependent TERS spectroscopy; the catalytic rate can be controlled well by the distance of the gap, and the localized experimental temperature can be estimated by TERS spectra of Stokes and anti-Stokes. We also performed a surface catalysis reaction of time-dependent remote excitation; the catalytic reaction of DMAB generated by 4NBT through a plasma waveguide, at atomic level, had smooth surfaces. Then, absorption spectroscopy revealed plasmon-exciton coupling on AgNPs-TiO_2 film hybridization, such as ultraviolet-visible (UV-vis) absorption, ultrafast transient absorption, SERS spectroscopy, and so on. Furthermore, we propose a uniform voltage-controlled plasmon co-driven oxidation and reduction reaction treatment method. On the one hand, the proposed mechanism could promote better understanding of co-driven chemical reactions of plasmon-exciton coupling. On the other hand, it may accelerate the development of plasma excitons in different fields. Lastly, we noted that plasma-driven double reduction reactions had been successfully observed in experiments in electrochemical environments.

Plasmon-induced chemical reactions caused increasing interest in photocatalytic reactions. For instance, a molecule can be adsorbed on metal nanostructures. However, this mechanism is still controversial because it is difficult to directly observe the chemical reaction in the plasma field, which is located near the metal surface. Surface plasmons in nanoscale-assisted molecular synthesis is an attractive topic in the field of surface catalytic reactions [5–8]. DMAB has been the subject of plasma-assisted catalytic reaction studies, but they are all plasma-assisted synthesis reactions. Nevertheless, since the hot electrons required for dissociation (produced by plasma decay) are mainly quenched in the environmental SERS, it is desirable to find a new technique for the dissociation process. Therefore, it is hoped to find a novel technology to carry out the dissociation process.

Hydrogen dissociation is a key step in many hydrogenation reactions for industrial chemical production. Catalytic hydrogenation is one of the largest branches of heterogeneous catalysis and is very important for numerous industrial reactions, such as the hydrogenation of unsaturated hydrocarbons and hydride chlorination. Supported precious metals have been widely used as catalysts because their favorable electronic structure allows hydrogen molecules to easily adsorb onto the metallic surface and dissociate. Currently, plasma metal nanoparticles have become new photocatalysts for small molecule activation and related general chemical reactions. In previous work we demonstrated the use of Au nanoparticles for plasmon-enabled hydrogen dissociation despite their intrinsically inert d-band configuration [4].

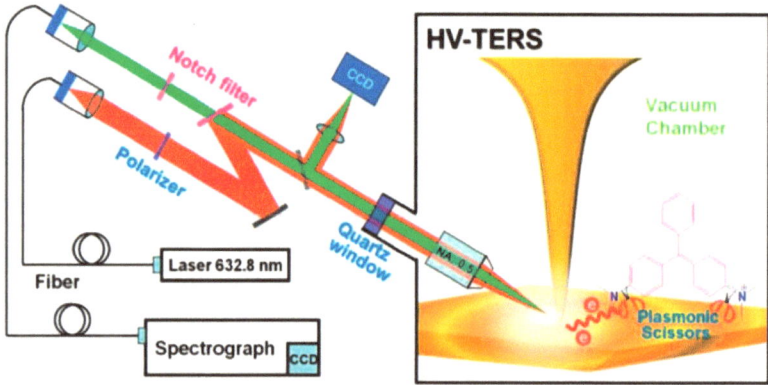

Figure 2. Home-made setup of high-vacuum tip-enhanced Raman spectroscopy (HV-TERS) and mechanism of resonant dissociation by plasmonic scissors. Hot electrons are excited by laser light in the nanogap of a TERS setup to cut molecular bonds in the substrate-adsorbed Malachite Green (MG) molecules. Adapted from [9].

In this review, we first introduced the resonance dissociation process with time-dependent measurements by using tip-enhanced Raman spectroscopy under high vacuum conditions. We demonstrate the process by which hot electrons dissociate Malachite Green (MG) and sensitively control the dissociation rate and probability by adjusting the thermal electron density. The hot electrons are generated by plasma decay in the nanogap of the HV-TERS setting (as demonstrated in Figure 2). Second, we recommend selective bond cleavage of DMAB by using thermal electrons as plasma scissors, especially by changing the pH to control the dissociation products. Third, we report the first experimental evidence for photocatalytic decomposition of hydrogen at room temperature. Then, we show that chemically synthesized aluminum nanocrystals (Al NCs) are capable of photocatalyzing hydrogen dissociation under laser illumination without the use of transition metal dopants, deep UV photons, or extreme reaction conditions such as elevated temperature or pressure. In addition, we prove the plasma enhancement of photocatalytic water decomposition under visible light irradiation by combining strong plasma Au nanoparticles with strongly catalyzed TiO_2. Lastly, we describe real-world and real-time observations of plasmon-induced general chemical reactions of individual molecules.

2. Molecular Resonant Dissociation of Surface-Adsorbed Molecules by Plasmonic Nanoscissors

Developing a method that enables pattern-specific chemistry was a very significant goal for chemists and an active area of research in the field of molecular reaction dynamics. The proof of resonance dissociation of MG was dependent on several complementary theoretical and experimental studies. To begin with, authors measured the light absorption spectrum of Malachite Green. The implication of Figure 3a is that there are strong absorption peaks, weak absorption peaks, and no absorption peaks at 632.8 nm, 514.5 nm, and 785 nm, respectively. In addition, the implication of Figure 3b is that the SERS spectra of Malachite Green in Ag sol were marked at these three frequencies. The normal Raman excitation is at 785 nm, the Raman excitation wavelength is 514 nm, similar to the 785 nm excitation Raman; and the resonance Raman excitation at 632.8 nm, where the Raman peak A–D was resonant electron transition selectivity at 632.8 nm excitation, at which point peak E for comparison was considered to be the normalized peak. Thirdly, our theoretical calculations determine that the peak A–D associated with the vibration mode in Malachite Green is related to $-NC_2H_6$ fragments (see Figure 3c). Malachite Green exhibits a C_2 symmetric mode with two vibration modes, a and b, respectively, and five selective enhanced vibration modes, which are recorded as a_{39}, a_{45}, b_{49}, b_{61} and a_{58} from low frequencies to high frequencies, respectively.

As demonstrated in Figure 3b, Peak E was a vibration mode associated with benzenyl C-C stretching mode. Consequently, their analysis exhibited that the resonant excitation energy at 632.8 nm was concentrated on the A–D mode associated with the $-NC_2H_6$ fragments. These modes could be excited by selective resonance. It is to be noted that the peaks B and C were superimposed due to the width at half maximum in Figure 3b, but obviously they were separated in the HV-TERS (see Figure 4a). By comparing the tip-enhanced resonance Raman spectroscopy (TERRS) excitation at 632.8 nm and the SERS excitation at 785 nm, we clearly see that the A–D peak could be selectively excited.

Then, authors used TERRS to verify the resonance decomposition process through time-dependent measurements under high vacuum conditions, and the wavelength of the incident light is 632.8 nm. Figure 4 gives the time sequence TERRS of the MG, and the MG in a high vacuum was adsorbed on the Ag surface in the presence of 632.8 nm of incident light. In the beginning, the implication of Figure 4a was that the Raman peaks of A–D, in the TERRS spectrum, were strongly enhanced by comparing the non-resonant SERS spectra excited at 785 nm, when the laser was irradiated on the sample at $t = 0$ minutes; in contrast, the E peak at 1592 cm^{-1} was considered to be a normalized Raman peak. This indicates that the resonant excitation energy during the electronic transition mainly excites four vibration modes selectively. The implication of Figure 4d,e is that the simulated Raman spectra of the dissociated fragments from the MG, as well as their vibrational modes, are also assigned. The fragment of HN(CH$_3$) exhibited C_{2v} symmetry; these four vibration modes are labeled a_1, a_2, b_1 and b_2, respectively. The large fragment in Figure 4d exhibited D_3 symmetry; however, when it was attached to a metal (as shown in Figure 4f), it exhibited C_2 symmetry and the two vibration modes are labeled a and b, respectively. The vibrational modes F–M (seen in Figure 4c) can be seen in Figure 4f. The temporal evolution of the TERRS spectrum (as demonstrated in Figure 4) provides evidence of the dissociation chemical reactions that occur in the MG. The simulation of the Raman spectrum due to dissociation debris is clearly demonstrated. The implication of Figure 4c is that the ultimate TERRS spectrum contains features from two dissociated fragments.

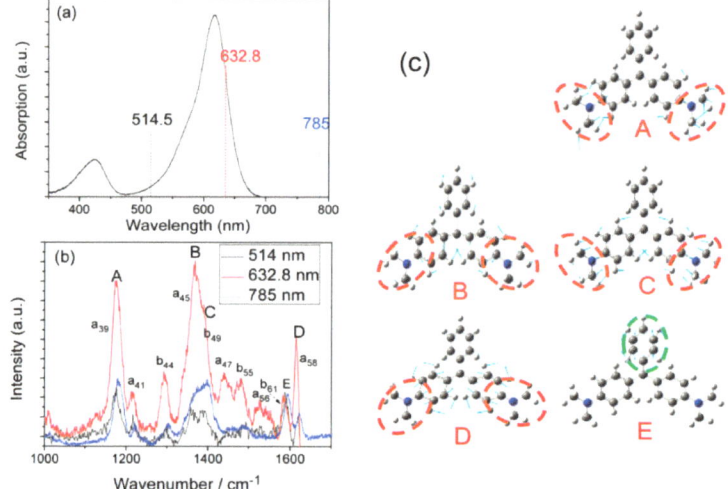

Figure 3. Absorption and Raman spectra and calculated normal modes of MG. (**a**) Absorption spectrum of MG in water, (**b**) surface-enhanced Raman spectroscopy (SERS) and surface-enhanced resonance Raman spectroscopy (SERRS) spectra of MG in Ag sol, and (**c**) vibrational modes A–E of MG. Adapted from [9].

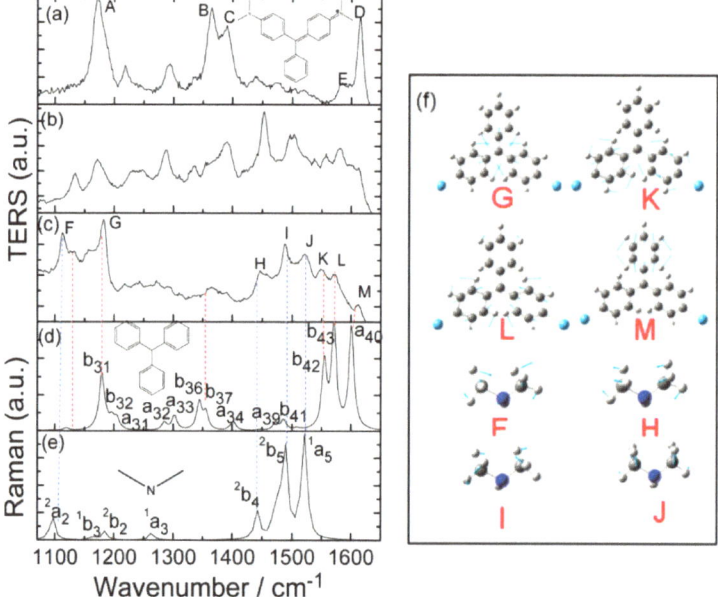

Figure 4. Time sequential tip-enhanced resonance Raman spectroscopy (TERRS) and the vibrational modes of MG. (**a**) The initial, (**b**) intermediate (20 min after continuous radiation using a laser), and (**c**) the final spectra (40 min after continuous radiation using a laser). The tunneling current and the bias voltage are 1 nA and 1 V, respectively. (**d**) and (**e**) Simulated Raman spectra of fragments (see the insets). (**f**) The vibrational modes of dissociated fragments of MG (corresponding to the experimental peaks in (**d**) and (**e**), as calculated by density functional theory). Adapted from [9].

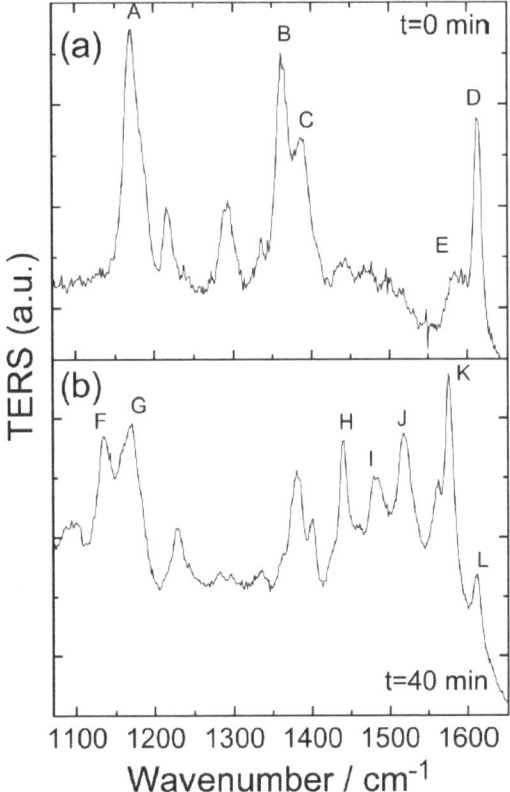

Figure 5. Time sequential TERRS and the vibrational modes of MG adsorbed on the Au film. (**a**) The initial and (**b**) the spectra 40 min after continuous radiation using a laser. The tunneling current and the bias voltage are 1 nA and 1 V, respectively. Adapted from [9].

The implication of Figure 5 is that the authors made a similar observation of Malachite Green adsorbed on the gold film; meanwhile, the authors concluded that the general chemical reaction on the silver film reacts faster than on the gold substrate. Generally speaking, the reaction rate and probability of a chemical reaction under high vacuum conditions can be controlled by a number of factors, such as bias voltage, laser intensity, tunnel current variation, and so on. The implication of Figure 6a,b is that laser power reduction of 10% of total laser power produces a slower dissociation rate—after 40 min, as demonstrated in Figure 6d, it was far from the stable configuration shown in Figure 6c. Obviously, the implication of Figure 6b is that, although a new Raman peak had appeared, the strongly enhanced Raman peaks A–C were still visible. Thus, the energy and tunneling current provided by the laser are insufficient to dissociate the molecules unless the laser illuminates the sample continuously to offer a continuous supply of large-density hot electrons.

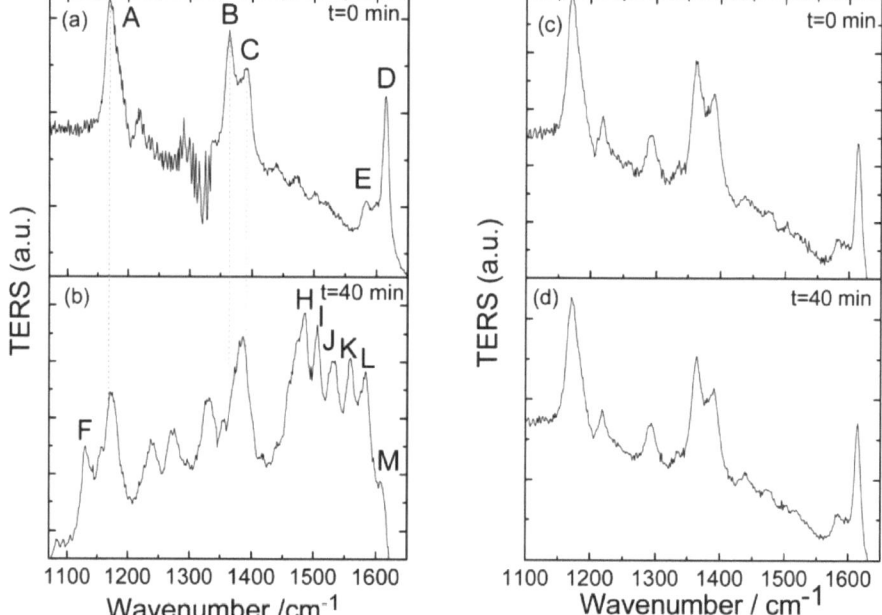

Figure 6. TERRS of MG adsorbed on the Ag film at $t = 0$ and $t = 40$ min. (**a**,**b**) Sample irradiated with 10% lower laser power compared to the sample analyzed in Figure 3a,c shows evidence of a slower dissociation rate; (**c**,**d**) without continuous laser irradiation of the sample shows that the tunneling current alone is not sufficient to induce dissociation. Adapted from [9].

By irradiating the nanogap in the HV-TERS between the tip and the substrate by a laser, large-density plasma was first generated. As demonstrated in Figure 7, the high kinetic energy of the hot electrons and the laser beam induced the dissociation process, and the hot electrons were generated by plasma decay. Compared with general Raman scattering, the resonance electron transition increases the lifetime of Raman scattering, and the resonant excitation of the laser causes the excitation energy to concentrate on the vibration mode associated with the Malachite Green $-NC_2H_6$ fragments. Therefore, we turn to selectively reducing the energy of these bonds, as demonstrated by process A in Figure 7. The thermal electrons generated by plasma decay had two purposes; to begin with, when hot electrons strike metal-adsorbing molecules, due to the high energy of hot electrons, they induce changes in the molecular excited-state potential energy surface (ESPES) from a neutral to excited state of transient negative ion (ESPES$^-$), as demonstrated by process B in Figure 7. The redox state caused by the hot electrons is a transient negative state [10]. In addition, during their interaction, the kinetic energy was transferred from thermionic to vibrational dissociation energy of the intramolecular, as demonstrated by Process C in Figure 7.

Malachite Green can be applied as a vibration mode for various chemical reactions, which can effectively reduce the activation energy of the reaction and increase the reaction rate.

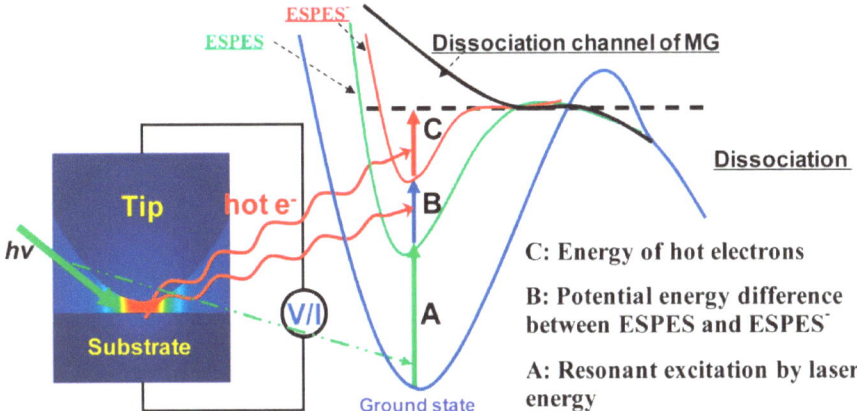

Figure 7. The mechanism of resonant dissociation by plasmon scissors. Three main energetic components drive the chemical reaction: (**A**) resonant absorption of the laser light enhances the vibrational modes associated with $-NC_2H_6$ and excites the MG molecules to an excited-state potential energy surface (ESPES); (**B**) hot electrons temporarily change the molecules' ESPES from a neutral to a negative ion excited state (ESPES$^-$); (**C**) the kinetic energy of the hot electrons is converted into intramolecular vibrational thermal energy. Adapted from [9].

3. Plasmonic Scissors for Molecular Design

Previously, plasma-assisted synthesis of DMAB revealed by surface enhanced Raman scattering spectroscopy from PATP and 4NBT had been achieved. DMAB had become a research project for plasma-assisted catalytic reactions, but they were all plasma-assisted synthesis reactions. Generally speaking, it was theoretically possible to implement the decomposition process by SERS spectroscopy. The implication of Figure 8a is that the TER spectrum was adsorbed on the silver film for DMAB. As shown in Figure 8b, this was the same as the normal Raman spectrum of the DMAB powder [5]. In particular, the ag_{17} vibrational mode at 1433 cm^{-1} of DMAB amounts to an N=N stretching mode. When DMAB is selectively decomposed by plasma scissors, this peak is reduced or even disappears in the TER spectrum. The author tried to observe the signal of decomposition by continuous emission, while its TER spectrum remains the same as was demonstrated in Figure 8a. In previous studies, in the molecular synthesis of DMAB from PATP and 4NBT, dissociation fragments could be rapidly dimerized into DMAB by the N=N bonds. Obviously, dissociation could not be observed during the competition between dissociation and synthesis; therefore, new approaches were needed.

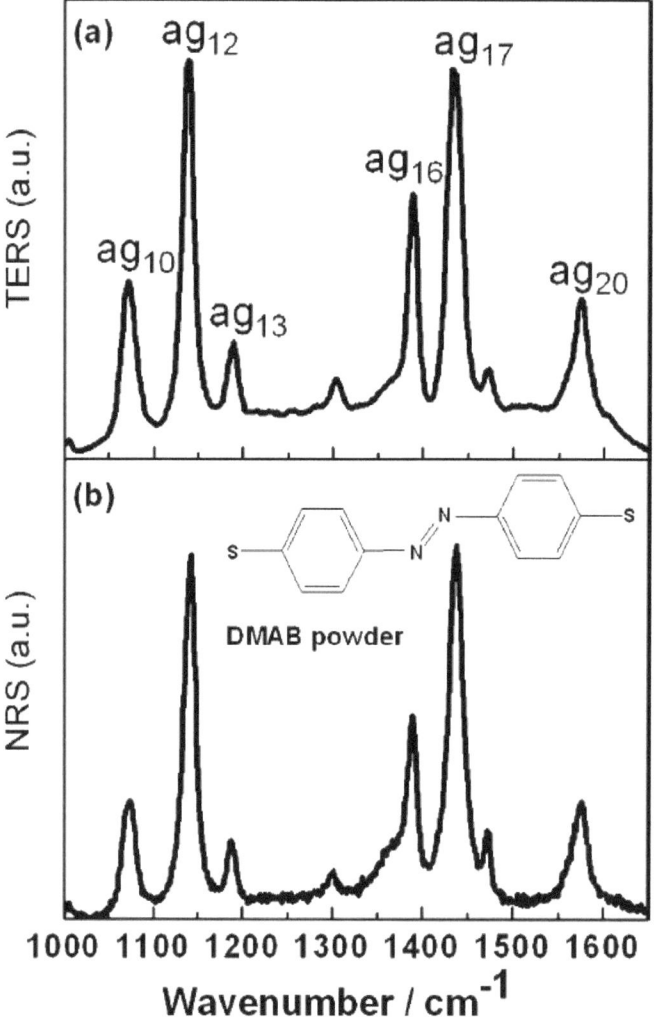

Figure 8. TER and nuclear resonance spectra (NRS) of DMAB. (**a**) TER spectrum of DMAB adsorbed on Ag film at tunneling current of 1 nA and bias voltage of 1 V. (**b**) NRS of DMAB powder. The excitation wavelength was 632.8 nm. Adapted from [11].

The authors manipulated the pH under HV-TERS conditions to control the reaction pathway in order to achieve dimerization of the free radical fragments that prevent dissociation. Under acidic conditions, DMAB was first excited by a 100% laser in the nanogap. Then, the TER spectrum was measured over a period of 0 to 60 min. If the N=N bond of DMAB is selectively destroyed in some systems, such as molecule, tip, substrate system and so on, the dissociated fragments will be set in the tip or substrate. The implication of Figure 9a,b is that there was no significant change in the intense stimulation over 60 min. As demonstrated in Figure 9c, a TER spectrum with weak plasma intensity was measured at the tip, and the results were completely different from Figure 9a. Obviously, the implication of Figure 9c is that the complete disappearance of the ag_{17} mode of DMAB indicated that its N=N bond was selectively dissociated. Under acidic conditions, because of the adsorption of a large amount of hydrogen ions on the substrate, the radical fragments will be translated into PATP via the adsorbed hydrogen ions after the decomposition process. The implication of Figure 9d is that the above results are verified by measuring the nuclear resonance spectra (NRS) of the PATP powder. Since Figure 9c,d are similar, it is shown that PATP was translated into DMAB again in the nanogap of the tip and the substrate, so it could not be observed in the slit. This indicates that PATP could not be easily converted to DMAB at relatively low plasma intensities [12].

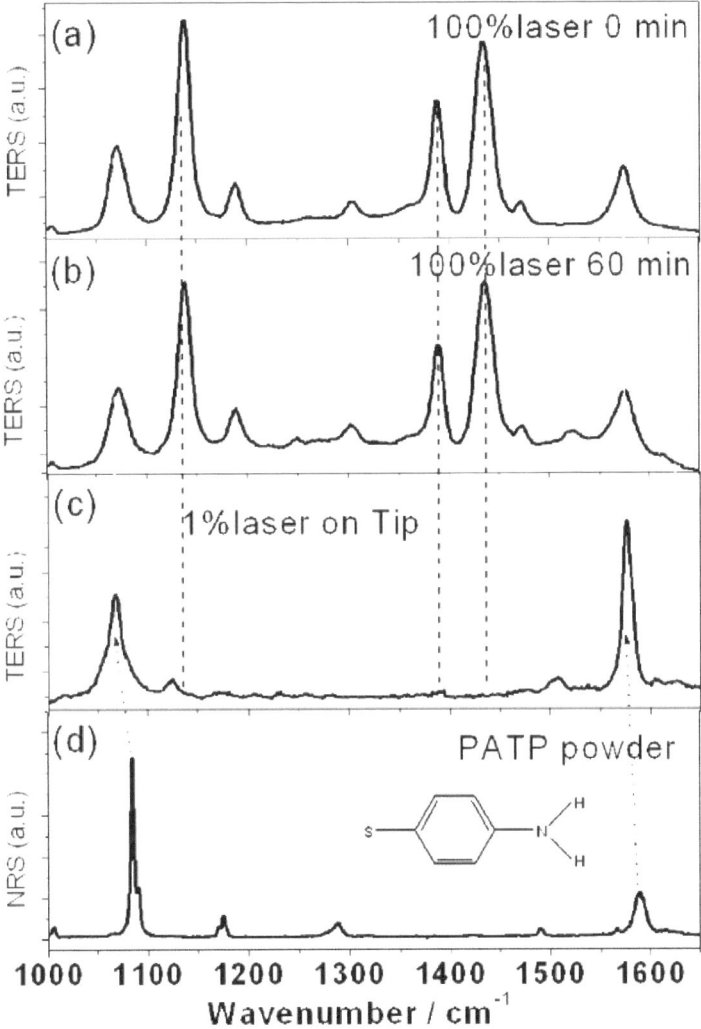

Figure 9. Experimental TER (acidic conditions) and NR spectra. TER spectra of DMAB at (**a**) 0 and (**b**) 60 min. (**c**) TER spectrum on tip with weak excitation after strong excitation for 60 min. (**d**) NRS of PATP powder. Adapted from [11].

At the same time, under alkaline conditions, the author also measured the TER spectrum of DMAB. These spectra in Figure 10a–c were measured at 0 min, 10 min, and 60 min, respectively. After 10 min of irradiation with 100% laser, a weak vibration peak at 1336 cm^{-1} is evident in Figure 10b. Obviously, the implication of Figure 10d is that the TER spectrum was measured at the tip with a laser intensity of 1%, and the result was completely different from that of Figure 10a. As demonstrated in Figure 10d, the vibration peak almost disappeared at 1433 cm^{-1}, while three Raman peaks were observed. This is similar to the NRS spectrum of the 4NBT powder (see Figure 10e). Therefore, under alkaline conditions, due to the adsorption of a large amount of oxygen ions on the substrate, the dissociated radical fragments were converted into 4NBT by the adsorbed oxygen ions.

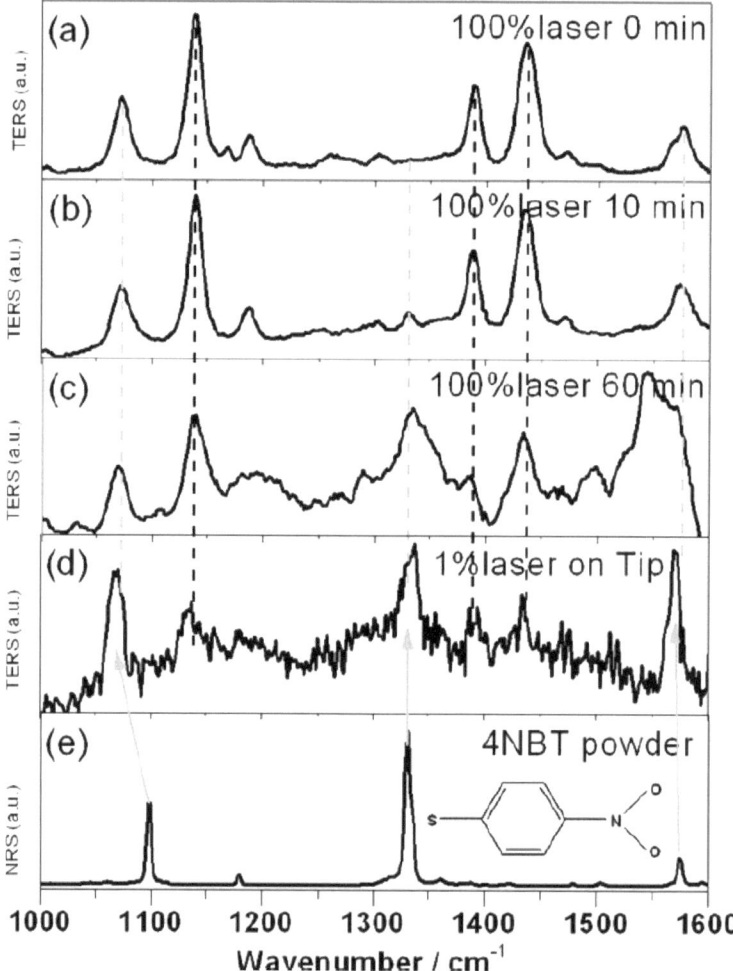

Figure 10. Experimental TER (alkaline conditions) and NR spectra. TER spectra of DMAB at (**a**) 0, (**b**) 10, and (**c**) 60 min. (**d**) TER spectrum on tip with weak excitation after strong excitation for 60 min. (**e**) NRS of 4NBT powder. Adapted from [11].

Figure 11 summarizes the physical source of the pH control product, the product of which was dissociated by plasma scissors in HV-TERS. To begin with, the hot electrons generated by the decay of the plasma are temporarily attached to the potential energy surface (PES) of the DMAB, which adds PES to the PES$^-$ and reduced the reaction barrier. In addition, the laser could excite the reaction energy above or near the dissociation energy. Last but not least, the dissociated free radicals absorb the hydrogen ions on the substrate under acidic conditions and, conversely, absorb the oxygen ions on the substrate under alkaline conditions. With this reaction mechanism, PATP or 4NBT could be produced by changing the pH to control the respective reaction pathways.

To sum up, for overcoming potential obstacles and providing sufficient energy, the products PATP and 4NBT can be converted to DMAB. Furthermore, DMAB, PATP, and 4NBT could exist simultaneously during the reaction. Compared with 4NBT, PATP had a faster reaction rate during the reaction, required less energy, and had lower reaction barriers. Because 4NBT had a high potential barrier, its reaction rate was slower and more difficult. Therefore, we know that the kinetics of the reaction process could be achieved under alkaline conditions, but cannot be achieved under acidic conditions, as shown in Figures 9 and 10.

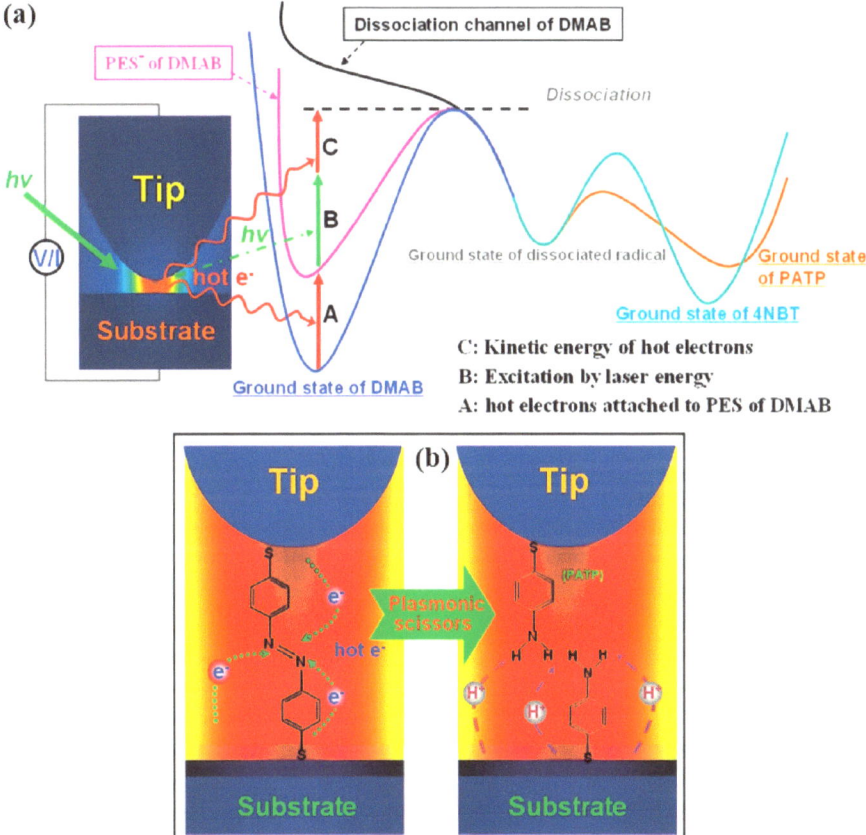

Figure 11. The process of pH control of products of the dissociation by plasmonic scissors in HV-TERS. (**a**) Scheme of reaction paths, in accordance with experimental results. (**b**) Dynamic process of chemical reaction under acidic conditions. Adapted from [11].

As mentioned above, we selectively dissociated DMAB with the N=N bond while obtaining a stable signal from the dissociated product. These three conditions were necessary. To begin with, this produces strong continuous plasma scissors from strong surface plasmons; in addition, in order to produce a stable reactant, it was necessary to adsorb oxygen ions and hydrogen ions because these ions provided a lower potential barrier. Last but not least, to avoid triggering the backward synthesis process, weak plasma was required to collect the general Raman signal of the product.

We conclude that the N=N selective bond cleavage of DMAB is obtained by using hot electrons as plasma scissors. Obviously, we can explain some physical mechanisms, for instance, bond-selective dissociation, pH-controlled products, and so on. Therefore, the dissociation product could be controlled by the change of pH. Under acidic conditions, the hydrogen ions adsorbed by PATP were generated from radical fragments. In contrast, 4NBT was generated by oxygen ions adsorbed under alkaline conditions.

4. Real-Space and Real-Time Observation of a Plasmon-Induced Chemical Reaction of a Single Molecule

The plasmon-induced chemical reaction of the molecule adsorbed on the surface of metal nanostructures has attracted extensive attention to photocatalytic reactions. Since the plasma was located near the metal surface, it was difficult to directly observe the chemical reaction in the plasma field. Here, the authors used scanning tunneling microscope (STM) to observe the plasmon-induced chemical reactions for the single molecule level.

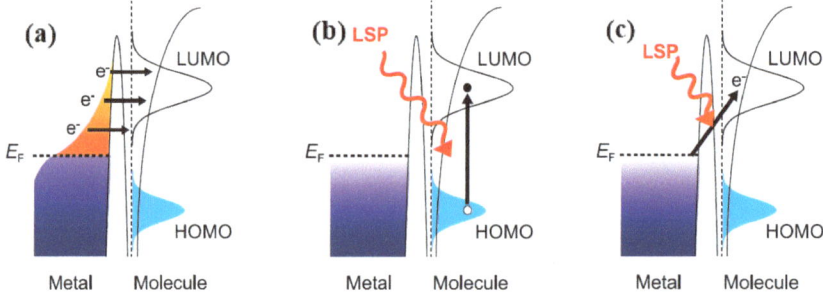

Figure 12. Excitation mechanisms for plasmon-induced chemical reactions. (**a**) Indirect hot-electron transfer mechanism. Hot electrons (e⁻) generated via non-radiative decay of an LSP transferred to form the transient negative state (TNI) states of the molecule. (**b**) Direct intramolecular excitation mechanism. The LSP induces direct excitation from the occupied state to the unoccupied state of the adsorbate. (**c**) Charge transfer mechanism. The electrons are resonantly transferred from the metal to the molecule. Adapted from [13].

The LSPR of the metal nanostructure surface concentrates light near the metal surface, making it smaller than the diffraction limit [14], while producing a local electric field that can be used for the near-field optics [15,16]. Moreover, LSPs helped to increase the conversion rate of solar energy in photovoltaics [17,18] and photocatalysts [19,20]. In particular, the plasmon-induced chemical reaction of the molecule adsorbed on the surface of metal nanostructures has attracted attention as a photocatalyst to form bonds or dissociate it. As demonstrated in Figure 12a, there is an indirect thermal electron transfer mechanism [21,22]. In the metal nanostructures, electron-hole pairs were produced by local surface plasma non-radiative decay [22–25], and the transferred hot electrons formed the transient negative ion state of a adsorbed molecule [22]. In previous experimental studies, the plasmon-induced decomposition of hydrogen and oxygen molecules was observed. This was due to the decomposition reaction that is excited by vibration after the thermal electron transfer produces transient anions.

Figure 12b reveals the direct intramolecular excitation mechanism that was proposed based on a chemical reaction of the single molecule induced by plasma. In the electronic structure of the adsorbate, the LSP excites electrons from an occupied state to an unoccupied state. Individual molecules in a strong localized plasma field near a metal surface could be observed in real space and real time.

Obviously, the implication of Figure 13a,b is that the experimental protocols for studying plasmon-induced chemical reactions and corresponding STM images are sound. Here, the authors chose $(CH_3S)_2$ as a target molecule for plasmon-induced chemical reactions. In order to excite the local surface plasmon, the gold tip was placed on the bare metal surface and the radius of curvature of the tip was ~60, while the bias voltage V_s was set to 20 mV and the tunnel voltage I_t was set to 0.2 nA. It is worth noting that tunneling electrons did not excite vibrational modes associated with any type of reaction at a bias voltage of 20 mV (for example, desorption, dissociation, or rotation of a molecule [26,27]).

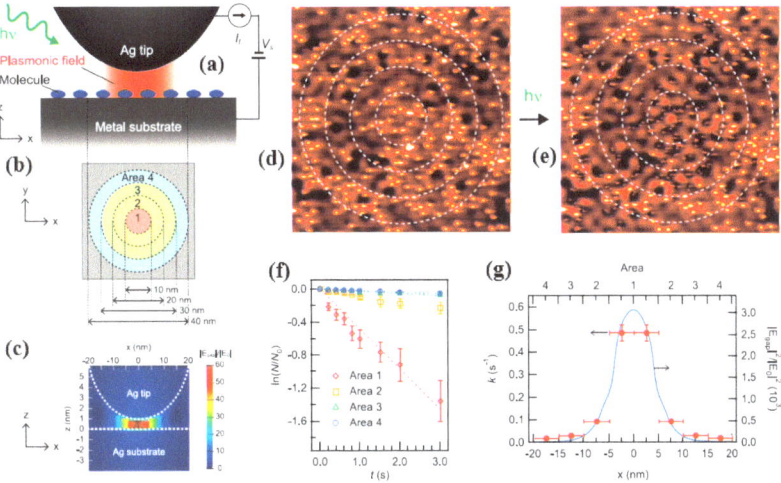

Figure 13. Real-space investigations of the plasmon-induced chemical reaction. (**a**) Schematic illustration of the experiment for the real-space investigation of the plasmon-induced chemical reaction in the nanogap between an Ag tip and a metal substrate. The tip was positioned over the metal surface during light irradiation with the feedback loop turned on to maintain the gap distance. The V_s and I_t were kept at 20 mV and 0.2 nA. $h\nu$, Planck's constant (h) multiplied by frequency (ν). (**b**) Division of the s scanning tunneling microscope (STM) image into four areas depicted with 10-nm wide concentric rings for analysis. (**c**) Simulated spatial distribution of the electric field at the 1-nm gap under p-polarized light at 532 nm. E_0 was the incident electric field. Topographic STM images of $(CH_3S)_2$ molecules on Ag(111) (**d**) before and (**e**) after irradiation with p-polarized light at 532 nm (~7.6 × 10^{17} photons cm^{-2} s^{-1}, 2 s) (V_s = 20 mV, I_t = 0.2 nA, 43 nm by 43 nm). The tip was positioned at the center of area one during light irradiation. (**f**) Time dependence of the dissociation ratio (N/N_0) under irradiation with p-polarized light at 532 nm (~5.9 × 10^{16} photons cm^{-2} s^{-1}) in the four areas shown in (**b**). Each data point represents the average of results from six trials. The dotted lines denote single exponential functions fitted to the data points [$\ln(N/N_0) = -kt$]. Error bars indicate SD. (**g**) The rate constant k obtained at areas one through four and the calculated lateral profile of electric field intensity at 0.1 nm above the substrate surface (z = 0.1 nm) under 533-nm light. x = 0 nm corresponds to the center of the tip. Adapted from [13].

Figure 13c shows that a local surface plasma in the nanogap generated the strong electric field. Figure 13d reveals the spatial distribution of the separated molecules, such as $(CH_3S)_2$, on Ag (111) before the LSP was excited with 532 nm of p-polarized light. Figure 13e displays the distribution after the p-polarized light excited the LSP. The dissociation ratio is N/N_0, where N was the number of the $(CH_3S)_2$ molecule after localized surface plasmon excitation with p-polarized light with a wavelength of 532 nm; N_0 is the number of the pre-adsorbed molecule. This ratio is used for quantitative analysis of experimental studies. Figure 13f shows the time-dependent dissociation ratio curve for p-polarized light excitation conditions at a wavelength of 532 nm, corresponding to the four regions in Figure 13b. As shown, the ratio was linearly related to time, indicating that dissociation from $(CH_3S)_2 \rightarrow 2CH_3S$ was the first-order reaction. The dissociation constant, whose value was k, could be determined by the oblique line in Figure 13f. Figure 13g reveals the magnitude of the dissociation constants in four different regions. As seen in Figure 13g, the value reached a maximum in region 1 and decreased as the distance from the tip increased. Comparing k and E_{gap}, the authors found that plasmon-induced decomposition had a very strong correlation with the electric field strength of photoexcited local surface plasmons.

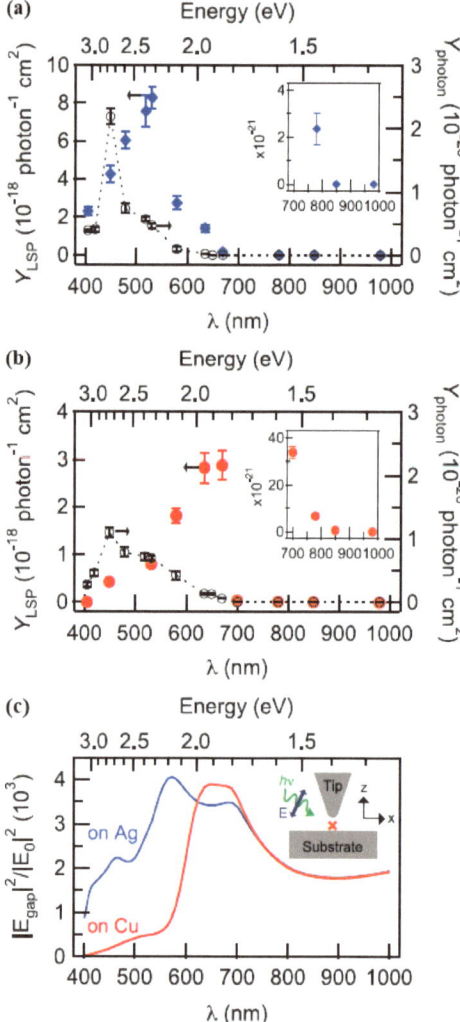

Figure 14. Wavelength dependence of the plasmon-induced chemical reaction and of the plasmonic electric field. Wavelength dependence of Y_{LSP} of $(CH_3S)_2$ molecules on (**a**) Ag(111) (blue diamonds) and (**b**) Cu(111) (red circles). Insets in (**a**) and (**b**) show Y_{LSP} for 700 to 980 nm. Each data point represents the average of results from six trials. The photodissociation yields without the excitation of the LSP [Y_{photon}, previously reported in [26]] (black circles) are also shown. The yield is determined from k divided by the number of incident photons per second (~6.0×10^{16} to 6.5×10^{16} photons $cm^{-2} \cdot s^{-1}$ for both plasmon-induced dissociation and photodissociation). Error bars indicate standard deviation (SD). (**c**) Calculated electric field intensity for a 1-nm gap between an Ag tip and the metal substrates under p-polarized light. The simulated point is z = 0.1 nm above the substrate surfaces and x = 0 nm. Adapted from [13].

The dissociation rate constant divided by the number of incident photons per second could be used to calculate the productivity of plasma-induced dissociation, Y_{LSP}. The wavelength at a maximum of a plasma-induced decomposition yield was about 532 nm with an energy value of about 2.33 eV. The threshold wavelength of Y_{LSP} in region 1 was λ_{Th}, which was 780 nm, and the energy value was 1.59 eV, as displayed in Figure 14a. The wavelength dependence of the plasmon-induced decomposition yield of the $(CH_3S)_2$ molecules on the Cu(111) substrate is displayed in Figure 14b. It can be seen from Figure 14b that it had different plasma characteristics and electronic structures.

In the indirect thermal electron transfer mechanism (shown in Figure 14a), the value of Y_{LSP} was confirmed by the energy distribution of the hot electrons and the density of state (DOS) of the lowest unoccupied molecular orbital (LUMO). The energy distribution of holes and hot electrons generated by local surface plasma decay to metal electronic band structure was sensitive [20,21]. Figure 14c displays the calculated electric field strength of the gap between the metal substrate and the Ag tip under the condition of p-polarized light excitation, wherein the gap width was 1 nm.

By using STM for real-time observation, the authors were able to measure the rate of plasmon-induced chemical reactions for single molecules while facilitating more in-depth studies of basic reaction pathways that were not available through spectroscopic analysis of $Y_{LSP}\lambda$ and traditional spectroscopy. I_t had a strong sensitivity to changes in gap distance, so the authors could collect real-time information by tracking I_t under illumination, as displayed in Figure 15a. However, Figure 15b reveals the current trace when a STM tip is located on the target molecule adsorbed on the Ag(111) under the irradiation conditions of p-polarized light having a wavelength of 532 nm. It is worth noting that the value of I_t drops suddenly, reflecting d changing from d_1 to d_2; as shown in Figure 15a, this might be due to molecular dissociation. The rate of dissociation was determined by the reciprocal of the time needed for plasma-induced dissociation. It is worth noting that at a voltage of 20 mV, the tunneling electrons did not induce a chemical reaction. Since the tunnel current under light irradiation conditions was relatively stable on the metal surface, the thermal expansion at the tip was negligible. The dissociation rate $1/t_R$ varies with the gap distance under p-polarized light with a wavelength of 532 nm. In the indirect thermo electron transfer mechanism, the reaction occurred through the inelastic electron tunneling process, as shown in Figure 15a, it transferred from a transient negative state (TNI) state shaped by the hot electrons to the molecule [20]. In addition, the authors could gain a deeper understanding of the reaction pathways triggered by the TNI state. The TNI state was formed by electron transfer from metal to molecule by the IET process, and having the STM [22,23]. As demonstrated in Figure 15c, based on a non-dissociated potential energy surface, the energy of the state of the transient anion was dissipated to the excited state of the vibration, and the transient negative ion state was formed by electrons being transferred from a metal to a molecule. These reactions led to dissociation and rotation of $(CH_3S)_2$, and rotation occurred before the dissociation process. As displayed in Figure 14, the authors could see from a $Y_{LSP}\lambda$ spectrum plasma-induced dissociation of $(CH_3S)_2$ by a direct dissociation reaction path that occurred from the neutral excited state; this neutral excited state was produced through direct intramolecular excitation, as revealed in Figure 15d. The LUMO of the $(CH_3S)_2$ molecule was weakly hybridized with a metal substrate [26,28]. Since the weak hybridization inhibited the excited state relaxation [29,30], it could enter the energy surface of the dissociation potential from a neutral excited state, which predicts the photodissociation for $(CH_3S)_2$ molecules in theory, in the gas phase, as shown in Figure 15d [31,32]. The results indicate that a plasmon-induced chemical reaction of molecules accompanying electronic states was likely to be hybridized by direct intermolecular excitation mechanisms rather than indirect thermal electron transfer mechanisms. These findings allowed us to further understand the interaction between local surface plasmons and metal surface molecules, so that we could drive plasmon-induced photocatalysis more efficiently.

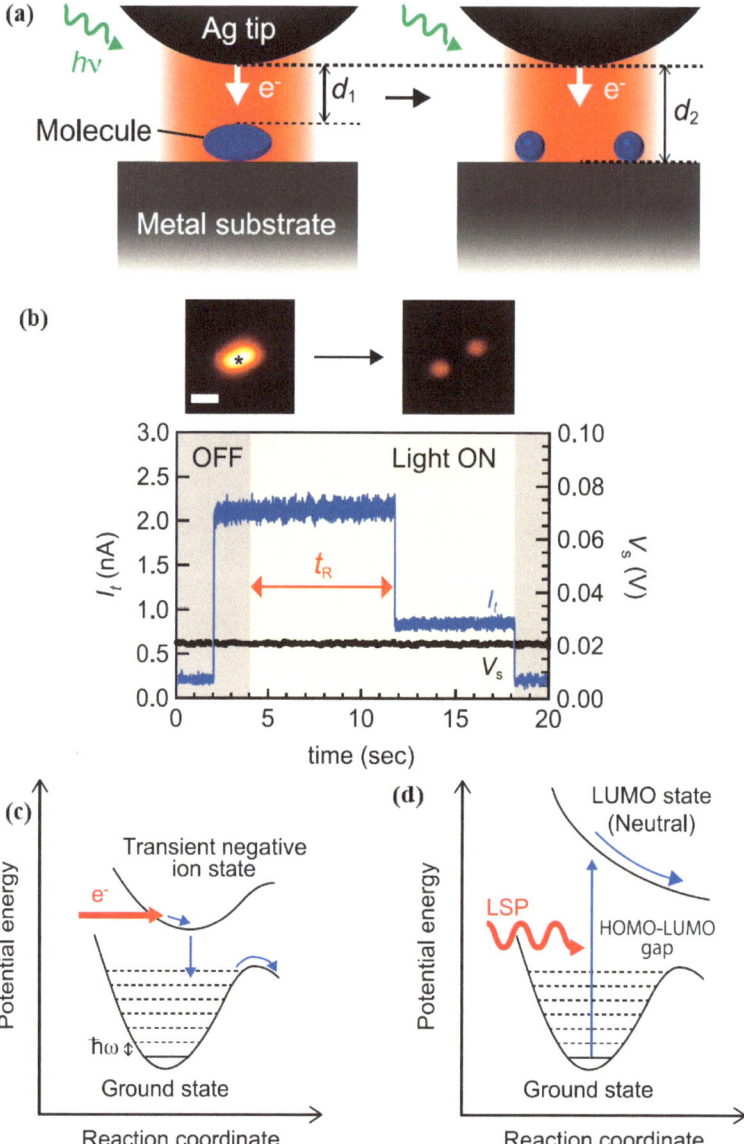

Figure 15. Real-time STM results for the plasmon-induced chemical reaction and transient negative state (IET)-induced reactions of a single molecule. (**a**) Schematic illustration of the real-time observation of the plasmon-induced chemical reaction. (**b**) Current trace for detecting the dissociation event for the target molecule (STM images) on Ag(111) induced by the LSP excited with p-polarized light at 532 nm (~2.7 × 10^{15} photons cm^{-2} s^{-1}). Schematic illustrations of the potential energy surface for the plasmon-induced chemical reactions on the basis of (**c**) the indirect hot-electron transfer mechanism and (**d**) the direct intramolecular excitation mechanism are shown. $\hbar\omega$, Planck's constant h divided by 2π ($\hbar$) multiplied by the angular frequency (ω). Adapted from [13].

5. Summary and Outlook

In summary, our research shows that plasma scissors are an effective tool for resonance control surface adsorption molecular dissociation; simultaneously, an important new area of plasmonic hot-electron mediated chemistry has been established to extend the range of photochemistry to chemical reactions. This also indicates that HV-TERS is a promising nanoscale in situ surface chemical analysis tool and manipulation technique. Furthermore, we clearly explain the physical mechanisms and principles of the bond-selective dissociation and the pH control products. In parallel, our results prove the quite important application of the plasmon in the field of heterogeneous photocatalysis, allowing hydrogen to carry out room-temperature dissociation reactions on gold nanoparticles. Finally, theoretical calculations combined with real-world and real-time STM examination showed that the dissociation of S-S bonds for a single molecule $(CH_3S)_2$ on the plasmon-induced silver and copper surfaces was mostly accomplished by direct intramolecular excitation to the lowest unoccupied molecular orbital state of antibonding S–S (σ^*ss) orbital between the silver tip and a metal surface by the decay of the optically excited localized surface plasma in the nanogap.

Author Contributions: Conceptualization, J.W. and X.M.; Investigation, J.W., X.W. and X.M.; Writing-Original Draft Preparation, J.W., X.W.; Writing-Review & Editing, J.W., X.W. and X.M.; Supervision, J.W.; Project Administration, X.M.; Funding Acquisition, J.W.

Funding: This research was funded by the National Natural Science Foundation of China, grant number [91436102 and 11374353], and The APC was funded by National Natural Science Foundation of China (grant number [91436102].

Conflicts of Interest: The authors declare no conflict of interest.

References

1. Sun, M.; Zhang, Z.; Wang, P.; Li, Q.; Ma, F.; Xu, H. Remotely excited Raman optical activity using chiral plasmon propagation in Ag nanowires. *Light Sci. Appl.* **2013**, *2*, e112. [CrossRef]
2. Sun, M.T.; Xu, H.X. A novel application of plasmonics: Plasmon-driven surface-catalyzed reactions. *Small* **2012**, *8*, 2777–2786. [CrossRef] [PubMed]
3. Zhang, Z.; Xu, P.; Yang, X.; Liang, W.; Sun, M. Surface plasmon-driven photocatalysis in ambient, aqueous and high-vacuum monitored by SERS and TERS. *J. Photochem. Photobiol. C Photochem. Rev.* **2016**, *27*, 100–112. [CrossRef]
4. Mukherjee, S.; Libisch, F.; Large, N.; Neumann, O.; Brown, L.V.; Cheng, J.; Lassiter, J.B.; Carter, E.A.; Nordlander, P.; Halas, N.J. Hot electrons do the impossible: Plasmon-Induced dissociation of H_2 on Au. *Nano Lett.* **2013**, *13*, 240–247. [CrossRef] [PubMed]
5. Fang, Y.R.; Li, Y.Z.; Xu, H.X.; Sun, M.T. Ascertaining p,p'-dimercaptoazobenzene produced from p-aminothiophenol by selective catalytic coupling reaction on silver nanoparticles. *Langmuir* **2010**, *26*, 7737–7746. [CrossRef] [PubMed]
6. Christopher, P.; Xin, H.L.; Linic, S. Visible-light-enhanced catalytic oxidation reactions on plasmonic silver nanostructures. *Nat. Chem.* **2011**, *3*, 467–472. [CrossRef] [PubMed]
7. Wang, J.; Lin, W.; Xu, X.; Ma, F.; Sun, M. Plasmon-Exciton Coupling Interaction for Surface Catalytic Reactions. *Chem. Rec.* **2018**, *18*, 481–490. [CrossRef] [PubMed]
8. Dong, B.; Fang, Y.R.; Chen, X.W.; Xu, H.X.; Sun, M.T. Substrate-, wavelength-, and time-dependent plasmon-assisted surface catalysis reaction of 4-nitrobenzenethiol dimerizing to p,p'-dimercaptoazobenzene on Au, Ag, and Cu films. *Langmuir* **2011**, *27*, 10677–10682. [CrossRef] [PubMed]
9. Zhang, Z.; Sheng, S.; Zheng, H.; Xu, H.; Sun, M. Molecular resonant dissociation of surface adsorbed molecules by plasmonic nanoscissors. *Nanoscale* **2014**, *6*, 4903–4908. [CrossRef] [PubMed]
10. Kim, K.H.; Watanabe, K.; Mulugeta, D.; Freund, H.-J.; Menzel, D. Enhanced Photoinduced Desorption from Metal Nanoparticles by Photoexcitation of Confined Hot Electrons Using Femtosecond Laser Pulses. *Phys. Rev. Lett.* **2011**, *107*, 047401. [CrossRef] [PubMed]
11. Sun, M.; Zhang, Z.; Kim, Z.H.; Zheng, H.; Xu, H. Plasmonic scissors for molecular design. *Chem. Eur. J.* **2013**, *19*, 14958–14962. [CrossRef] [PubMed]

12. Cao, E.; Guo, X.; Zhang, L.; Shi, Y.; Lin, W.; Liu, X.; Fang, Y.; Zhou, L.; Sun, Y.; Song, Y.; et al. Electrooptical Synergy on Plasmon–Exciton-Codriven Surface Reduction Reactions. *Adv. Mater. Interfaces* **2017**, *4*, 1700869. [CrossRef]
13. Kazuma, E.; Jung, J.; Ueba, H.; Trenary, M.; Kim, Y. Real-space and real-time observation of a plasmon-induced chemical reaction of a single molecule. *Science* **2018**, *360*, 521–526. [CrossRef] [PubMed]
14. Schuller, J.A.; Barnard, E.S.; Cai, W.; Jun, Y.C.; White, J.S.; Brongersma, M.L. Plasmonics for extreme light concentration and manipulation. *Nat. Mater.* **2010**, *9*, 193–204. [CrossRef] [PubMed]
15. Fang, Y.; Zhang, Z.; Sun, M. High vacuum tip-enhanced Raman spectroscope based on a scanning tunneling microscope. *Rev. Sci. Instrum.* **2016**, *87*, 033104. [CrossRef] [PubMed]
16. Anker, J.N.; Hall, W.P.; Lyandres, O.; Shah, N.C.; Zhao, J.; van Duyne, R.P. Biosensing with plasmonic nanosensors. *Nat. Mater.* **2008**, *7*, 442–453. [CrossRef]
17. Atwater, H.A.; Polman, A. Plasmonics for improved photovoltaic devices. *Nat. Mater.* **2010**, *9*, 205–213. [CrossRef]
18. Clavero, C. Plasmon-induced hot-electron generation at nanoparticle/metal-oxide interfaces for photovoltaic and photocatalytic devices. *Nat. Photonics* **2014**, *8*, 95–103. [CrossRef]
19. Lin, W.; Cao, Y.; Wang, P.; Sun, M. Unified Treatment for Plasmon−Exciton Co-driven Reduction and Oxidation Reactions. *Langmuir* **2017**, *33*, 12102–12107. [CrossRef]
20. Linic, S.; Aslam, U.; Boerigter, C.; Morabito, M. Photochemical transformations on plasmonic metal nanoparticles. *Nat. Mater.* **2015**, *14*, 567–576. [CrossRef]
21. Brongersma, M.L.; Halas, N.J.; Nordlander, P. Plasmon-Induced Hot Carrier Science and Technology. *Nat. Nanotechnol.* **2015**, *10*, 25–34. [CrossRef] [PubMed]
22. Sundararaman, R.; Narang, P.; Jermyn, A.S.; Goddard, W.A., 3rd; Atwater, H.A. Theoretical predictions for hot-carrier generation from surface plasmon decay. *Nat. Commun.* **2014**, *5*, 5788. [CrossRef] [PubMed]
23. Zhang, Z.; Sheng, S.; Wang, R.; Sun, M. Tip-Enhanced Raman Spectroscopy. *Anal. Chem.* **2016**, *88*, 9328–9346. [CrossRef] [PubMed]
24. Ohara, M.; Kim, Y.; Yanagisawa, S.; Morikawa, Y.; Kawai, M. Role of Molecular Orbitals Near the Fermi Level in the Excitation of Vibrational Modes of a Single Molecule at a Scanning Tunneling Microscope Junction. *Phys. Rev. Lett.* **2008**, *100*, 136104. [CrossRef] [PubMed]
25. Motobayashi, K.; Kim, Y.; Ohara, M.; Ueba, H.; Kawai, M. The role of thermal excitation in the tunneling-electron-induced reaction: Dissociation of dimethyl disulfide on Cu(111). *Surf. Sci.* **2016**, *643*, 18–22. [CrossRef]
26. Kazuma, E.; Jung, J.; Ueba, H.; Trenary, M.; Kim, Y. A direct pathway of molecular photodissociation with visible light on metal surfaces. *J. Am. Chem. Soc.* **2017**, *139*, 3115–3121. [CrossRef] [PubMed]
27. Lin, W.; Cao, E.; Zhang, L.; Xu, X.; Song, Y.; Liang, W.; Sun, M. Electrically enhanced hot hole driven oxidation catalysis at the interface of a plasmon–exciton hybrid. *Nanoscale* **2018**, *10*, 5482–5488. [CrossRef] [PubMed]
28. Lin, W.; Shi, Y.; Yang, X.; Li, J.; Cao, E.; Xu, X.; Pullerits, T.; Liang, W.; Sun, M. Physical mechanism on exciton-plasmon coupling revealed by femtosecond pump-probe transient absorption spectroscopy. *Mater. Today Phys.* **2017**, *3*, 33–40. [CrossRef]
29. Ding, Q.; Li, R.; Chen, M.; Sun, M. Ag nanoparticles-TiO$_2$ film hybrid for plasmon-exciton co-driven surface catalytic reactions. *Appl. Mater. Today* **2017**, *9*, 251–258. [CrossRef]
30. Lin, W.; Ren, X.; Cui, L.; Zonga, H.; Sun, M. Electro-optical tuning of plasmon-driven double reduction interface catalysis. *Appl. Mater. Today* **2018**, *11*, 189–192. [CrossRef]
31. Stipe, B.C.; Rezaei, M.A.; Ho, W. Single-molecule vibrational spectroscopy and microscopy. *Science* **1998**, *280*, 1732–1735. [CrossRef] [PubMed]
32. Sun, M.; Zhang, Z.; Zheng, H.; Xu, H. In-situ plasmon-driven chemical reactions revealed by high vacuum tip-enhanced Raman spectroscopy. *Sci. Rep.* **2013**, *2*, 647. [CrossRef] [PubMed]

© 2019 by the authors. Licensee MDPI, Basel, Switzerland. This article is an open access article distributed under the terms and conditions of the Creative Commons Attribution (CC BY) license (http://creativecommons.org/licenses/by/4.0/).

Review

Morphology-Governed Performance of Plasmonic Photocatalysts

Zhishun Wei [1,*], Marcin Janczarek [2], Kunlei Wang [3,4], Shuaizhi Zheng [5] and Ewa Kowalska [3,*]

1. Hubei Provincial Key Laboratory of Green Materials for Light Industry, Hubei University of Technology, Wuhan 430068, China
2. Institute of Chemical Technology and Engineering, Faculty of Chemical Technology, Poznan University of Technology, 60-965 Poznan, Poland; marcin.janczarek@put.poznan.pl
3. Institute for Catalysis, Hokkaido University, Sapporo 001-0021, Japan; kunlei@cat.hokudai.ac.jp
4. Northwest Research Institute, Co. Ltd. of C.R.E.C., Lanzhou 730000, China
5. Key Laboratory of Low Dimensional Materials and Application Technology of Ministry of Education, School of Materials Science and Engineering, Xiangtan University, Xiangtan 411105, China; xishuai423@hotmail.com
* Correspondence: wei.zhishun@hbut.edu.cn (Z.W.); kowalska@cat.hokudai.ac.jp (E.K.)

Received: 11 August 2020; Accepted: 12 September 2020; Published: 17 September 2020

Abstract: Plasmonic photocatalysts have been extensively studied for the past decade as a possible solution to energy crisis and environmental problems. Although various reports on plasmonic photocatalysts have been published, including synthesis methods, applications, and mechanism clarifications, the quantum yields of photochemical reactions are usually too low for commercialization. Accordingly, it has been proposed that preparation of plasmonic photocatalysts with efficient light harvesting and inhibition of charge carriers' recombination might result in improvement of photocatalytic activity. Among various strategies, nano-architecture of plasmonic photocatalysts seems to be one of the best strategies, including the design of properties for both semiconductor and noble-metal-deposits, as well as the interactions between them. For example, faceted nanoparticles, nanotubes, aerogels, and super-nano structures of semiconductors have shown the improvement of photocatalytic activity and stability. Moreover, the selective deposition of noble metals on some parts of semiconductor nanostructures (e.g., specific facets, basal or lateral surfaces) results in an activity increase. Additionally, mono-, bi-, and ternary-metal-modifications have been proposed as the other ways of performance improvement. However, in some cases, the interactions between different noble metals might cause unwanted charge carriers' recombination. Accordingly, this review discusses the recent strategies on the improvements of the photocatalytic performance of plasmonic photocatalysts.

Keywords: plasmonic photocatalysis; vis-responsive photocatalysts; morphology; faceted particles; nanotubes; gold; silver; copper; platinum

1. Introduction

The consumption of fossil fuels has been significantly increasing in recent years because of the fast industry development and world population growth. In the consequence, continuous environmental pollution and shortage of renewable energy resources have become major challenges faced by humanity. Therefore, the development of environment-friendly, clean, safe, and sustainable energy sources is one of the biggest challenges for the whole world. Among all available options, solar energy is one of the most prospective, especially for the developing countries with usually high intensity and duration (daily and yearly) of solar radiation, and serious energy and environmental problems. Accordingly, the development of environment-friendly technologies using solar energy is an important and useful way to solve humanity problems.

For example, heterogeneous photocatalysis under solar radiation has been considered as environment-friendly and probable candidate to overcome these challenges [1–4]. In general, photocatalyst (semiconductor) is activated under light irradiation with energy equal to or larger than its bandgap, resulting in transfer of electrons from valence band (VB) to conduction band (CB), thus forming charge carriers, i.e., electrons in CB and holes in VB, as shown in Figure 1a. Charge carriers might either react with some reagents adsorbed on the photocatalyst surface or recombine in the bulk or on the surface, i.e., bulk and surface recombination [5], respectively. The recombination results in an obvious decrease in photocatalytic efficiency, reaching quantum yields much lower than expected 100%. Accordingly, various methods have been proposed to limit this recombination [6], including properties' improvements, e.g., higher crystallinity, less defects' content, perfect crystal morphology (faceted particles) [7–11], larger content of shallow than deep electron traps (ETs) [12]), and various modifications, e.g., doping [13–15], surface modifications [16–19], heterojunction formation (coupled semiconductors) [20–22]. One of the most efficient methods to inhibit the charge carriers' recombination is the surface modification of photocatalyst with deposits of noble metals, because noble metals work as an electrons' scavenger (Figure 1b) because of larger work function of metals than the electron affinity of oxide semiconductors, as first found for platinum-modified titania almost half century ago by Bard [23].

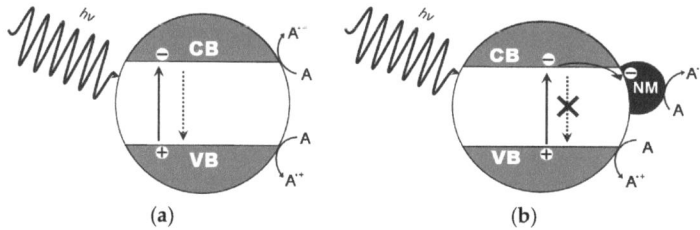

Figure 1. The schematic drawings of: (**a**) Semiconductor excitation, charge carriers' formation (↑), and following redox reactions on the semiconductor surface or charge carriers: recombination (↓); and (**b**) electron scavenging by noble metal (NM) deposits; A—reagent adsorbed on the semiconductor surface; - electron; + hole.

Another problem of heterogeneous photocatalysts is their wide bandgaps (despite being positive for high redox activity), and thus low activity under solar radiation, as they must be excited with UV irradiation, being only ca. 4% of solar light. Accordingly, various strategies have been proposed to obtain vis-active materials, including also the methods used for the inhibition of charge carriers' recombination, e.g., doping, surface modification, and heterojunction formation [24–28]. Interestingly, noble metals, known for activity enhancement under UV irradiation for many years, have also been found to activate wide-bandgap semiconductors toward vis and even NIR range of solar spectrum, because of plasmonic properties [29–34], and thus noble-metal-modified semiconductors with activity under vis/NIR have been named as plasmonic photocatalysts.

Three main mechanisms of plasmonic photocatalysis have been considered, i.e., energy transfer (Figure 2a), electron transfer (Figure 2b) and plasmonic heating. Moreover, the light scattering on the deposits of noble metals has been proposed to enhance the overall photocatalytic performance by efficient light harvesting. In the case of energy transfer, the respective energy levels should overlap, and thus more probable is energy transfer for Ag/TiO_2 with similar energies, i.e., ca. 2.8–3.0 for Ag deposits (depending on the shape and size) and 3.0–3.2 eV for titania than that for Au/TiO_2 with much different energy of gold localized surface plasmon resonance (LSPR; ca. 2.2–2.5 eV for spherical Au NPs) than that of titania. Therefore, only pre-modified titania with vis-response has been reported as able for energy transfer from Au excitation, e.g., titania modified with nitrogen [35] and defect-rich titania, i.e., with crystal defects [36] and amorphous titania with disorders—the localized states inside bandgap [37].

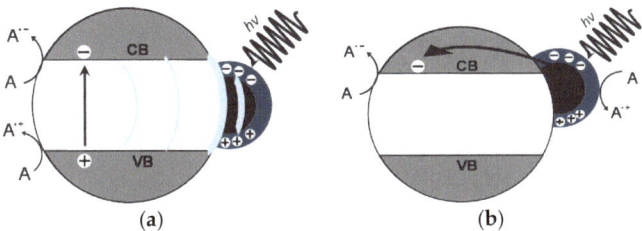

Figure 2. The schematic drawings of the proposed mechanism of plasmonic photocatalysis: (**a**) energy transfer, and (**b**) electron transfer; A—reagent adsorbed on the semiconductor surface.

In the case of electron transfer, noble metals work as sensitizers (being even named as "plasmonic sensitizers"), and thus "hot" electron transfer from noble metal to CB of semiconductor has been proposed as the first step of plasmonic photocatalysis [30]. Next, the surface reactions between "hot" electrons and adsorbed species are considered to take place on the semiconductor surface, e.g., with molecular oxygen, resulting in the formation of reactive oxygen species (ROS). At the same time, to keep noble metal at zero-valent state, oxidation of some compounds on the surface of electron-deficient noble metal must proceed [38]. Although, there are no direct proof for electron transfer (as detected electrons might also origin from energy transfer in some cases), there are many studies suggesting this mechanism as the most probable via various experimental approaches, including femtosecond transient absorption spectroscopy (electrons transferred from Au to TiO_2) [39,40], EPR spectroscopy (different oxidation species under UV and vis irradiation) [41,42], time-resolved microwave conductivity (TRMC) method (conductivity under vis irradiation only for titania modified with noble metals: Au [43] and Ag [44]), electrochemical study (shift of electrode potential to negative or positive values and generation of anodic or cathodic photocurrent depending on electrode configuration, i.e., ITO/TiO_2/Au or ITO/Au/TiO_2, respectively, as shown in Figure 3) [45].

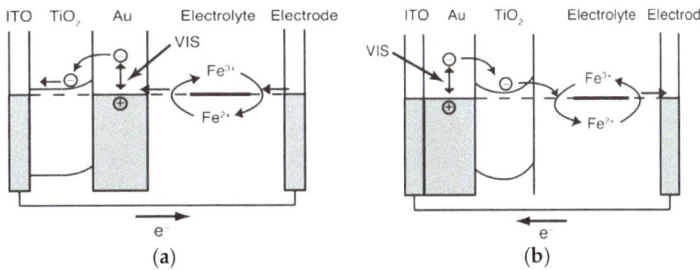

Figure 3. The schematic drawings of the photoelectrochemical cell with: (**a**) an ITO/TiO_2/Au electrode; and (**b**) an ITO/Au/TiO_2 electrode; adapted from [45]. Copyright 2009 Wiley.

The plasmonic heating has also been proposed as the main mechanism of some reactions, i.e., local heating around noble-metal-deposits resulting in the cleavage of some chemical bonds [46]. The mechanism might depend on the semiconductor type. For example, both Au/TiO_2 and Au/ZnO are active under vis irradiation, but only Au/ZnO is active during heating in the dark [32,47]. Accordingly, plasmon-assisted photocatalysis and plasmon-assisted catalysis (plasmonic heating) might be expected for Au/TiO_2 and Au/ZnO, respectively. Although some studies have proposed plasmonic heating as the main mechanism of plasmonic photocatalysis [46–49], plenty of reports have rejected this possibility, because of the inactivity of noble-modified insulators or unsupported noble-metals in comparison to highly active noble-metal-modified semiconductors [31,36,50–53]. Moreover, studies on activation energy have also excluded plasmonic heating as the main mechanism for photocurrent generation [34] and decomposition of organic compounds [54].

Although the final mechanism has not been clarified yet, and might depend on the properties/morphology/composition of the photocatalysts and studied reactions, it has been proven that plasmonic photocatalysts are active for various photocatalytic reactions under vis range of solar spectrum, e.g., solar energy conversion (into electricity [45] and fuels [55]), degradation of organic compounds [56] and microorganisms [57], synthesis of organic compounds [58], and cancer treatment [59]. However, the overall activity is usually much lower than that under UV, and thus various procedures have been proposed to improve the photocatalytic performance. Among these methods, the nanoarchitecture of morphology has been proposed as efficient one for activity enhancement, as discussed in the next sections.

2. Morphology Design of Plasmonic Photocatalysts

The most typical plasmonic photocatalysts (like all other heterogeneous photocatalysts) are semiconductors in the form of particles (nano or micro) with deposits of noble metals (usually in the form of NPs). In the case of these photocatalysts, the activity under vis irradiation is mainly governed by the properties of noble-metal deposits. Accordingly, it has been proposed that higher polydispersity of noble metals (both in the size and the shape) results in higher photocatalytic activity as a result of efficient light harvesting [38,56,60], i.e., different size/shape of noble metal causes the absorption of photons with different energy, as LSPR depends on the properties of noble metals (longer wavelengths for larger NPs and with more complex morphology than spherical NPs). Although usually particles of semiconductors are highly polydisperse (in both composition and the particles' morphology), even for commercial and famous photocatalysts, e.g., titania P25 (anatase/rutile/amorphous phases) [61], the morphology-controlled particles have also been used for photocatalysis. Among these particles, faceted ones, i.e., with well-observed crystalline facets, have recently been intensively studied because of high photocatalytic activities [8,9,62–66]. Obviously, faceted particles have also been used as supports for noble-metal deposits, as shortly discussed below.

2.1. Faceted Semiconductors as Supports for Noble Metals

The most famous faceted titania photocatalysts are anatase particles of either octahedral (same as anatase crystals in nature) or decahedral shapes, i.e., with eight equivalent {101} facets or eight {101} facets and two additional {001} facets on the top/bottom of anatase crystal, respectively [8,10–12,62,63,65–68]. Additionally, other faceted materials have also been investigated, such as rutile [69,70] and brookite [71] titania, iridium [72], $AgNbO_3$ [73], $TiOF_2$ [74], $MgAl_2O_4$ [75], BiOBr [76], BiOCl [77,78], Ag_2MO_4 [79], $BiVO_4$ [80], ZnO [81], and Cu_2WS_4 [82]. Since faceted photocatalysts show usually much higher activity than other particles of irregular shape, they have been proposed for various applications in the photocatalysis field, including also plasmonic photocatalysis. One of the first reports on faceted plasmonic photocatalysts have been shown by Janczarek et al. for decahedral anatase particles (DAP) modified with mono- and bi-metal NPs of silver and copper by photodeposition method, i.e., under UV irradiation in the presence of methanol as hole scavenger and in the absence of oxygen to avoid capture of electrons (being main reducer for respective metal cations) [83]. It has been found that the sequence of metal deposition on DAP has been detrimental for the resultant properties (Figure 4), and thus final activities. For example, in the case of sequential deposition, the second metal is also deposited on the surface of the first deposited metal since under UV irradiation, photogenerated electrons are scavenged by the first metal, and thus the electron-rich metal is attractive for cations' adsorption of second metal (Figure 4c,d). Accordingly, in the case of bi-metal samples prepared by sequential deposition both monometallic and bi-metallic deposits are obtained. In contrast, in the case of co-deposition, the competition between metals for photogenerated electrons results in the formation of fine NPs of Cu and large NPs of Ag (Figure 4b). Although, all samples show high enhancement of photocatalytic activity under UV irradiation (3–4× than that by bare DAP), only sample with mono-metallic silver exhibits high activity under vis irradiation (>455 nm), probably since only this sample keeps plasmonic properties as copper is easily oxidized in air than silver.

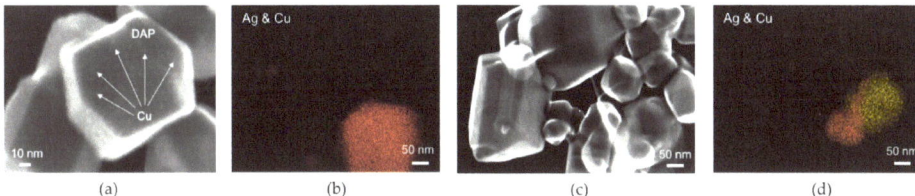

Figure 4. SEM (**a**,**c**) and EDS (**b**,**d**) images of: (**a**) decahedral anatase particles (DAP) with monometallic Cu, (**b**) DAP with large Ag NPs and fine Cu NPs, prepared by co-deposition, (**c**,**d**) DAP with bimetallic deposits prepared by deposition of Cu on Ag/DAP; red—Ag, yellow-Cu; adapted with permission from [83]. Copyright 2017 Creative Commons Attribution.

These findings have been confirmed with another faceted anatase sample, i.e., octahedral anatase particles (OAP) with eight equivalent (101) facets by Wei et al. [84]. It is found that vis activity decreases in the following order: Au/OAP > Ag/OAP > Cu/OAP, correlating with the content of zero-valent state of noble metals. Next, Au/OAP samples with different sizes of Au NPs are investigated by changing the photodeposition conditions, i.e., from anaerobic (usually used) to initially aerobic and from methanol to 2-propanol (as hole scavenger) [85]. Interestingly, it is found that an increase in the size of Au NPs results in an increase in the photocatalytic activity under vis irradiation, probably because of larger areas with field enhancement, as shown in Figure 5.

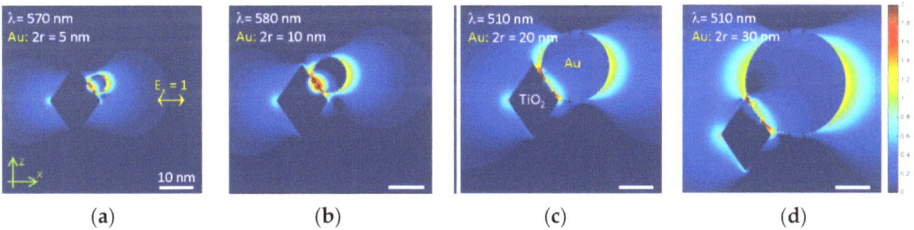

Figure 5. Simulated light intensity enhancement for different size of Au NPs on OAP, i.e., gold NPs diameter of: (**a**) 5 nm, (**b**) 10 nm, (**c**) 20 nm, and (**d**) 30 nm. The intensity color map is logarithmic and shows up to 10^2 times enhancement localized mainly at the interface between gold and titania; the same scale for all images; adapted with permission from [85]. Copyright 2017 Creative Commons Attribution.

Similar findings have been obtained for Ag-modified OAP samples, i.e., an increase in the activity with an increase in the size of Ag NPs [44]. Moreover, the participation of the mobile electrons in the mechanism of 2-propanol oxidation under vis irradiation has been confirmed by time-resolved microwave conductivity (TRMC). The decay of TRMC signal correlates with the activity, suggesting that more "hot" electrons are formed in the case of larger Ag deposits, as shown in Figure 6.

The continuation of these studies results in the comparison between noble-metal-modified DAP and OAP [86]. It should be pointed out that DAP shows the highest photocatalytic activity among various titania samples (commercial and self-synthesized, including OAP) despite not having the best surface properties (specific surface area of only ca. 5–20 m^2g^{-1} [10]), because of intrinsic properties of two-kind facets, i.e., the natural separation of charge carriers when electrons migrate to (101), whereas holes to (001) facets [63]. Indeed, under UV irradiation DAP samples (bare and modified with NPs of Ag, Cu, and Au) show higher activity than that by respective OAP samples. However, the opposite trend has been observed under vis irradiation where OAP samples are more active than DAP ones. Interestingly, the comparison between faceted and commercial titania samples shows that Au/OAP exhibits ca. one order in magnitude higher activity than that by Au-modified commercial titania samples (of similar properties; Figure 7a), suggesting fast charge carriers' separation (Figure 7b), which is quite reasonable considering the preferential distribution of shallow than deep electron traps (ETs) in OAP [12]. In contrast, Au/DAP shows the worst performance among 16 tested samples, possibly because of intrinsic properties of DAP (as discussed above), i.e., back electron transfer to (101) facets (Au → TiO$_2$(101) → Au), as gold is mainly deposited on these facets, as shown in Figure 7c.

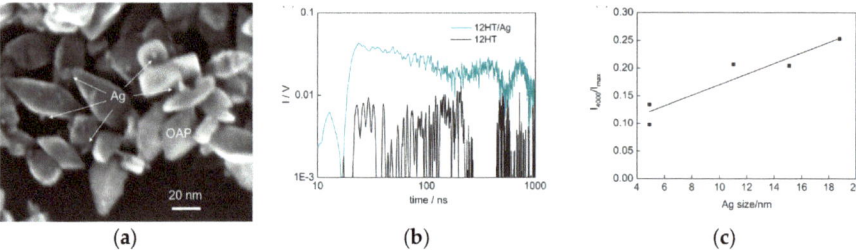

Figure 6. Characteristics of Ag/OAP samples: (**a**) SEM image, (**b**) time-resolved microwave conductivity (TRMC) signal (OAP-12HT; Ag/OAP-12HT/Ag). (**c**) The correlation between the decay of TRMC signal and the size of Ag NPs; adapted with permission from [44]. Copyright 2018 Creative Commons Attribution.

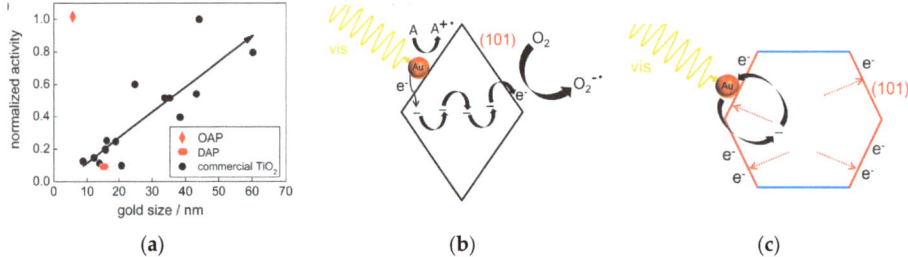

Figure 7. (**a**) The comparison of vis activity of Au-modified commercial titania samples with that by Au/OAP and Au/DAP, and (**b**,**c**) the schematic drawings of possible electron transfer/recombination under vis excitation for: (**b**) Au/OAP and (**c**) Au/DAP; adapted with permission from [86]. Copyright 2018 Creative Commons Attribution.

Interestingly, quite different performance has been observed against microorganisms, where Cu/DAP exhibits much higher activity than Cu/OAP under vis irradiation, probably because of larger content of positively charged Cu [87]. It should be pointed out that for antimicrobial action, positively charged noble metals are much more active than zero-valent ones, because of both intrinsic action of noble-metal cations (in dark) and easier adsorption on negatively charged bacteria [88]. Accordingly, Ag/TiO$_2$ and Cu/TiO$_2$ are usually much more active than Au/TiO$_2$ and Pt/TiO$_2$ samples. Indeed, only Cu- and Ag-modified DAP and OAP show higher activity under vis than that in the dark [87]. Moreover, a decrease in activity is observed under UV irradiation, which should be caused by opposite direction of electron transfer (from TiO$_2$ to noble metal), resulting in negatively charged noble metal, and thus causing a repulsion between noble metal and bacteria.

Other groups have also intensively studied plasmonic photocatalysts using faceted semiconductors. For example, Mao et al. have obtained Pt-modified DAP samples, and tested them for photocatalytic CO$_2$ reduction [89]. It has been shown that without Pt-loading, {010} facets adsorb more CO$_2$

molecules, showing longer lifetime of photogenerated charges than {001} facets, and thus causing higher photocatalytic activity. Interestingly, the presence of small Pt NPs loaded on {010} facets enhances the separation of photogenerated carriers more efficiency than Pt particles (aggregates) deposited on {001} facets. Unfortunately, the study has been performed only under UV irradiation, and thus strong plasmonic effect is not expected. In another work [90], DAP has been modified with Pt NPs by different methods and platinum precursors, which influences the size, distribution, and chemical states of platinum deposits. For example, photodeposition method results in facet-selective deposition, because of photo-driven spatial separation of photogenerated charges (as discussed above). The selective deposition of Pt NPs is also attributed to adsorption behavior of the metal precursors on different crystal facets. The uniform deposition of Pt on {101} facets by chemical reduction of H_2PtCl_6 precursor (different than that for $Pt(NH_3)_4Cl_2$) might be explained by the good adsorption of $PtCl_6^{2-}$ ions on the {101} facets. The adsorption of $Pt(NH_3)_4^{2+}$ is weaker, causing the phenomenon that platinum particles reduced from $Pt(NH_3)_4^{2+}$ are agglomerated easily before being deposited on the titania facets. Moreover, Kobayashi et al. have shown that adsorption of noble metals on DAP facets depends on the surface charges and pH value of the reaction mixture [91].

Similar issue has been examined by Zielinska-Jurek et al. where Pt NPs have been deposited on DAP by different methods [92]. The selective deposition of Pt NPs on DAP results from the fact that the surface of {101} facets is electron-rich, whereas the oxidized {001} surface is electron-deficient. Therefore, Pt NPs are deposited preferably on {101} facets with oxygen vacancies. This phenomenon occurs in the case of thermal reduction method—electrons located on the oxygen vacancy states are released, reducing the adsorbed platinum ions. In the case of chemical reduction method (with $NaBH_4$), a solution-dominated reduction pathway promotes the mechanism of seed-mediated growth of Pt deposits on titania facets (Figure 8). The surface of seeds simultaneously acts as both deposition and reduction site, which influences further growth of Pt NPs and causes the formation of larger particles. In contrast to OAP and commercial titania samples modified with noble metals by photodeposition, it has been found that a decrease in the size of Pt NPs results in activity enhancement under both UV and vis irradiation.

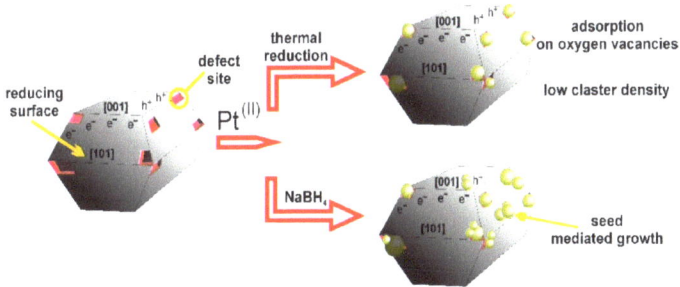

Figure 8. Proposed mechanism of Pt NPs deposition on DAPs; adapted with permission from [92]. Copyright 2019 Creative Commons Attribution.

Application of noble metals to faceted anatase particles might be performed also as a metal–metal alloy to improve the photocatalytic performance, e.g., selective photoconversion of CO_2 into hydrocarbons. For example, Au-Pd alloys, deposited selectively on {101} facets of DAP, have solved the problem of desorption of reaction intermediates (such as CO molecules) [93]. Moreover, the adjustment of the stoichiometric ratio between Au and Pd allows to change the selectivity of the reaction, i.e., CO/hydrocarbons ratio. However, those studies have been performed also only under UV irradiation, and thus could not be considered as plasmonic photocatalysis.

Interesting work about interactions between Ag NPs and different facets of rutile photoanodes has been performed by Ballestas-Barrientos et al. [70]. The special combination of experimental techniques: SEM and electron backscatter diffraction (EBSD) detector with electrochemical impendence spectroscopy has been applied to determine the phenomenon of the Ag NPs photodeposition on the grain boundary defects of preferentially oriented {100} rutile nanorod films. Accordingly, it has been reported that the presence of grain boundary defects results in larger Schottky barriers and better charge carrier separation efficiency and is beneficial for the "plasmonic enhancement" during water splitting. However, similar to other studies, UV light has not been excluded (solar simulator), and thus titania excitation should be considered as the main mechanism of action rather than plasmonic photocatalysis. There are many interesting reports on noble-metal modified faceted semiconductors but tested only under UV/vis irradiation [71,82], and thus not further discussed in this review on plasmonic photocatalysis (activity under vis due to LSPR).

The promising strategy for design of plasmonic photocatalysts based on faceted supports is the composition of coupled systems. For example, Bian et al. prepared $Au/TiO_2/BiVO_4$ plasmonic nanocomposites, at the first stage from $BiVO_4$ 2D-structured nanoflakes via ion exchange process from BiOCl nanosheets, followed by coupling (001) facet exposed anatase 2D-structured nanosheets and finally modified with Au nanorods [94]. The important issue of this study is to match different components of nanocomposite for the construction of effective heterojunction system. The {001} facet-exposed TiO_2 is considered as an efficient acceptor of photogenerated electrons from $BiVO_4$ nanoflakes and plasmonic Au nanorods. The proposed system has been active for photocatalytic CO_2 conversion ensuring wide visible-light response in the range from 400 nm to 660 nm (78.9% of whole visible light spectrum).

2.2. Other Particulate Plasmonic Photocatalysts with Advanced Morphology

Besides mono-metallic plasmonic photocatalysts, multi-metallic plasmonic photocatalysts have also been investigated, including bi- (many examples [60,83,95–97]) and tri- [98] metal-modified semiconductors. It has been suggested that preparation of multi-metallic plasmonic photocatalysts should result in enhanced activity because of efficient light harvesting, as the position of LSPR band depends on the kind of noble metal. Indeed, high enhancement of activity has been observed in some cases [95], but other reports suggest that second metal causes charge carriers' recombination instead of separation. For example, in the case of titania modification with Ag and Au, only samples with mono-metallic deposits (Ag NPs and Au NPs deposited separately on the support) show enhanced activity, whereas bi-metallic deposits (e.g., core-shell NPs) causes a decrease in the activity (Figure 9a) [97]. Interestingly, strong field enhancement has been observed for core-shell NPs (Figure 9b–d), and thus obtained data indirectly support the mechanism of charge transfer rather than energy transfer of plasmonic photocatalysis since enhanced field should cause an activity enhancement (in contrast to experimental results) in the case of energy transfer.

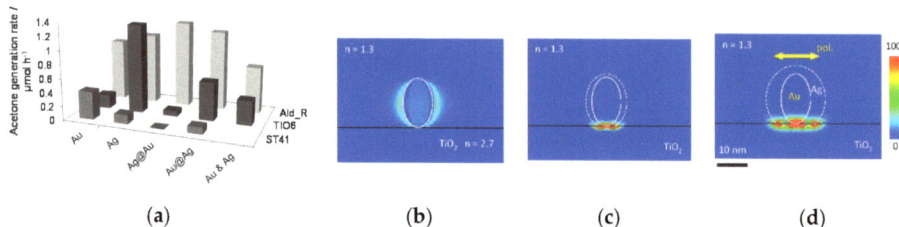

Figure 9. (a) The comparison of photocatalytic activities of mono- and bi-metal modified commercial titania samples (Ald_R—large rutile particles; TIO6—fine rutile NPs; ST41—large anatase NPs; Ag@Au/Ald_R and Au@Ag/Ald_R samples with separately deposited Ag NPs and Au NPs.); (b–d) The cross-sectional patterns of the near-field intensity enhancement factor of mono-(Au/TiO_2) (b) and bi- ($Ag/Au/TiO_2$) (c,d) metallic samples; for: (b) t = 0 at 510 nm wavelength; (c) t = 2 nm at 600 nm, (d) t = 5 nm at 570 nm; t -width of silver layer on gold NP; adapted with permission from [97]. Copyrights 2014 Elsevier.

Titania aerogels modified with NPs of noble metals (gold and copper) have also been proposed as efficient plasmonic photocatalysts [99–101]. Interestingly, it has been found that the localization of gold is decisive for the resultant photocatalytic activity, i.e., the photocatalyst with gold NPs incorporated inside titania network exhibits much higher activity under vis irradiation than that with gold NPs deposited in the porosity, as shown in Figure 10. Although, authors suggest that both mechanisms, i.e., energy and electron transfer, are possible, it seems that such large difference in the vis activity might support the mechanism of electron transfer since the larger interface between titania and gold should facilitate charge migration, whereas same energy transfer (if any) would be expected for similar structures (with similar physicochemical properties) [100]. Moreover, incorporation of less noble metals inside titania aerogel network might also stabilize them against oxidation in air, as plasmonic photocatalysts with zero-valent copper incorporated inside titania aerogel network has been successfully prepared [99].

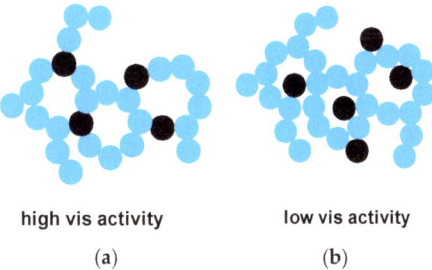

Figure 10. The schematic drawings showing titania aerogels (blue color) with Au NPs (black color): (**a**) Au NPs incorporated inside titania network, and (**b**) Au NPs deposited in the porosity; drawn based on [100].

Janus particles have also been proposed as efficient plasmonic photocatalyst for hydrogen evolution under vis irradiation [37]. It has been shown that Janus particles composed of amorphous titania (one site) and 50-nm Au NPs (another site) possess much higher activity than respective core(Au)-shell(TiO_2) NPs, because of strong localization of LSPR at the interface between Au and TiO_2, as shown in Figure 11 (white area). The vis activities (>400 nm) of core/shell NPs, bare TiO_2 and bare Au NPs has been much lower than that by Janus particles, reaching only 58%, 1%, and <<1%, respectively.

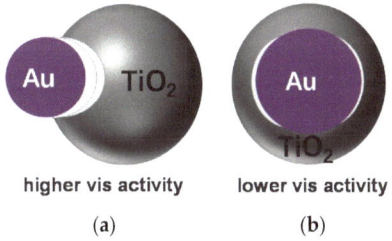

Figure 11. The schematic drawings of: (**a**) Janus particle, and (**b**) core/shell particles; violet-Au; grey-TiO_2; white/l. grey-plasmonic near-field distribution; drawn based on [37].

Although, usually noble metals are deposited on the surface of semiconductors, the reverse structures have been investigated. For example, Horiguchi et al. have proposed Au nanorods (NRs) covered with silver layer and then with TiO_2 [102]. Interestingly, it has been thought that predominant mechanism might change depending on the width of titania layer, i.e., from electron transfer to energy transfer with an increase in the width, as shown in Figure 12. The electron transfer for the structure with a thin titania layer (<10 nm) has been proposed because the release of silver cations has been observed (silver oxidation after "hot" electron transfer to titania). However, energy transfer has been suggested for broader titania layers (>10 nm) because the best photocatalytic activity has been obtained for the optimal width of titania layer (10 nm). The concept of optimal titania layer has been suggested as inconsistent with an electron transfer mechanism since an increase in titania width might result in activity decrease (high probability of electrons' trapping before their arrival at the photocatalyst surface).

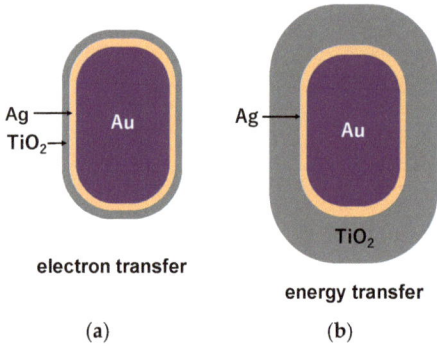

Figure 12. The schematic drawings showing tri-layered photocatalyst composed of Au NR as a core, silver as an interlayer and TiO$_2$ as a shell with: (**a**) a thin titania layer with the mechanism of an electron transfer, and (**b**) a thick titania layer with the mechanism of an energy transfer; drawn based on [102].

Interesting approach has been proposed by Bielan et al. for plasmonic photocatalysts (TiO$_2$ with Pt and Cu/Cu$_x$O) supported on magnetic core for easy separation of photocatalyst after reaction (Figure 13) [103,104]. The photocatalysts exhibit good reusability with almost same reaction rate for fresh and reused (after recovery by a magnet) photocatalysts during consequent photocatalytic cycles. Although, Pt-modified titania samples show vis activity due to LSPR, the magnetic photocatalysts lose the vis response, probably because of "hot" electron trapping by magnetic core (Pt→TiO$_2$→Fe$_3$O$_4$/SiO$_2$) [103], similarly as suggested for bimetallic [97] and hybrid (co-modified with ruthenium dye [105–107]) plasmonic photocatalysts. Interestingly, self-doped titania (titania with defects) significantly improves the photocatalytic activity under vis irradiation for both supported (on magnetic core) and unsupported plasmonic photocatalysts [104].

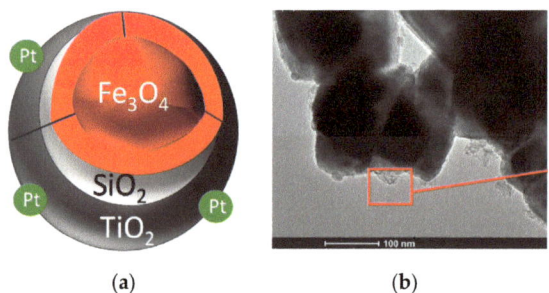

Figure 13. Pt-modified core(Fe$_3$O$_4$)/interlayer(SiO$_2$)/shell(TiO$_2$) photocatalysts magnetically separable and recyclable: (**a**) the schematic drawing; and (**b**) TEM image (Pt NP in red rectangle); adapted from [104]. Copyright 2020 Creative Commons Attribution.

2.3. One-Dimensional Plasmonic Photocatalysts

One-dimensional nanomaterials are considered as ideal candidates for direct transport of electrons because of their orientation characteristics along a certain direction, such as nanowires, nanofibers, nanotubes, nanorods, nanobelts, and so on. Obviously, they have also been applied as a support for noble metals to obtained plasmonic photocatalysts.

Similar to faceted plasmonic photocatalysts (discussed in Section 2.1.), the size and distribution of noble-metal NPs on the 1D titania is also important. Accordingly, uniform distribution of Ag NPs on one-dimensional anatase titania results in fast decomposition of 2,4-dichlorophenol under vis irradiation and high reusability (almost same activity after 10 cycles) [108].

Bimetallic NPs have also been applied on 1D titania structures. For example, titania nanowires (TNWs) have been modified with Au/Ag alloyed NPs by a facile hydrothermal and photodeposition method [109]. The synergistic effect between Au and Ag has been obtained for selective CO_2 reduction with H_2 to CO and hydrocarbons under vis irradiation. Interestingly, it has been proposed that bimetallic deposits might have double function, i.e., as a plasmonic sensitizer (under vis excitation) and as an electron sink (same as under UV irradiation), as shown in Figure 14, and thus charge carriers' recombination (main shortcoming of plasmonic photocatalysis) might be hindered.

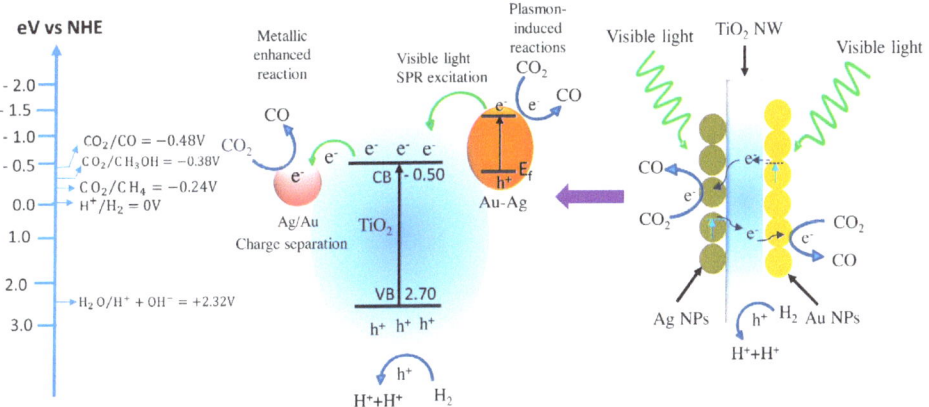

Figure 14. Proposed mechanism showing a photocatalytic process for CO_2 reduction by H_2 to CO on Au/Ag-NPs/TiO_2 NWs under vis irradiation; adapted with permission from [109]. Copyrights 2017 Elsevier.

Another structure using 1D titania has been proposed by Nasir et al. in which branched titania (Figure 15a) has been coupled with g-C_3N_4 quantum dots (QDs) and surface modified with Au NPs [110]. The composite has exhibited very high quantum efficiency (19.5%) for hydrogen generation under vis irradiation (420 nm). It has been proposed that both Au NPs and g-C_3N_4 QDs might absorb visible light, resulting in an electron transfer to CB of titania, as shown in Figure 15. However, another possibility should be also considered, i.e., an electron transfer from g-C_3N_4 QDs via CB of TiO_2 to Au NPs (similar to that under UV irradiation for Au/TiO_2 photocatalyst) since it is well-known that noble metals work as co-catalyst for hydrogen evolution (negligible activity of bare titania).

Titania nanofibers modified with Au NPs, synthesized by one-step electrospinning of TiO_2 and Au precursors, and polyvinylpyrrolidone (PVP) in DMF solution followed by pyrolysis, have also been used for the performance enhancement of dye-sensitized solar cells (DSSCs) [111]. It has been found that 0.5 wt% is the optimal content of Au (varied by the concentration of $HAuCl_4 \cdot 3H_2O$ in the electrospinning solution).

Besides the solid 1D plasmonic photocatalysts (nanowires, nanorods, nanofibers, and so on), hollow 1D photocatalysts (nanotubular arrays or various length of nanotubes) have attracted extensive interests as a support for plasmonic NPs. For example, the photocatalytic activity of Au-modified titania nanotubes (TNTs) have been studied for CO_2 conversion to CH_4 [112]. It has been found that the activity of TNTs has been significantly improved by Au decoration. Although authors suggest plasmonic photocatalysis with an electron transfer from Au to TiO$_2$, the activity has been tested under UV/vis, and thus titania excitation could not be excluded (as also observed by vis activity of bare TNTs), and thus Au NPs might work predominantly as an electron acceptor rather than plasmonic sensitizer.

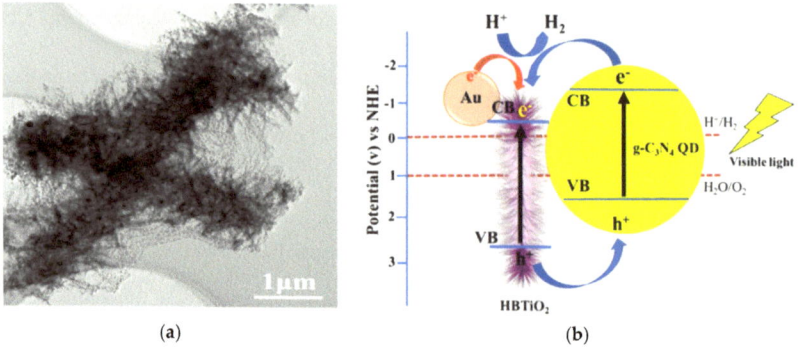

Figure 15. (a) TEM of Au/TiO$_2$/g-C$_3$N$_4$ sample after photocatalytic reaction, (b) the schematic diagram for an electron transfer mechanism under vis excitation; adapted with permission from [110]. Copyrights 2020 Elsevier.

Interesting study has been presented by Fu et al. for single-wall TiO$_2$ nanotube array (TNTA) with in situ deposition of Au NPs [113]. The "hot" electron transfer under vis irradiation has been confirmed by photoelectrochemical study, as shown in Figure 16. The Au/TNTA has been successfully used for DSSCs, and the best performance has reached high power conversion efficiency of 8.93%, which is 190.4% that of the pristine TNTA-based DSSC.

Besides gold, other noble metals have also been applied for modifications of 1D nanostructures. For example, Ag-TNTA films with the preferential orientation of crystals have been prepared on ITO glass by magnetron sputtering and anodization [114]. It has been found that Ag/TNTA exhibits superior separation/transfer of photo-induced charges, resulting in high photocatalytic activities, i.e., 99.1% removal of Cr (VI) during 90 min of vis illumination. However, in the case of less noble metals, their oxidation should be also considered, and thus the coupled photocatalysts, e.g., TiO$_2$/Ag$_2$O/Ag are usually obtained [83,84,86]. Accordingly, the mechanism of vis response might be quite complex, containing both plasmonic photocatalysis and heterojunction between two semiconductors.

Interesting approach for the preparation of Ag/TNT has been proposed by Mazierski et al. via one-step synthesis by anodization of a Ti–Ag alloy [115]. It has been shown that both zero-valent and oxidized forms of silver (Ag and Ag$_2$O) have been obtained. Silver NPs have been formed during the in situ generation of Ag ions and are (i) embedded in the TNT walls, (ii) stuck on the external TNT walls, and (iii) placed inside the TNTs. It has been proposed that the enhanced photocatalytic activity under UV irradiation originates from the optimal content of Ag$_2$O/Ag NPs, responsible for decreasing of the number of recombination centers. In contrast, under vis irradiation the activity increases with an increase in the content of both Ag$_2$O and Ag^0, probably because of more efficient light harvesting by both forms of silver. The electron transfer from Ag$_2$O and Ag (plasmonic excitation) to CB of titania has been proposed as the first step of photocatalytic reaction. However, zero-valent silver could also work as an electron sink, as already proposed/discussed.

The comparison between 0D and 1D plasmonic photocatalysts have been performed by Wei et al. for OAP (faceted anatase titania) and titanate nanowires (TNWs; precursor used for OAP synthesis) modified with NPs of Au, Ag, and Pt [116]. It has been confirmed that the modifications with noble metals have improved the photocatalytic performance of both semiconductors significantly under UV and vis irradiation. However, the photocatalytic activities of bare and modified OAP samples are much higher than those by TNW samples, probably because of anatase presence, higher crystallinity, and fast electron mobility in faceted NPs. Interestingly, noble metals show different influence on the activity depending on the semiconductor support, i.e., gold-modified TNW and platinum-modified OAP exhibit the highest activity for acetic acid decomposition under UV irradiation, whereas silver- and gold-modified samples are the most active under vis irradiation, respectively.

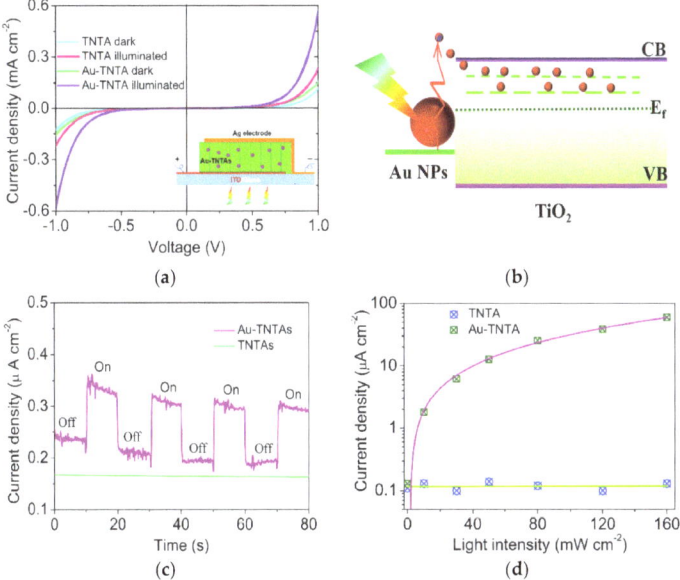

Figure 16. Hot electron injection induced current response measurement: (**a**) J-V curves of conductivity test of TNTA (titania nanotube array) and Au-TNTA films in the dark and under illumination (100 mW cm^{-2}). The inset shows the structure of the tested electrode; (**b**) schematic illustration of hot electron injection process from Au-NPs into TNTA; (**c**) photocurrent response of TNTA and Au-TNTA film toward on/off of the input laser light (520 nm, 20 mW cm^{-2}); (**d**) photocurrent variation of TNTA and Au-TNTA film under 520 nm laser illumination with different light intensity. The applied bias is 1.0 V; adapted with permission from [113]. Copyrights 2019 Elsevier.

2.4. Two- and Three-Dimensional Plasmonic Photocatalysts

To improve photocatalytic performance of plasmonic photocatalysts, various advanced nanostructures have been proposed. For example, "advanced superstructure system" composed of titania mesocrystals and NPs of noble metals has been designed to avoid charge carriers' recombination [117]. It has been found that localization of noble metals is crucial for the overall activity, i.e., Au NPs on the basal surface results in much higher activity than that when Au NPs are placed on the lateral one since "hot" electrons easily migrate via titania mesocrystals (similar charge migration as that in faceted titania—DAP), as shown in Figure 17. Because of this anisotropic electron flow, which hinders electron/hole recombination in Au NPs, electrons migrate to the edges of mesocrystals where they might be stored for further reactions. Additionally, in the case of hydrogen evolution reaction, it is known that for plasmonic photocatalysis second metal should be deposited as hydrogen evolution takes place on the metal surface (negligible activity of bare titania and mono-metallic plasmonic photocatalysts because of an electron-deficient noble metal). Accordingly, the deposition of Au NPs on the basal surface and Pt NPs on the lateral surface has caused the high photocatalytic activity under vis irradiation, even for hydrogen evolution (Usually plasmonic photocatalysts are almost inactive for hydrogen evolution due to the opposite direction of electron transfer, i.e., from noble metal to titania, than that under UV irradiation).

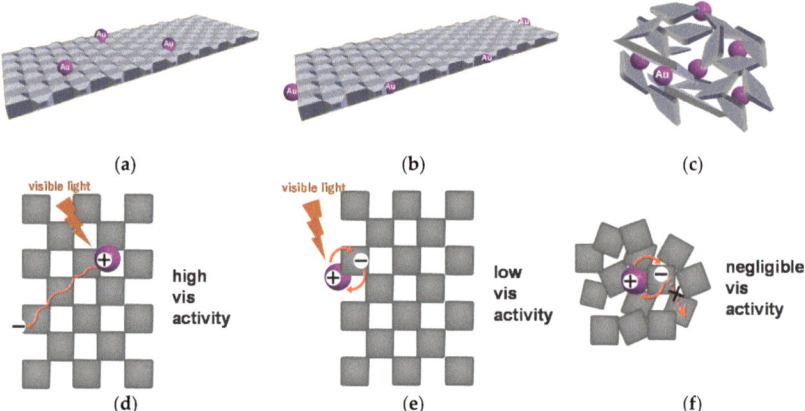

Figure 17. The schematic drawings of: (**a,b,d,e**) the superstructure composed of titania mesocrystals and Au NPs deposited on basal (**a,d**) and lateral (**b,e**) surfaces, and (**c,f**) titania mesocrystals with Au NPs; (**a**–**c**) side view, (**e,f**) top view.

The faceted titania nanosheets (2D) have also been used as plasmonic photocatalysts [118]. It has been found that the photocatalytic efficiency of Au/TiO$_2$ (Au NPs selectively deposited on {001} facets) for vis-induced H$_2$ evolution depends on the content of {001} facets, i.e., activity increase with an increase in the content of {001}. Besides titania, other 2D materials have been examined, e.g., tin(IV) oxide nanoflakes (2D) modified with Ag NPs [119]. It has been proposed that the activity under vis irradiation is caused by electron transfer from Ag NPs, resulting from plasmonic excitation. However, bare titania is also active, as experiments have been performed for dye discoloration, and thus photocatalyst sensitization by dye should be considered in the overall mechanism.

Moreover, 3D materials have been proposed for plasmonic photocatalysis. Li et al. have decorated TiO$_2$ nanorods with Au NPs via one-step approach to provide a dense coverage of titania nanorods' surfaces with Au NPs, resulting in high quality of Au-TiO$_2$ interface, as shown in Figure 18 [120]. This structure provides high activity in vis-induced photoelectrochemical water splitting reaction. FDTD simulations have indicated that the enhanced photoactivity resulting from Au presence is mainly caused by the electric-field amplification effect and hot-electron generation upon LSPR excitation in the visible light.

Interesting comparison between 2D and 3D plasmonic photocatalysts have been provided by Song et al. for rods (Au/RR) and 3D hierarchical structure (Au/3DR), respectively, as shown in Figure 19 [121]. In the case of visible light ($\lambda > 455$ nm) induced photocatalytic degradation of p-nitrosodimethylaniline, the reaction efficiency is higher for Au/3DR than that for Au/RR sample. After white-light irradiation on a single particle, a substantial red shift in the LSPR peak and amplified plasmonic coupling in Au/3DR is observed (Figure 19a). To verify the advantage of 3D structure, plasmonic coupling of two Au NPs depending on the irradiated light angle has been calculated by FDTD simulation (Figure 19c). When the tilted angle is lower than 45°, the transverse mode becomes more dominant than the longitudinal mode and a blue shift with decreased intensity of electromagnetic (EM) field in the LSPR peak occurs. When the tilted angle is higher than 45°, the longitudinal mode is formed and the red shift in the LSPR peak occurs, reaching the maximum at 90°. Rods from Au/RR are randomly oriented in the photocatalytic system and plasmonic coupling effects in all angles are negatively averaged (Figure 19d). On the other hand, rods in the Au/3DR are oriented in an orderly manner and most of Au NP array can be irradiated by light in the vertical direction (longitudinal plasmonic coupling), which leads to the LSPR peak of Au/3DR to become red-shift with high intensity.

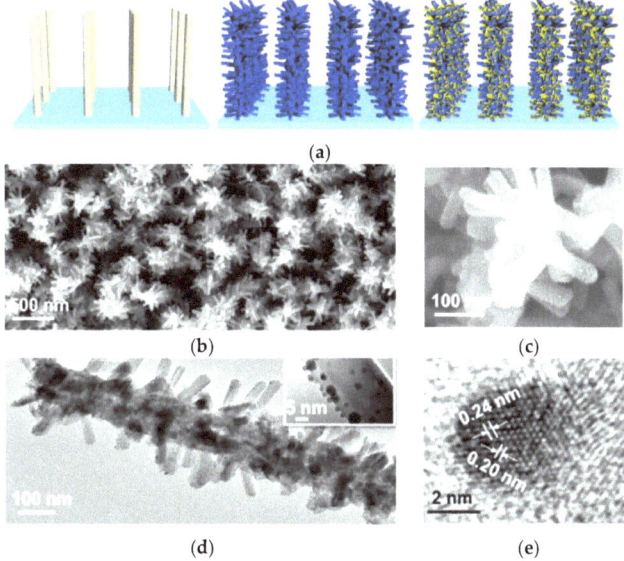

Figure 18. (**a**) Schematic representation of the fabrication of Au-NP-decorated 3D branched TiO$_2$ NR architectures; (**b**) top-viewed SEM image of Au NP–3D TiO$_2$ NR arrays; (**c**) SEM image showing homogeneously distributed Au NPs on the entire TiO$_2$ surface; (**d**) TEM image of an individual branched TiO$_2$ NR array. Inset: Single TiO$_2$ NR decorated with Au NPs; (**e**) HRTEM image of a Au NP residing on the TiO$_2$ surface; adapted with permission from [120]. Copyright 2017 ACS.

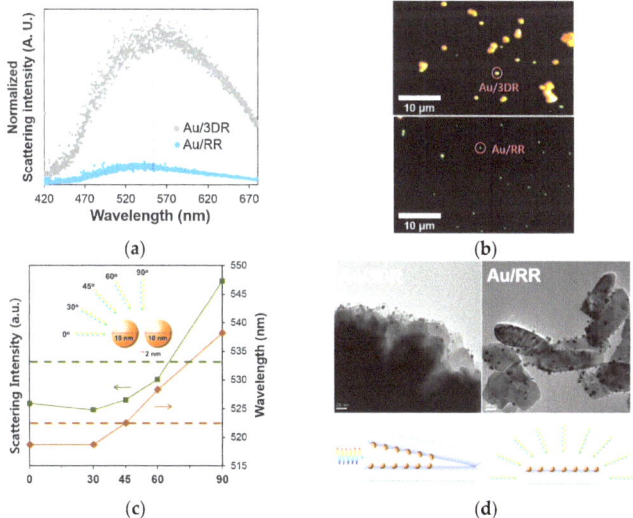

Figure 19. (**a**,**b**) Dark field images of Au/3DR and Au/RR (typical particles—red circles); (**c**) FDTD simulations performed according to inset scheme of coupled Au NPs with light irradiation in different directions; the values of original 10 nm Au NP indicated by dotted lines; (**d**) TEM images (top) of Au/3DR and Au/RR with simplified schemes including light irradiation (bottom); adapted with permission from [121]. Copyright 2016 Elsevier.

For activity improvements of any photocatalytic materials, two main strategies are mainly considered, i.e., inhibition of charge carriers' recombination and efficient light harvesting. Accordingly, various strategies for charge carriers' separation have been proposed, as already discussed in this review. In the case of efficient light harvesting, the preparation of highly polydisperse NPs of noble metals (broad LSPR [38,56]) and co-modifications with other metals [98], metal oxides [96,122] and metal complexes [105–107] have been investigated. Recently, another strategy has been proposed, i.e., using semiconductors in the form of photonic crystals with photonic bandgap and slow photon effect. Usually titania inverse opal, prepared from respective opal (e.g., silica or polystyrene (PS) sacrificial template), is used as photonic crystal for support of noble metals, as shown in Figure 20. It has been proposed that the overlapping of photonic bandgap with LSPR of noble metal causes the significant enhancement of the photocatalytic activity (photons are efficiently used by multi-scattering effect), as already comprehensively presented in the review paper on the photonic crystals for plasmonic photocatalysis [123]. Although, efficient light harvesting and high photocatalytic activity have been achieved in some cases [124–126], there are various challenges facing the photonic crystal-based plasmonic photocatalysts, including complex synthesis, delicate nature (easy destruction of opal/inverse opal structure), and still unknown key factors of the high photocatalytic performance. Besides that, it seems that probably this is the hottest topic of plasmonic photocatalysis in recent years, and more studies/reports are expected in the nearest future, based on the excellent light harvesting ability, resulting from extraordinary optical properties in such nanostructures (similar to metal films and sub-wavelength hole arrays [127,128]).

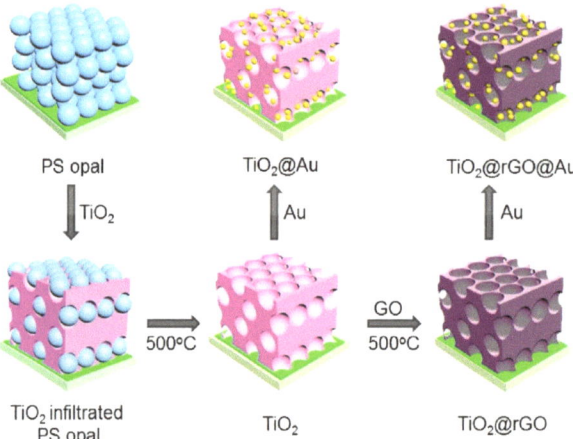

Figure 20. The schematic drawings of preparation of noble-metal-modified photonic crystals (TiO2 inverse opal and rGO-modified TiO2 inverse opal), adapted from [129]. Copyright 2017 ACS.

2.5. Other Plasmonic Photocatalysts

Another recent research on new plasmonic photocatalysts focuses on ferroelectrics, which have been applied in photocatalysis reactions, since the polarization of ferroelectrics might improve the charge carriers' separation, and thus the photocatalytic activities [130]. Accordingly, a range of ferroelectrics has attracted interests in photocatalysis, whereas their photocatalytic performance still needs to be improved. Ferroelectrics usually have a large bandgap, a low absorption coefficient, and a low photon conversion efficiency when used alone. Therefore, various methods have been tried to enhance their photocatalytic efficiency, among which the modification with NPs of noble metals turns out to be efficient. For example, Su et al. have enhanced the activity of ferroelectric $BaTiO_3$ after its modification with Ag [131]. Besides the ferroelectricity and specific charge-transfer kinetics in the $Ag/BaTiO_3$ hybrid, LSPR of Ag has been suggested to contribute to the enhancement as well. Wu et al.

have prepared a ternary Z-scheme heterojunction photocatalyst of $BaTiO_3/Au/g-C_3N_4$, in which Au NPs besides acting as an electron mediator, could absorb visible light by LSPR effect to inject "hot" electrons into CB of $g-C_3N_4$ to participate in the photocatalytic reactions [132]. Recently, a dual-modified photocatalyst of $BaTiO_3$ has been synthesized, i.e., $Ag/BaTiO_3/MnO_x$, where LSPR of Ag NPs on $BaTiO_3$ is noticed, but its influence on the photocatalytic reactions has not been discussed [133]. Lan et al. have observed an enhanced photocatalytic activity in one-dimensional $KNbO_3$ nanowires (1D) modified with Au NPs due to LSPR and interband transitions on Au NPs, depending on the size of Au NPs, i.e., an increase in the activity with an increase in the size of Au NPs from 5 to 10 nm [134] (which might suggest more important LSPR effect for the overall activity). In another study, it has been proposed that the improved photocatalytic activity and stability of $Ag/AgNbO_3$ is caused by LSPR effect, ferroelectric polarization, specific exposed facets, and high crystallinity at the heterogeneous interface [73]. Meanwhile, $BiFeO_3$ (one of the most promising multiferroic materials) decorated with NPs of Au and Ag has proven to be active under vis irradiation due to strong field enhancement especially near Ag NPs [135]. Similarly, as mentioned in former paragraphs, the presence of LSPR does not guarantee its participation in promoting the photocatalytic activities. For example, Cui et al. have coated $BaTiO_3$ with nanostructured Ag, resulting in vis absorption due to LSPR, but photo-decolorization of RhB (under both UV- and visible-light-blocking filters) indicates that "hot" electrons do not contribute to the photocatalytic activity [136]. Besides, it has been proposed that the geometry of metal deposits on ferroelectric substrate plays an important role in enlarging or inhibiting the polarization effect, and thus the metal deposits should not be too thick to block the effect from polarization [137]. Chao et al. have indicated that in the case of $Au/BaTiO_3$ the "hot" electrons (LSPR of Au) have only little effect on the improvement of vis photocatalytic performance [138].

Summarizing, as pointed by Kumar et al. it is important to realize that the integration of plasmonic metals with ferroelectrics might be beneficial for two aspects: (1) The injected charge carriers are effectively driven to spatially separate over the ferroelectric particles; and (2) tuning the Schottky barrier height, e.g., by changing the chemical composition of noble-metal/semiconductor, thus providing an interfacial contact that favors the hot charge injection [139]. Both effects can prolong the lifetimes of both the excited electrons and holes. Therefore, for effective utilization of LSPR in ferroelectric-based photocatalysts, more efforts are needed to reveal the critical parameters and achieve higher photocatalytic efficiency.

3. Summary

Plasmonic photocatalysts have been intensively investigated in recent years, because of tunable vis activity and broad possibilities of applications, including environmental purification (decomposition of organic compounds and inactivation of microorganisms) and solar energy conversion (water splitting, photocurrent generation, CO_2 conversion—artificial photosynthesis). Although, plasmonic photocatalysts have shown vis response and high potential for various applications, the activity under vis irradiation is usually too low for commercialization. Accordingly, morphology design has been intensively studied to improve the photocatalytic performance. Faceted nanoparticles, aerogels, core/shells, Janus particles, photonic crystals, 1D, 2D, and 3D nanostructures have been proposed as excellent support for noble metals. However, it should be pointed out that some morphology might result in a decrease of photocatalytic activity, e.g., highly UV-active decahedral anatase with noble-metal NPs deposited on {101} facets accelerates charge carriers' recombination under vis irradiation.

The comparison between various nanostructures is quite difficult as different irradiation sources, photoreactors, tested compounds, and reaction conditions have been used, and only several studies show apparent quantum yields. Moreover, solar radiation or solar simulators are commonly used for activity testing, and thus different mechanism is expected, i.e., excitation of semiconductor rather than plasmon resonance. Accordingly, even if "plasmonic photocatalysis" term is used in the title of some reports, it does not guarantee that really plasmonic photocatalysis has been investigated.

It should be reminded that noble metals have been known as an electron sink for ca. 50 years, and thus improving photocatalytic activity of wide-bandgap semiconductors under their excitation. Moreover, for some reactions noble metals are necessary as co-catalyst, e.g., hydrogen evolution, as bare semiconductors are hardly active. Additionally, activity tests should not be performed for dye discoloration, as photocatalyst sensitization by dye might be the predominant mechanism of their degradation rather than plasmonic photocatalysis.

Another important issue is the mechanism clarifications, as three main mechanisms have been postulated, i.e., energy transfer, electron transfer and plasmonic heating. Although, plasmonic heating has not been considered as the main mechanism in majority of studies because of the inefficiency to cleave the chemical bonds, it might not be excluded for antimicrobial activity as microorganisms are highly sensitive to any environmental changes. Although, it is hard to find a direct proof for any mechanism, and only indirect proofs have been shown, it might be concluded that the mechanism depends on the morphology.

The role of morphology is undoubtedly an important issue, which should be taken into consideration during design of new plasmonic photocatalysts. Combined with the presented state-of-the-art, future efforts to prepare morphologically controlled plasmonic materials should focus, among others, on the following points:

(i) Better understanding of the destination photocatalytic reaction mechanism and synergistic effects between LSPR phenomenon and particle morphology is necessary—the usage of advanced characterization techniques to perform a detailed analysis of physicochemical properties of prepared materials coupled with providing theoretical simulations and calculations to deeply understand these synergistic interactions.

(ii) The issue of the reaction selectivity by adjusting morphological properties of plasmonic photocatalysts (e.g., understanding the role of titania crystal facets configuration for the course of photocatalytic CO_2 reduction) should be considered in more detail.

Although, many aspects of plasmonic photocatalysts are still unknown, unclear, or even inconsistent, it is thought that plasmonic photocatalysts might be successfully commercialized in the nearest future.

Funding: This research was financially supported by National Natural Science Foundation of China (NSFC; contract No. 51802087) and "Yugo-Sohatsu Kenkyu" for an Integrated Research Consortium on Chemical Sciences (IRCCS) project from Ministry of Education and Culture, Sport, Science and Technology-Japan (MEXT).

Conflicts of Interest: The authors declare no conflict of interest.

References

1. Balcerski, W.; Ryu, S.Y.; Hoffmann, M.R. Visible-Light Photoactivity of Nitrogen-Doped TiO_2: Photo-oxidation of HCO_2H to CO_2 and H_2O. *J. Phys. Chem. C* **2007**, *111*, 15357–15362. [CrossRef]
2. Hoffmann, M.R.; Martin, S.T.; Choi, W.Y.; Bahnemann, D.W. Environmental applications of semiconductor photocatalysis. *Chem. Rev.* **1995**, *95*, 69–96. [CrossRef]
3. Fujishima, A.; Honda, K. Electrochemical photolysis of water at a semiconductor electrode. *Nature* **1972**, *238*, 37–38. [CrossRef] [PubMed]
4. Abe, R.; Shinmei, K.; Koumura, N.; Hara, K.; Ohtani, B. Visible-light-induced water splitting based on Two-step photoexcitation between dye-sensitized layered niobate and tungsten oxide photocatalysts in the presence of a triiodide/iodide shuttle redox mediator. *J. Am. Chem. Soc.* **2013**, *135*, 16872–16884. [CrossRef] [PubMed]
5. Herrmann, J.M. Heterogeneous photocatalysis: State of the art and present applications. *Top. Catal.* **2005**, *34*, 49–65. [CrossRef]
6. Wang, K.L.; Janczarek, M.; Wei, Z.S.; Raja-Mogan, T.; Endo-Kimura, M.; Khedr, T.M.; Ohtani, B.; Kowalska, E. Morphology- and crystalline composition-governed activity of titania-based photocatalysts: Overview and perspective. *Catalysts* **2019**, *9*, 1054. [CrossRef]

7. Murakami, N.; Katayama, S.; Nakamura, M.; Tsubota, T.; Ohno, T. Dependence of photocatalytic activity on aspect ratio of shape-controlled rutile titanium(IV) oxide nanorods. *J. Phys. Chem. C* **2011**, *115*, 419–424. [CrossRef]
8. Amano, F.; Yasumoto, T.; Mahaney, O.O.P.; Uchida, S.; Shibayama, T.; Terada, Y.; Ohtani, B. Highly active titania photocatalyst particles of controlled crystal phase, size, and polyhedral shapes. *Top. Catal.* **2010**, *53*, 455–461. [CrossRef]
9. Amano, F.; Prieto-Mahaney, O.O.; Terada, Y.; Yasumoto, T.; Shibayama, T.; Ohtani, B. Decahedral single-crystalline particles of anatase titanium(IV) oxide with high photocatalytic activity. *Chem. Mater.* **2009**, *21*, 2601–2603. [CrossRef]
10. Janczarek, M.; Kowalska, E.; Ohtani, B. Decahedral-shaped anatase titania photocatalyst particles: Synthesis in a newly developed coaxial-flow gas-phase reactor. *Chem. Eng. J.* **2016**, *289*, 502–512. [CrossRef]
11. Wei, Z.S.; Kowalska, E.; Ohtani, B. Enhanced photocatalytic activity by particle morphology: Preparation, characterization, and photocatalytic activities of octahedral anatase titania particles. *Chem. Lett.* **2014**, *43*, 346–348. [CrossRef]
12. Wei, Z.; Kowalska, E.; Verrett, J.; Colbeau-Justin, C.; Remita, H.; Ohtani, B. Morphology-dependent photocatalytic activity of octahedral anatase particles prepared by ultrasonication-hydrothermal reaction of titanates. *Nanoscale* **2015**, *7*, 12392–12404. [CrossRef] [PubMed]
13. Zaleska, A. Doped-TiO_2: A review. *Recent Pat. Eng.* **2008**, *2*, 157–164. [CrossRef]
14. Asahi, R.; Morikawa, T.; Ohwaki, T.; Aoki, K.; Taga, Y. Visible-light photocatalysis in nitrogen-doped titanium oxides. *Science* **2001**, *293*, 269–271. [CrossRef]
15. Ohno, T.; Akiyoshi, M.; Umebayashi, T.; Asai, K.; Mitsui, T.; Matsumura, M. Preparation of S-doped TiO_2 photocatalysts and their photocatalytic activities under visible light. *Appl. Catal. A Gen.* **2004**, *265*, 115–121. [CrossRef]
16. Beranek, R.; Kisch, H. Surface-modified anodic TiO_2 films for visible light photocurrent response. *Electrochem. Commun.* **2007**, *9*, 761–766. [CrossRef]
17. Janus, M.; Tryba, B.; Inagaki, M.; Morawski, A.W. New preparation of a carbon-TiO_2 photocatalyst by carbonization of n-hexane deposited on TiO_2. *Appl. Catal. B Environ.* **2004**, *52*, 61–67. [CrossRef]
18. Sclafani, A.; Mozzanega, M.N.; Pichat, P. Effect of silver deposits on the photocatalytic activity of titanium dioxide samples for the dehydrogenation or oxidation of 2-propanol. *J. Photochem. Photobiol. A Chem.* **1991**, *59*, 181–189. [CrossRef]
19. Wang, K.L.; Endo-Kimura, M.; Belchi, R.; Zhang, D.; Habert, A.; Boucle, J.; Ohtani, B.; Kowalska, E.; Herlin-Boime, N. Carbon/graphene-modified titania with enhanced photocatalytic activity under UV and vis irradiation. *Materials* **2019**, *12*, 4158. [CrossRef]
20. Di Paola, A.; Palmisano, L.; Venezia, A.M.; Augugliaro, V. Coupled semiconductor systems for photocatalysis. Preparation and characterization of polycrystalline mixed WO_3/WS_2 powders. *J. Phys. Chem. B* **1999**, *103*, 8236–8244. [CrossRef]
21. Endo-Kimura, M.; Janczarek, M.; Bielan, Z.; Zhang, D.; Wang, K.; Markowska-Szczupak, A.; Kowalska, E. Photocatalytic and antimicrobial properties of Ag_2O/TiO_2 heterojunction. *ChemEngineering* **2019**, *3*, 3. [CrossRef]
22. Janczarek, M.; Wang, K.L.; Kowalska, E. Synergistic effect of Cu_2O and urea as modifiers of TiO_2 for enhanced visible light activity. *Catalysts* **2018**, *8*, 240. [CrossRef]
23. Kraeutler, B.; Bard, A.J. Heterogeneous photocatalytic preparation of supported catalysts. Photodeposition of platinum on TiO_2 powder and other substrates. *J. Am. Chem. Soc.* **1978**, *100*, 4317–4318. [CrossRef]
24. Xu, A.-W.; Gao, Y.; Liu, H.-Q. The preparation, characterization, and their photocatalytic activities of rare-earth-doped TiO_2 nanoparticles. *J. Catal.* **2002**, *207*, 151–157. [CrossRef]
25. Morikawa, T.; Asahi, R.; Ohwaki, T.; Aoki, K.; Taga, Y. Band-gap narrowing of titanium dioxide by nitrogen doping. *Jpn. J. Appl. Phys.* **2001**, *40*, L561–L563. [CrossRef]
26. Anpo, M. Use of visible light. Second-generation titanium oxide photocatalysts prepared by the application of an advanced metal ion-implantation method. *Pure Appl. Chem.* **2000**, *72*, 1787–1792. [CrossRef]
27. Wang, Y.Q.; Cheng, H.M.; Zhang, L.; Hao, Y.Z.; Ma, J.M.; Xu, B.; Li, W.H. The preparation, characterization, photoelectrochemical and photocatalytic properties of lanthanide metal-ion-doped TiO_2 nanoparticles. *J. Mol. Catal. A Chem.* **2000**, *151*, 205–216. [CrossRef]

28. Hirano, K.; Suzuki, E.; Ishikawa, A.; Moroi, T.; Shiroishi, H.; Kaneko, M. Sensitization of TiO_2 particles by dyes to achieve H_2 evolution by visible light. *J. Photochem. Photobiol. A* **2000**, *136*, 157–161. [CrossRef]
29. Tian, Y.; Liu, H.; Zhao, G.; Tatsuma, T. Shape-controlled electrodeposition of gold nanostructures. *J. Phys. Chem. B* **2006**, *110*, 23478–23481. [CrossRef]
30. Tian, Y.; Tatsuma, T. Mechanisms and applications of plasmon-induced charge separation at TiO_2 films loaded with gold nanoparticles. *J. Am. Chem. Soc.* **2005**, *127*, 7632–7637. [CrossRef]
31. Liu, Z.; Hou, W.; Pavaskar, P.; Aykol, M.; Cronin, S.B. Plasmon resonance enhancement of photocatalytic water splitting under visible illumination. *Nano Lett.* **2011**, *11*, 1111–1116. [CrossRef] [PubMed]
32. Hou, W.B.; Hung, W.H.; Pavaskar, P.; Goeppert, A.; Aykol, M.; Cronin, S.B. Photocatalytic conversion of CO_2 to hydrocarbon fuels via plasmon-enhanced absorption and metallic interband transitions. *ACS Catal.* **2011**, *1*, 929–936. [CrossRef]
33. Ueno, K.; Misawa, H. Surface plasmon-enhanced photochemical reactions. *J. Photochem. Photobiol. C* **2013**, *15*, 31–52. [CrossRef]
34. Nishijima, Y.; Ueno, K.; Yokata, Y.; Murakoshi, K.; Misawa, H. Plasmon-assisted photocurrent generation from visible to near-infrared wavelength using a Au-nanorods/TiO_2 electrode. *J. Phys. Chem. Lett.* **2010**, *1*, 2031–2036. [CrossRef]
35. Bouhadoun, S.; Guillard, C.; Dapozze, F.; Singh, S.; Amans, D.; Boucle, J.; Herlin-Boime, N. One step synthesis of N-doped and Au-loaded TiO_2 nanoparticles by laser pyrolysis: Application in photocatalysis. *Appl. Catal. B Environ.* **2015**, *174*, 367–375. [CrossRef]
36. Hou, W.; Liu, Z.; Pavaskar, P.; Hsuan Hung, W.; Cronin, S.B. Plasmonic enhancement of photocatalytic decomposition of methyl orange under visible light. *J. Catal.* **2011**, *277*, 149–153. [CrossRef]
37. Seh, Z.W.; Liu, S.W.; Low, M.; Zhang, S.-Y.; Liu, Z.; Mlayah, A.; Han, M.-Y. Janus Au-TiO_2 photocatalysts with strong localization of plasmonic near fields for efficient visible-light hydrogen generation. *Adv. Mater.* **2012**, *24*, 2310–2314. [CrossRef]
38. Kowalska, E.; Abe, R.; Ohtani, B. Visible light-induced photocatalytic reaction of gold-modified titanium(IV) oxide particles: Action spectrum analysis. *Chem. Commun.* **2009**, 241–243. [CrossRef] [PubMed]
39. Du, L.; Furube, A.; Hara, K.; Katoh, R.; Tachiya, M. Plasmon induced electron transfer at gold-TiO_2 interface under femtosecond near-IR two photon excitation. *Thin Solid Films* **2009**, *158*, 861–864. [CrossRef]
40. Du, L.C.; Furube, A.; Yamamoto, K.; Hara, K.; Katoh, R.; Tachiya, M. Plasmon-induced charge separation and recombination dynamics in gold-TiO_2 nanoparticle systems: Dependence on TiO_2 particle size. *J. Phys. Chem. C* **2009**, *113*, 6454–6462. [CrossRef]
41. Caretti, I.; Keulemans, M.; Verbruggen, S.W.; Lenaerts, S.; Van Doorslaer, S. Light-induced processes in plasmonic gold/TiO_2 photocatalysts studied by electron paramagnetic resonance. *Top. Catal.* **2015**, *58*, 776–782. [CrossRef]
42. Priebe, J.B.; Radnik, J.; Lennox, A.J.J.; Pohl, M.M.; Karnahl, M.; Hollmann, D.; Grabow, K.; Bentrup, U.; Junge, H.; Beller, M.; et al. Solar hydrogen production by plasmonic Au-TiO_2 catalysts: Impact of synthesis protocol and TiO_2 phase on charge transfer efficiency and H_2 evolution rates. *ACS Catal.* **2015**, *5*, 2137–2148. [CrossRef]
43. Mendez-Medrano, M.G.; Kowalska, E.; Lehoux, A.; Herissan, A.; Ohtani, B.; Rau, S.; Colbeau-Justin, C.; Rodriguez-Lopez, J.L.; Remita, H. Surface modification of TiO_2 with Au nanoclusters for efficient water treatment and hydrogen generation under visible light. *J. Phys. Chem. C* **2016**, *120*, 25010–25022. [CrossRef]
44. Wei, Z.; Janczarek, M.; Endo, M.; Colbeau-Justin, C.; Ohtani, B.; Kowalska, E. Silver-modified octahedral anatase particles as plasmonic photocatalyst. *Catal. Today* **2018**, *310*, 19–25. [CrossRef] [PubMed]
45. Sakai, N.; Fujiwara, Y.; Takahashi, Y.; Tatsuma, T. Plasmon-resonance-based generation of cathodic photocurrent at electrodeposited gold nanoparticles coated with TiO_2 films. *Chem. Phys. Chem.* **2009**, *10*, 766–769. [CrossRef]
46. Chen, X.; Zhu, H.-Y.; Zhao, J.-C.; Zheng, Z.-F.; Gao, X.-P. Visible-light-driven oxidation of organic contaminants in air with gold nanoparticle catalysts on oxide supports. *Angew. Chem. Int. Ed.* **2008**, *47*, 5353–5356. [CrossRef]
47. Wang, C.J.; Ranasingha, O.; Natesakhawat, S.; Ohodnicki, P.R.; Andio, M.; Lewis, J.P.; Matranga, C. Visible light plasmonic heating of Au-ZnO for the catalytic reduction of CO_2. *Nanoscale* **2013**, *5*, 6968–6974. [CrossRef]

48. Mohamed, R.M.; Aazam, E.S. Characterization and catalytic properties of nano-sized Au metal catalyst on titanium containing high mesoporous silica (Ti-HMS) synthesized by photo-sssisted deposition and impregnation methods. *Int. J. Photoenergy* **2011**. [CrossRef]
49. Trammell, S.A.; Nita, R.; Moore, M.; Zabetakis, D.; Chang, E.; Knight, D.A. Accelerating the initial rate of hydrolysis of methyl parathion with laser excitation using monolayer protected 10 nm Au nanoparticles capped with a Cu(bpy) catalyst. *Chem. Commun.* **2012**, *48*, 4121–4123. [CrossRef] [PubMed]
50. Mukherjee, S.; Libisch, F.; Large, N.; Neumann, O.; Brown, L.V.; Cheng, J.; Lassiter, J.B.; Carter, E.A.; Nordlander, P.; Halas, N.J. Hot electrons do the impossible: Plasmon-induced dissociation of H_2 on Au. *Nano Lett.* **2013**, *13*, 240–247. [CrossRef] [PubMed]
51. Cushing, S.K.; Li, J.T.; Meng, F.K.; Senty, T.R.; Suri, S.; Zhi, M.J.; Li, M.; Bristow, A.D.; Wu, N.Q. Photocatalytic activity enhanced by plasmonic resonant energy transfer from metal to semiconductor. *J. Am. Chem. Soc.* **2012**, *134*, 15033–15041. [CrossRef] [PubMed]
52. Silva, C.G.; Juarez, R.; Marino, T.; Molinari, R.; Garcia, H. Influence of excitation wavelength (UV or visible light) on the photocatalytic activity of titania containing gold nanoparticles for the generation of hydrogen or oxygen from water. *J. Am. Chem. Soc.* **2011**, *133*, 595–602. [CrossRef]
53. Son, M.S.; Im, J.E.; Wang, K.K.; Oh, S.L.; Kim, Y.R.; Yoo, K.H. Surface plasmon enhanced photoconductance and single electron effects in mesoporous titania nanofibers loaded with gold nanoparticles. *Appl. Phys. Lett.* **2010**, *96*, 023115. [CrossRef]
54. Kominami, H.; Tanaka, A.; Hashimoto, K. Gold nanoparticles supported on cerium(IV) oxide powder for mineralization of organic acids in aqueous suspensions under irradiation of visible light of λ = 530 nm. *Appl. Catal. A Gen.* **2011**, *397*, 121–126. [CrossRef]
55. Ilie, M.; Cojocaru, B.; Parvulescu, V.I.; Garcia, H. Improving TiO_2 activity in photo-production of hydrogen from sugar industry wastewaters. *Int. J. Hydrogen Energy* **2011**, *36*, 15509–15518. [CrossRef]
56. Kowalska, E.; Mahaney, O.O.P.; Abe, R.; Ohtani, B. Visible-light-induced photocatalysis through surface plasmon excitation of gold on titania surfaces. *Phys. Chem. Chem. Phys.* **2010**, *12*, 2344–2355. [CrossRef] [PubMed]
57. Endo, M.; Wei, Z.S.; Wang, K.L.; Karabiyik, B.; Yoshiiri, K.; Rokicka, P.; Ohtani, B.; Markowska-Szczupak, A.; Kowalska, E. Noble metal-modified titania with visible-light activity for the decomposition of microorganisms. *Beilstein J. Nanotech.* **2018**, *9*, 829–841. [CrossRef]
58. Mkhalid, I.A. Visible light photocatalytic synthesis of aniline with an $Au/LaTiO_3$ nanocomposites. *J. Alloy Compd.* **2015**, *631*, 298–302. [CrossRef]
59. Seo, J.H.; Jeon, W.I.; Dembereldorj, U.; Lee, S.Y.; Joo, S.W. Cytotoxicity of serum protein-adsorbed visible-light photocatalytic $Ag/AgBr/TiO_2$ nanoparticles. *J. Hazard. Mater.* **2011**, *198*, 347–355. [CrossRef]
60. Zielińska-Jurek, A.; Kowalska, E.; Sobczak, J.W.; Lisowski, W.; Ohtani, B.; Zaleska, A. Preparation and characterization of monometallic (Au) and bimetallic (Ag/Au) modified-titania photocatalysts activated by visible light. *Appl. Catal. B Environ.* **2011**, *101*, 504–514. [CrossRef]
61. Wang, K.L.; Wei, Z.S.; Ohtani, B.; Kowalska, E. Interparticle electron transfer in methanol dehydrogenation on platinum-loaded titania particles prepared from P25. *Catal. Today* **2018**, *303*, 327–333. [CrossRef]
62. Yang, H.G.; Liu, G.; Qiao, S.Z.; Sun, C.H.; Jin, Y.G.; Smith, S.C.; Zou, J.; Cheng, H.M.; Lu, G.Q. Solvothermal synthesis and photoreactivity of anatase TiO_2 nanosheets with dominant {001} facets. *J. Am. Chem. Soc.* **2009**, *131*, 4078–4083. [CrossRef] [PubMed]
63. Tachikawa, T.; Yamashita, S.; Majima, T. Evidence for crystal-face-dependent TiO_2 photocatalysis from single-molecule imaging and kinetic analysis. *J. Am. Chem. Soc.* **2011**, *133*, 7197–7204. [CrossRef] [PubMed]
64. Fazio, G.; Ferrighi, L.; Di Valentin, C. Spherical versus faceted anatase TiO_2 nanoparticles: A model study of structural and electronic properties. *J. Phys. Chem. C* **2015**, *119*, 20735–20746. [CrossRef]
65. Wei, Z.; Kowalska, E.; Wang, K.; Colbeau-Justin, C.; Ohtani, B. Enhanced photocatalytic activity of octahedral anatase particles prepared by hydrothermal reaction. *Catal. Today* **2017**, *280*, 29–36. [CrossRef]
66. Wei, Z.; Kowalska, E.; Ohtani, B. Influence of post-treatment operations on structural properties and photocatalytic activity of octahedral anatase titania particles prepared by an ultrasonication-hydrothermal reaction. *Molecules* **2014**, *19*, 19573–19587. [CrossRef]
67. Zhu, J.A.; Wang, S.H.; Bian, Z.F.; Xie, S.H.; Cai, C.L.; Wang, J.G.; Yang, H.G.; Li, H.X. Solvothermally controllable synthesis of anatase TiO_2 nanocrystals with dominant {001} facets and enhanced photocatalytic activity. *CrystEngComm* **2010**, *12*, 2219–2224. [CrossRef]

68. Yang, H.G.; Sun, C.H.; Qiao, S.Z.; Zou, J.; Liu, G.; Smith, S.C.; Cheng, H.M.; Lu, G.Q. Anatase TiO$_2$ single crystals with a large percentage of reactive facets. *Nature* **2008**, *453*, 638–641. [CrossRef]
69. Murakami, N.; Kurihara, Y.; Tsubota, T.; Ohno, T. Shape-controlled anatase titanium(IV) oxide particles prepared by hydrothermal treatment of peroxo titanic acid in the presence of polyvinyl alcohol. *J. Phys. Chem. C* **2009**, *113*, 3062–3069. [CrossRef]
70. Ballestas-Barrientos, A.; Li, X.B.; Yick, S.; Yuen, A.; Masters, A.F.; Maschmeyer, T. Interactions of plasmonic silver nanoparticles with high energy sites on multi-faceted rutile TiO$_2$ photoanodes. *ChemCatChem* **2020**, *12*, 469–477. [CrossRef]
71. Li, K.; Peng, T.Y.; Ying, Z.H.; Song, S.S.; Zhang, J. Ag-loading on brookite TiO$_2$ quasi nanocubes with exposed {210} and {001} facets: Activity and selectivity of CO$_2$ photoreduction to CO/CH$_4$. *Appl. Catal. B Environ.* **2016**, *180*, 130–138. [CrossRef]
72. Bryl, R.; Olewicz, T.; de Bocarme, T.V.; Kruse, N. Thermal faceting of clean and oxygen-covered Ir nanocrystals. *J. Phys. Chem. C* **2011**, *115*, 2761–2768. [CrossRef]
73. Yu, Z.Z.; Zhan, B.W.; Ge, B.H.; Zhu, Y.T.; Dai, Y.; Zhou, G.J.; Yu, F.P.; Wang, P.; Huang, B.B.; Zhan, J. Synthesis of high efficient and stable plasmonic photocatalyst Ag/AgNbO$_3$ with specific exposed crystal-facets and intimate heterogeneous interface via combustion route. *Appl. Surf. Sci.* **2019**, *488*, 485–493. [CrossRef]
74. Hou, C.T.; Liu, W.L.; Zhu, J.M. Synthesis of NaOH-modified TiOF$_2$ and its enhanced visible light photocatalytic performance on RhB. *Catalysts* **2017**, *7*, 243. [CrossRef]
75. Li, W.Z.; Kovarik, L.; Mei, D.H.; Liu, J.; Wang, Y.; Peden, C.H.F. Stable platinum nanoparticles on specific MgAl$_2$O$_4$ spinel facets at high temperatures in oxidizing atmospheres. *Nat. Commun.* **2013**, *4*, 2481. [CrossRef] [PubMed]
76. Li, J.Y.; Dong, X.A.; Sun, Y.J.; Cen, W.L.; Dong, F. Facet-dependent interfacial charge separation and transfer in plasmonic photocatalysts. *Appl. Catal. B Environ.* **2018**, *226*, 269–277. [CrossRef]
77. Zhang, L.; Wang, W.Z.; Sun, S.M.; Jiang, D.; Gao, E.P. Selective transport of electron and hole among {001} and {110} facets of BiOCl for pure water splitting. *Appl. Catal. B Environ.* **2015**, *162*, 470–474. [CrossRef]
78. Dong, F.; Xiong, T.; Yan, S.; Wang, H.Q.; Sun, Y.J.; Zhang, Y.X.; Huang, H.W.; Wu, Z.B. Facets and defects cooperatively promote visible light plasmonic photocatalysis with Bi nanowires@BiOCl nanosheets. *J. Catal.* **2016**, *344*, 401–410. [CrossRef]
79. Warmuth, L.; Ritschel, C.; Feldmann, C. Facet-, composition- and wavelength-dependent photocatalysis of Ag$_2$MoO$_4$. *RSC Adv.* **2020**, *10*, 18377–18383. [CrossRef]
80. Li, R.G.; Han, H.X.; Zhang, F.X.; Wang, D.G.; Li, C. Highly efficient photocatalysts constructed by rational assembly of dual-cocatalysts separately on different facets of BiVO$_4$. *Energy Environ. Sci.* **2014**, *7*, 1369–1376. [CrossRef]
81. Wang, X.W.; Wang, W.Y.; Miao, Y.Q.; Feng, G.; Zhang, R.B. Facet-selective photodeposition of gold nanoparticles on faceted ZnO crystals for visible light photocatalysis. *J. Colloid Interf. Sci.* **2016**, *475*, 112–118. [CrossRef] [PubMed]
82. Wang, B.; Liu, M.C.; Zhou, Z.H.; Guo, L.J. Surface activation of faceted photocatalyst: When metal cocatalyst determines the nature of the facets. *Adv. Sci.* **2015**, *2*, 1500153. [CrossRef] [PubMed]
83. Janczarek, M.; Wei, Z.; Endo, M.; Ohtani, B.; Kowalska, E. Silver- and copper-modified decahedral anatase tiania particles as visible light-responsive plasmonic photocatalyst. *J. Photonics Energy* **2017**, *7*, 1–16.
84. Wei, Z.; Endo, M.; Wang, K.; Charbit, E.; Markowska-Szczupak, A.; Ohtani, B.; Kowalska, E. Noble metal-modified octahedral anatase titania particles with enhanced activity for decomposition of chemical and microbiological pollutants. *Chem. Eng. J.* **2017**, *318*, 121–134. [CrossRef]
85. Wei, Z.; Rosa, L.; Wang, K.; Endo, M.; Juodkazi, S.; Ohtani, B.; Kowalska, E. Size-controlled gold nanoparticles on octahedral anatase particles as efficient plasmonic photocatalyst. *Appl. Catal. B Environ.* **2017**, *206*, 393–405. [CrossRef]
86. Wei, Z.; Janczarek, M.; Endo, M.; Wang, K.L.; Balcytis, A.; Nitta, A.; Mendez-Medrano, M.G.; Colbeau-Justin, C.; Juodkazis, S.; Ohtani, B.; et al. Noble metal-modified faceted anatase titania photocatalysts: Octahedron versus decahedron. *Appl. Catal. B Environ.* **2018**, *237*, 574–587. [CrossRef]
87. Endo, M.; Janczarek, M.; Wei, Z.; Wang, K.; Markowska-Szczupak, A.; Ohtani, B.; Kowalska, E. Bactericidal properties of plasmonic photocatalysts composed of noble-metal nanoparticles on faceted ana-tase titania. *J. Nanosci. Nanotechnol.* **2019**, *19*, 442–452. [CrossRef]

88. Endo-Kimura, M.; Kowalska, E. Plasmonic photocatalysts for microbiological applications. *Catalysts* **2020**, *10*, 824. [CrossRef]
89. Mao, J.; Ye, L.Q.; Li, K.; Zhang, X.H.; Liu, J.Y.; Peng, T.Y.; Zan, L. Pt-loading reverses the photocatalytic activity order of anatase TiO_2 {001} and {010} facets for photoreduction of CO_2 to CH_4. *Appl. Catal. B Environ.* **2014**, *144*, 855–862. [CrossRef]
90. Xiong, Z.; Lei, Z.; Chen, X.X.; Gong, B.G.; Zhao, Y.C.; Zhang, J.Y.; Zheng, C.G.; Wu, J.C.S. CO_2 photocatalytic reduction over Pt deposited TiO_2 nanocrystals with coexposed {101} and {001} facets: Effect of deposition method and Pt precursors. *Catal. Commun.* **2017**, *96*, 1–5. [CrossRef]
91. Kobayashi, K.; Takashima, M.; Takase, M.; Ohtani, B. Mechanistic study on facet-dependent deposition of metal nanoparticles on decahedral-shaped anatase titania photocatalyst particles. *Catalysts* **2018**, *8*, 542. [CrossRef]
92. Zielinska-Jurek, A.; Wei, Z.S.; Janczarek, M.; Wysocka, I.; Kowalska, E. Size-controlled synthesis of Pt particles on TiO_2 surface: Physicochemical characteristic and photocatalytic activity. *Catalysts* **2019**, *9*, 940. [CrossRef]
93. Chen, Q.; Chen, X.J.; Fang, M.L.; Chen, J.Y.; Li, Y.J.; Xie, Z.X.; Kuang, Q.; Zheng, L.S. Photo-induced Au-Pd alloying at TiO_2 {101} facets enables robust CO_2 photocatalytic reduction into hydrocarbon fuels. *J. Mater. Chem. A* **2019**, *7*, 1334–1340. [CrossRef]
94. Bian, J.; Qu, Y.; Zhang, X.L.; Sun, N.; Tang, D.Y.; Jing, L.Q. Dimension-matched plasmonic $Au/TiO_2/BiVO_4$ nanocomposites as efficient wide-visible-light photocatalysts to convert CO_2 and mechanistic insights. *J. Mater. Chem. A* **2018**, *6*, 11838–11845. [CrossRef]
95. Verbruggen, S.W.; Keulemans, M.; Goris, B.; Blommaerts, N.; Bals, S.; Martens, J.A.; Lenaerts, S. Plasmonic 'rainbow' photocatalyst with broadband solar light response for environmental applications. *Appl. Catal. B Environ.* **2016**, *188*, 147–153. [CrossRef]
96. Mendez-Medrano, M.G.; Kowalska, E.; Lehoux, A.; Herissan, A.; Ohtani, B.; Bahena, D.; Briois, V.; Colbeau-Justin, C.; Rodriguez-Lopez, J.; Remita, H. Surface modification of TiO_2 with Ag nanoparticles and CuO nanoclusters for applications in photocatalysis. *J. Phys. Chem. C* **2016**, *120*, 5143–5154. [CrossRef]
97. Kowalska, E.; Janczarek, M.; Rosa, L.; Juodkazi, S.; Ohtani, B. Mono- and bi-metallic plasmonic photocatalysts for degradation of organic compounds under UV and visible light irradiation. *Catal. Today* **2014**, *230*, 131–137. [CrossRef]
98. Malankowska, A.; Kobylanski, M.P.; Mikolajczyk, A.; Cavdar, O.; Nowaczyk, G.; Jarek, M.; Lisowski, W.; Michalska, M.; Kowalska, E.; Ohtani, B.; et al. TiO_2 and $NaTaO_3$ decorated by trimetallic Au/Pd/Pt core-shell nanoparticles as efficient photocatalysts: Experimental and computational studies. *ACS Sustain. Chem. Eng.* **2018**, *6*, 16665–16682. [CrossRef]
99. DeSario, P.A.; Pietron, J.J.; Brintlinger, T.H.; McEntee, M.; Parker, J.F.; Baturina, O.; Stroud, R.M.; Rolison, D.R. Oxidation-stable plasmonic copper nanoparticles in photocatalytic TiO_2 nanoarchitectures. *Nanoscale* **2017**, *9*, 11720–11729. [CrossRef]
100. DeSario, P.A.; Pietron, J.J.; DeVantier, D.E.; Brintlinger, T.H.; Stroud, R.M.; Rolison, D.R. Plasmonic enhancement of visible-light water splitting with $Au-TiO_2$ composite aerogels. *Nanoscale* **2013**, *5*, 8073–8083. [CrossRef]
101. Panayotov, D.A.; DeSario, P.A.; Pietron, J.J.; Brintlinger, T.H.; Szymczak, L.C.; Rolison, D.R.; Morris, J.R. Ultraviolet and visible photochemistry of methanol at 3D mesoporous networks: TiO_2 and $Au-TiO_2$. *J. Phys. Chem. C* **2013**, *117*, 15035–15049. [CrossRef]
102. Horiguchi, Y.; Kanda, T.; Torigoe, K.; Sakai, H.; Abe, M. Preparation of gold/silver/titania trilayered nanorods and their photocatalytic activities. *Langmuir* **2014**, *30*, 922–928. [CrossRef] [PubMed]
103. Bielan, Z.; Kowalska, E.; Dudziak, S.; Wang, K.; Ohtani, B.; Zielinska-Jurek, A. Mono- and bimetallic (Pt/Cu) titanium(IV) oxide core–shell photocatalysts with UV/Vis light activity and magnetic separability. *Catal. Today* **2020**, in press. [CrossRef]
104. Bielan, Z.; Sulowska, A.; Dudziak, S.; Siuzdak, K.; Ryl, J.; Zielinska-Jurek, A. Defective TiO_2 core-shell magnetic photocatalyst modified with plasmonic nanoparticles for visible light-induced photocatalytic activity. *Catalysts* **2020**, *10*, 672. [CrossRef]
105. Zheng, S.; Wei, Z.; Yoshiiri, K.; Braumuller, M.; Ohtani, B.; Rau, S.; Kowalska, E. Titania modification with ruthenium(II) complex and gold nanoparticles for photocatalytic degradation of organic compounds. *Photochem. Photobiol. Sci.* **2016**, *15*, 69–79. [CrossRef]

106. Zheng, S.Z.; Wang, K.L.; Wei, Z.S.; Yoshiiri, K.; Braumuller, M.; Rau, S.; Ohtani, B.; Kowalska, E. Mono- and dual-modified titania with a ruthenium(II) complex and silver nanoparticles for photocatalytic degradation of organic compounds. *J. Adv. Oxid. Technol.* **2016**, *19*, 208–217. [CrossRef]
107. Kowalska, E.; Yoshiiri, K.; Wei, Z.; Zheng, S.; Kastl, E.; Remita, H.; Ohtani, B.; Rau, S. Hybrid photocatalysts composed of titania modified with plasmonic nanoparticles and ruthenium complexes for decomposition of organic compounds. *Appl. Catal. B Environ.* **2015**, *178*, 133–143. [CrossRef]
108. Jiang, L.M.; Zhou, G.; Mi, J.; Wu, Z.Y. Fabrication of visible-light-driven one-dimensional anatase TiO_2/Ag heterojunction plasmonic photocatalyst. *Catal. Commun.* **2012**, *24*, 48–51. [CrossRef]
109. Tahir, M.; Tahir, B.; Amin, N.A.S. Synergistic effect in plasmonic Au/Ag alloy NPs co-coated TiO_2 NWs toward visible-light enhanced CO_2 photoreduction to fuels. *Appl. Catal. B Environ.* **2017**, *204*, 548–560. [CrossRef]
110. Nasir, M.S.; Yang, G.R.; Ayub, I.; Wang, S.L.; Yan, W. In situ decoration of g-C_3N_4 quantum dots on 1D branched TiO_2 loaded with plasmonic Au nanoparticles and improved the photocatalytic hydrogen evolution activity. *Appl. Surf. Sci.* **2020**, *519*, 146208. [CrossRef]
111. Zheng, F.; Zhu, Z.T. Preparation of the Au@TiO_2 nanofibers by one-step electrospinning for the composite photoanode of dye-sensitized solar cells. *Mater. Chem. Phys.* **2018**, *208*, 35–40. [CrossRef]
112. Khatun, F.; Abd Aziz, A.; Sim, L.C.; Monir, M.U. Plasmonic enhanced Au decorated TiO_2 nanotube arrays as a visible light active catalyst towards photocatalytic CO_2 conversion to CH_4. *J. Environ. Chem. Eng.* **2019**, *7*, 103233. [CrossRef]
113. Fu, N.Q.; Jiang, X.Z.; Chen, D.C.; Duan, Y.D.; Zhang, G.G.; Chang, M.L.; Fang, Y.Y.; Lin, Y. Au/TiO_2 nanotube array based multi-hierarchical architecture for highly efficient dye-sensitized solar cells. *J. Power Sources* **2019**, *439*, 227076. [CrossRef]
114. Wang, S.; Zhang, Z.; Huo, W.; Zhu, K.; Zhang, X.; Zhou, X.; Fang, F.; Xie, Z.G.; Jiang, J. Preferentially oriented Ag-TiO_2 nanotube array film: An efficient visible-light-driven photocatalyst. *J. Hazard. Mater.* **2020**, *399*, 123016. [CrossRef]
115. Mazierski, P.; Malankowska, A.; Kobylanski, M.; Diak, M.; Kozak, M.; Winiarski, M.J.; Klimczuk, T.; Lisowski, W.; Nowaczyk, G.; Zaleska-Medynska, A. Photocatalytically active TiO_2/Ag_2O nanotube arrays interlaced with silver nanoparticles obtained from the one-step anodic oxidation of Ti-Ag alloys. *ACS Catal.* **2017**, *7*, 2753–2764. [CrossRef]
116. Wei, Z.S.; Endo-Kimura, M.; Wang, K.L.; Colbeau-Justin, C.; Kowalska, E. Influence of semiconductor morphology on photocatalytic activity of plasmonic photocatalysts: Titanate nanowires and octahedral anatase nanoparticles. *Nanomaterials* **2019**, *9*, 1447. [CrossRef]
117. Bian, Z.F.; Tachikawa, T.; Zhang, P.; Fujitsuka, M.; Majima, T. Au/TiO_2 superstructure-based plasmonic photocatalysts exhibiting efficient charge separation and unprecedented activity. *J. Am. Chem. Soc.* **2014**, *136*, 458–465. [CrossRef] [PubMed]
118. Yan, J.Q.; Wu, G.J.; Dai, W.L.; Guan, N.J.; Li, L.D. Synthetic design of gold nanoparticles on anatase TiO_2 {001} for enhanced visible light harvesting. *ACS Sustain. Chem. Eng.* **2014**, *2*, 1940–1946. [CrossRef]
119. Kumar, N.S.; Asif, M.; Reddy, T.R.K.; Shanmugam, G.; Ajbar, A. Silver quantum dot decorated 2D-SnO_2 nanoflakes for photocatalytic degradation of the water pollutant Rhodamine B. *Nanomaterials* **2019**, *9*, 1536. [CrossRef]
120. Li, H.X.; Li, Z.D.; Yu, Y.H.; Ma, Y.J.; Yang, W.G.; Wang, F.; Yin, X.; Wang, X.D. Surface-plasmon-resonance-enhanced photoelectrochemical water splitting from Au-nanoparticle-decorated 3D TiO_2 nanorod architectures. *J. Phys. Chem. C* **2017**, *121*, 12071–12079. [CrossRef]
121. Song, C.K.; Baek, J.; Kim, T.Y.; Yu, S.; Han, J.W.; Yi, J. Exploring crystal phase and morphology in the TiO_2 supporting materials used for visible-light driven plasmonic photocatalyst. *Appl. Catal. B Environ.* **2016**, *198*, 91–99. [CrossRef]
122. Mendez-Medrano, M.G.; Kowalska, E.; Endo, M.; Wang, K.; Bahena, D.; Rodriguez-Lopez, J.L.; Remita, H. Inhibition of fungal growth using modified TiO_2 with core@shell structure of Ag@CuO clusters. *ACS Appl. Bio Mater.* **2019**, *2*, 5626–5633. [CrossRef]
123. Raja-Mogan, T.; Ohtani, B.; Kowalska, E. Photonic crystals for plasmonic photocatalysis. *Catalysts* **2020**, *10*, 827. [CrossRef]

124. Lu, Y.; Yu, H.T.; Chen, S.; Quan, X.; Zhao, H.M. Integrating plasmonic nanoparticles with TiO_2 photonic crystal for enhancement of visible-light-driven photocatalysis. *Environ. Sci. Technol.* **2012**, *46*, 1724–1730. [CrossRef]
125. Zhang, X.; Liu, Y.; Lee, S.T.; Yang, S.H.; Kang, Z.H. Coupling surface plasmon resonance of gold nanoparticles with slow-photon-effect of TiO_2 photonic crystals for synergistically enhanced photoelectrochemical water splitting. *Energy Environ. Sci.* **2014**, *7*, 1409–1419. [CrossRef]
126. Zhang, S.S.; Peng, B.Y.; Yang, S.Y.; Wang, H.G.; Yu, H.; Fang, Y.P.; Peng, F. Non-noble metal copper nanoparticles-decorated TiO_2 nanotube arrays with plasmon-enhanced photocatalytic hydrogen evolution under visible light. *Int. J. Hydrogen Energy* **2015**, *40*, 303–310. [CrossRef]
127. Ebessen, T.W.; Lezec, H.J.; Ghaemi, H.F.; Thio, T.; Wolff, P.A. Extraordinary optical transmission through sub-wavelength hole arrays. *Nature* **1998**, *391*, 667669. [CrossRef]
128. Rodrigo, S.G.; de Leon-Perez, F.; Martin-Moreno, L. Extraordinary optical transmission: Fundamentals and applications. *Proc. IEEE* **2016**, *104*, 2288–2306. [CrossRef]
129. Boppella, R.; Kochuveedu, S.T.; Kim, H.; Jeong, M.J.; Mota, F.M.; Park, J.H.; Kim, D.H. Plasmon-sensitized graphene/TiO_2 inverse opal nanostructures with enhanced charge collection efficiency for water splitting. *ACS Appl. Mater. Interfaces* **2017**, *9*, 7075–7083. [CrossRef]
130. Kakekhani, A.; Ismail-Beigi, S.; Altman, E.I. Ferroelectrics: A pathway to switchable surface chemistry and catalysis. *Surf. Sci.* **2016**, *650*, 302–316. [CrossRef]
131. Su, R.; Shen, Y.J.; Li, L.L.; Zhang, D.W.; Yang, G.; Gao, C.B.; Yang, Y.D. Silver-modified nnosized frroelectrics as a nvel potocatalyst. *Small* **2015**, *11*, 202–207. [CrossRef] [PubMed]
132. Wu, M.; Ding, T.; Wang, Y.; Zhao, W.; Xian, H.; Tian, Y.; Zhang, T.; Li, X. Rational construction of plasmon Au assisted ferroelectric-$BaTiO_3$/Au/g-C_3N_4 Z-scheme system for efficient photocatalysis. *Catal. Today* **2019**. [CrossRef]
133. Cui, Y.F.; Sun, H.H.; Shen, G.D.; Jing, P.P.; Pu, Y.P. Effect of dual-cocatalyst surface modification on photodegradation activity, pathway, and mechanisms with highly efficient Ag/$BaTiO_3$/MnO_x. *Langmuir* **2020**, *36*, 498–509. [CrossRef] [PubMed]
134. Lan, J.Y.; Zhou, X.M.; Liu, G.; Yu, J.G.; Zhang, J.C.; Zhi, L.J.; Nie, G.J. Enhancing photocatalytic activity of one-dimensional KNbO3 nanowires by Au nanoparticles under ultraviolet and visible-light. *Nanoscale* **2011**, *3*, 5161–5167. [CrossRef]
135. Zhang, X.G.; Wang, B.; Wang, X.Z.; Xiao, X.H.; Dai, Z.G.; Wu, W.; Zheng, J.F.; Ren, F.; Jiang, C.Z. Preparation of M@$BiFeO_3$ nanocomposites (M = Ag, Au) bowl arrays with enhanced visible light photocatalytic activity. *J. Am. Ceram. Soc.* **2015**, *98*, 2255–2263. [CrossRef]
136. Cui, Y.F.; Briscoe, J.; Dunn, S. Effect of ferroelectricity on solar-light-driven photocatalytic activity of $BaTiO_3$-Influence on the carrier separation and stern layer formation. *Chem. Mater.* **2013**, *25*, 4215–4223. [CrossRef]
137. Cui, Y.F.; Goldup, S.M.; Dunn, S. Photodegradation of Rhodamine B over Ag modified ferroelectric $BaTiO_3$ under simulated solar light: Pathways and mechanism. *RSC Adv.* **2015**, *5*, 30372–30379. [CrossRef]
138. Chao, C.Y.; Zhou, Y.P.; Han, T.L.; Yang, Y.Y.; Wei, J.Y.; Li, H.; He, W.W. Ferroelectric polarization-enhanced photocatalytic properties and photo-induced charge carrier behavior of Au/$BaTiO_3$. *J. Alloy. Compd.* **2020**, *825*, 154060. [CrossRef]
139. Kumar, V.; O'Donnell, S.; Zoellner, B.; Martinez, J.; Wang, G.; Maggard, P.A. Interfacing plasmonic nanoparticles with ferroelectrics for hot-carrier-driven photocatalysis: Impact of Schottky barrier height. *ACS Appl. Energy Mater.* **2019**, *2*, 7690–7699. [CrossRef]

© 2020 by the authors. Licensee MDPI, Basel, Switzerland. This article is an open access article distributed under the terms and conditions of the Creative Commons Attribution (CC BY) license (http://creativecommons.org/licenses/by/4.0/).

MDPI AG
Grosspeteranlage 5
4052 Basel
Switzerland
Tel.: +41 61 683 77 34

Catalysts Editorial Office
E-mail: catalysts@mdpi.com
www.mdpi.com/journal/catalysts

Disclaimer/Publisher's Note: The title and front matter of this reprint are at the discretion of the Guest Editor. The publisher is not responsible for their content or any associated concerns. The statements, opinions and data contained in all individual articles are solely those of the individual Editor and contributors and not of MDPI. MDPI disclaims responsibility for any injury to people or property resulting from any ideas, methods, instructions or products referred to in the content.